Nutrient Cycling in Forest Ecosystems

Nutrient Cycling in Forest Ecosystems

Editor

Robert G. Qualls

MDPI • Basel • Beijing • Wuhan • Barcelona • Belgrade • Manchester • Tokyo • Cluj • Tianjin

Editor
Robert G. Qualls
Department of Natural
Resources and Environmental
Science, University of Nevada
USA

Editorial Office
MDPI
St. Alban-Anlage 66
4052 Basel, Switzerland

This is a reprint of articles from the Special Issue published online in the open access journal *Forests* (ISSN 1999-4907) (available at: https://www.mdpi.com/journal/forests/special_issues/NC).

For citation purposes, cite each article independently as indicated on the article page online and as indicated below:

LastName, A.A.; LastName, B.B.; LastName, C.C. Article Title. *Journal Name* **Year**, *Article Number*, Page Range.

ISBN 978-3-03936-800-6 (Hbk)
ISBN 978-3-03936-801-3 (PDF)

Cover image courtesy of Robert G. Qualls.

Contents

About the Editor

Robert G. Qualls is Professor Emeritus in the Department of Natural Resources and Environmental Science at the University of Nevada, USA. He received a B.S. in Biology, with Honors in Biology and Honors in Creative Writing at the University of North Carolina. He received an M.S.P.H. in Environmental Science and Engineering from the University of North Carolina and a Ph.D. in Ecology from the University of Georgia, while completing research at the Coweeta Hydrologic Laboratory. He was later an Assistant Research Professor at Duke University, working on biogeochemical cycles in the Everglades of Florida. He has taught courses in Microbial Ecology, Wetland Ecology and Management, Forest and Range Soils, Soil Genesis and Classification, Natural Resource Ecology, Biodiversity, Conservation and Humans, and Ecology of Flowing Waters. After some early research work in ultraviolet light disinfection of bacteria and viruses, and kinetics of reactions of oxidizing chlorine with humic substances, his research work has centered on the biogeochemistry of forests, wetlands, and streams. This work has included dissolved organic matter dynamics, primary succession, and soil organic matter formation, root production during succession, formation and decomposition of humic substances, protein content in dissolved organic matter, the bioavailability of N and P in rivers and lakes, and ecophysiology of invasive plants. Dr. Qualls has also been a Visiting Professor at Yokohama National University in Japan. He received the "Pioneer of Disinfection Award" from the International Water Association and the Water Environment Association. He currently serves as the Editor for the journal Forests. Outside the academic arena, he is the current (2019) U.S.A. Outdoor Track and Field Champion in the 5000 and 10,000 meter runs in his age group. He, and the three other members of his relay team, hold a current world record in the 65–69 year age group for the 4 × 800 meter relay.

Preface to "Nutrient Cycling in Forest Ecosystems"

The long-term productivity of forest ecosystems depends on the cycling of nutrients. The effect of carbon dioxide fertilization on forest productivity may ultimately be limited by the rate of nutrient cycling. Contemporary and future disturbances such as climatic warming, N-deposition, deforestation, short rotation sylviculture, fire (both wild and controlled), and the invasion of exotic species all place strains on the integrity of ecosystem nutrient cycling. Global differences in climate, soils, and species make it difficult to extrapolate even a single important study worldwide.

Despite advances in the understanding of nutrient cycling and carbon production in forests, many questions remain.

The chapters in this volume reflect many contemporary research priorities. The thirteen studies in this volume are arranged in the following subject groups:

- N and P resorption from foliage worldwide, along chronosequences and along elevation gradients;

- Litter production and decomposition;

- N and P stoichiometry as affected by N deposition, geographic gradients, species changes, and ecosystem restoration;

- Effects of N and P addition on understory biomass, litter, and soil;

- Effects of burning on soil nutrients;

- Effects of N addition on soil fauna.

Robert G. Qualls
Editor

 *forests*

Article

Nitrogen and Phosphorus Resorption in Planted Forests Worldwide

Dalong Jiang [1,2], Qinghong Geng [1], Qian Li [3], Yiqi Luo [4], Jason Vogel [2], Zheng Shi [1], Honghua Ruan [1] and Xia Xu [1,]*

[1] College of Biology and the Environment, Co-Innovation Center for Sustainable Forestry in Southern China, Nanjing Forestry University, Nanjing 210037, China; dalong.jiang1988@outlook.com (D.J.); gengqh0127@163.com (Q.G.); zheng.shi@ou.edu (Z.S.); hhruan@njfu.edu.cn (H.R.)

[2] School of Forest Resources and Conservation, University of Florida, Gainesville, FL 32608, USA; jvogel@ufl.edu

[3] Advanced Analysis and Testing Center, Nanjing Forestry University, Nanjing 210037, China; liqian.1987@outlook.com

[4] Center for Ecosystem Science and Society, Northern Arizona University, Flagstaff, AZ 86011, USA; yiqi.luo@nau.edu

* Correspondence: xuxia.1982@njfu.edu.cn; Tel.: +86-25-8542-7086

Received: 31 January 2019; Accepted: 25 February 2019; Published: 26 February 2019

Abstract: Nutrient resorption from senescing leaves is one of the plants' essential nutrient conservation strategies. Parameters associated with resorption are important nutrient-cycling constraints for accurate predictions of long-term primary productivity in forest ecosystems. However, we know little about the spatial patterns and drivers of leaf nutrient resorption in planted forests worldwide. By synthesizing results of 146 studies, we explored nitrogen (N) and phosphorus (P) resorption efficiency (NRE and PRE) among climate zones and tree functional types, as well as the factors that play dominant roles in nutrient resorption in plantations globally. Our results showed that the mean NRE and PRE were 58.98% ± 0.53% and 60.21% ± 0.77%, respectively. NRE significantly increased from tropical to boreal zones, while PRE did not significantly differ among climate zones, suggesting differential impacts of climates on NRE and PRE. Plant functional types exert a strong influence on nutrient resorption. Conifer trees had higher PRE than broadleaf trees, reflecting the adaptation of the coniferous trees to oligotrophic habitats. Deciduous trees had lower PRE than evergreen trees that are commonly planted in P-limited low latitudes and have long leaf longevity with high nutrient use efficiency. While non-N-fixing trees had higher NRE than N-fixing trees, the PRE of non-N-fixing trees was lower than that of N-fixing trees, indicating significant impact of the N-fixing ability on the resorption of N and P. Our multivariate regression analyses showed that variations in NRE were mainly regulated by climates (mean annual precipitation and latitude), while variations in PRE were dominantly controlled by green leaf nutrient concentrations (N and P). Our results, in general, suggest that the predicted global warming and changed precipitation regimes may profoundly affect N cycling in planted forests. In addition, green leaf nutrient concentrations may be good indicators for PRE in planted forests.

Keywords: nutrient resorption; nitrogen and phosphorous; planted forests; climate zones; plant functional types; precipitation; green leaf nutrient

---

1. Introduction

Nutrient availability is a critical constraint in plant productivity and carbon (C) stocks in terrestrial ecosystems [1,2]. Nutrient resorption, through which nutrients are reabsorbed before leaf senescence and reused for plant growth directly, is an essential component of nutrient conservation strategies [3,4].

It affects key ecosystem processes such as nutrient uptake [5,6], plant competition [7,8], C cycling and resource-use efficiency [9,10], and hence productivity [1,11]. From a physiological perspective, resorption efficiency is an important issue in nutrient conservation [6,12]. In ecosystem modeling, reliable estimates of resorption efficiency are of key importance for modeling nutrient cycling and for quantifying ecosystem productivity, particularly in the new generation of coupled global models [13,14]. While previous synthesis studies mainly focus on the resorption efficiency of natural forests [1,4,5,9], we know little about nutrient resorption and its associated drivers in planted forests worldwide.

Increasing demands for timber products have promoted more research on plantations, which require more nutrients for their rapid growth [15,16]. Nutrient resorption in planted forests is less studied, since many synthesis studies excluded data of nutrient resorption from plantations [5,7,17]. Additionally, few nutrient-resorption studies have separated planted forests from other forest types [18–21]. Planted forests may have different nutrient resorption efficiencies in comparison to natural forests, since they are commonly pioneer tree species and grow on nutrient-poor soils [4,22]. Recent studies show that plantations had significantly higher nitrogen resorption efficiency (NRE) than the adjacent natural forests in northeast China and inland Hokkaido, Japan [16,23]. Phosphorous resorption efficiency (PRE) was found to be lower in plantations than in natural forests [24]. Plant productivity is largely constrained by nutrient availability [25,26], making exploration of nutrient resorption efficiency of high necessity in planted forests worldwide.

Empirical and regional studies have undoubtedly advanced our understanding of nutrient resorption in planted forests [4,27–30]. A synthesis study showed that while phosphorous resorption efficiency (PRE) was higher in subtropical than in temperate zones, NRE did not differ between the two climate zones in China's planted forests [24]. In contrary, Zhang et al. found that NRE increased and PRE first increased and then decreased along latitude in *Metasequoia glyptostroboides* forests in the east coast of China [27]. In terms of plant functional types, while Lal et al. found that deciduous species had higher NRE and PRE than evergreen species in a dry tropical environment [31], Machado et al. showed that evergreen species presented higher PRE, but not NRE, than deciduous species in plantations in the Brazilian Amazon [29]. Few studies have explored NRE in N-fixing trees vs. non-N-fixing trees in planted forests, though Yuan et al. found lower NRE in N-fixing trees than in the non-N-fixing trees (39% vs. 50%) in a semi-arid region of northern China [32]. A site study in the Karst ecosystem of southwestern China showed higher NRE and lower PRE for the conifer *Pinus yunnanensis* Franch. than for the broadleaf *Eucalyptus maideni* F. Muell. [33]. Much uncertainty still surrounds the global patterns of nutrient resorption efficiency of plantations across different climate zones and functional groups. This lack of clear understanding limits the accuracy in predicting long-term primary productivity in forest ecosystems. Therefore, it is urgent to examine nutrient resorption of planted forests regarding spatial patterns and drivers.

Planted forests, accounting for 7% of forest areas globally in 2015, are one of the essential components of terrestrial forest ecosystems providing us economic and social benefits [34,35]. Moreover, plantations have been advocated as important C sinks to mitigate future climate change [36,37]. Nutrient resorption enables plants to store reabsorbed nutrients for rapid and sustained growth at the beginning of the next growing season [38,39]. Understanding the nutrient resorption patterns in plantations is thus beneficial for improving nutrient conservation and management of planted forests. In this systematic review, we aimed to: (1) explore the global patterns of nutrient resorption (N and P) in planted forests and (2) identify various factors that play important roles in nutrient resorption worldwide.

2. Materials and Methods

2.1. Data Description

In this study, we compiled a dataset of 149 independent sampling sites based on 146 peer-reviewed papers about NRE and PRE in planted forests worldwide (Figure 1, Supplementary materials 1 and

2). In total, 643 observations for NRE and 539 observations for PRE were included in the dataset. We utilized the searching tools of ISI Web of Science and Google Scholar for retrieving articles on nutrient resorption, and the China National Knowledge Infrastructure (CNKI) to search for papers published in Chinese. Both methods included combinations of the terms 'nutrient resorption' or 'nutrient retranslocation' or 'nutrient reabsorption' and 'concentration', 'forest', and 'tree' as the searching keywords. We selected articles and extracted data (using Graph Digitizer 2.24, http://getdata-graph-digitizer.com/) with the following criteria: (1) N and P concentrations based on dry mass in green and senesced leaves were directly available or could be calculated based on presented tables or graphs; (2) we only selected data of trees from plots identified as planted forests; (3) any data from greenhouse, nursery, fertilized, and polluted sites were excluded; and (4) any data from cases of possible 'premature senescence' (e.g., drought, pests, …) were also eliminated.

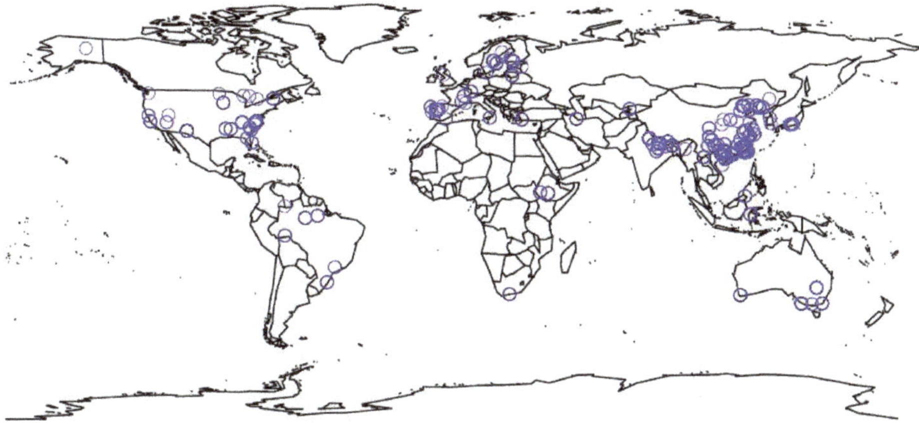

Figure 1. Global distribution of 149 independent sampling sites included in this global data synthesis.

NRE and PRE were estimated by percent reduction between green and senesced leaves [5,7,40,41]. However, to eliminate the underestimations of nutrient resorption, we used a mass loss correction factor (MLCF) with the calculation as follows [1,42]:

$$\text{NuRE } (\%) = \left(1 - \frac{\text{Nu}_{\text{senesced}}}{\text{Nu}_{\text{green}}} \text{MLCF} \right) \times 100\% \tag{1}$$

where NuRE is nutrient resorption efficiency; Nu_{green} and $\text{Nu}_{\text{senesced}}$ are nutrient concentrations in green and senesced leaves, respectively. The MLCF was calculated from the mass of green and senesced leaves or from the percentage of leaf mass loss during senescence. If green and senesced leaf mass and percentage of leaf mass loss were not available from a specific study, we used 0.745 for conifers, 0.780 for evergreen broad-leaved species, and 0.784 for deciduous broad-leaved species [1].

For studies that reported leaf nutrient concentrations during the growing season, we used maximum values for green leaf nutrient concentrations [5]. We also collected climate variables including latitude (°, LAT), longitude (°), mean annual temperature (°C, MAT), and mean annual precipitation (mm, MAP) and forest properties (e.g., tree species, functional groups, and stand ages) from those papers. Climate zones were subdivided into boreal, temperate, subtropical, and tropical regions [13]. Functional groups were separated into coniferous vs. broadleaf trees, deciduous vs. evergreen trees, and non-N-fixing vs. N-fixing trees.

2.2. Statistical Analysis

Before statistical analysis, the normality and homoscedasticity of the NRE and PRE were verified by the Kolmogorov–Smirnov test and Levene's test, respectively. One-way analysis of variance (ANOVA) was performed to analyze the differences between NRE and PRE. Linear mixed-effects models (LMEs) were applied to quantify the effects of climate zones and different functional groups on NRE and PRE worldwide. Climate zones and functional groups were treated as fixed factors, and studies were treated as a random factor. The significance of fixed effects terms was assessed via ANOVA with F tests. If the differences were significant, post hoc multiple comparisons were subsequently conducted using Duncan's test ($p = 0.05$). Stepwise regression analyses were used to explore the controls on the variation of NRE and PRE associated with climate variables, leaf nutrient status, and stand ages. All statistical analyses were performed using SPSS 22.0 for Windows (SPSS Inc., Chicago, IL, USA).

3. Results

3.1. Nutrient Resorption Patterns in Planted Forests

Globally, mean values of NRE and PRE in the planted forests were 58.98% ± 0.53% and 60.21% ± 0.77%, respectively, with no significant differences between them ($p > 0.05$, Figure 2). Generally, variations in both NRE and PRE showed remarkable differences among different climate zones and tree functional groups. While NRE significantly increased with increasing latitude ($p < 0.05$, Table 1, Figure 3a), PRE did not differ among climate zones ($p > 0.05$, Table 1, Figure 3b). Additionally, NRE did not differ between conifer and broadleaf trees ($p > 0.05$, Table 1, Figure 4a) and the PRE of conifer trees was higher than that of broadleaf trees ($p < 0.05$, Table 1, Figure 4b). Though NRE did not differ between deciduous and evergreen trees ($p > 0.05$, Table 1, Figure 4a), PRE was significantly higher in evergreen than deciduous trees ($p < 0.001$, Table 1, Figure 4b). The NRE of N-fixing trees was much lower than that of non-N-fixing trees ($p < 0.001$, Table 1, Figure 4a) and the PRE of N-fixing trees was much higher than that of non-N-fixing trees ($p < 0.05$, Table 1, Figure 4b).

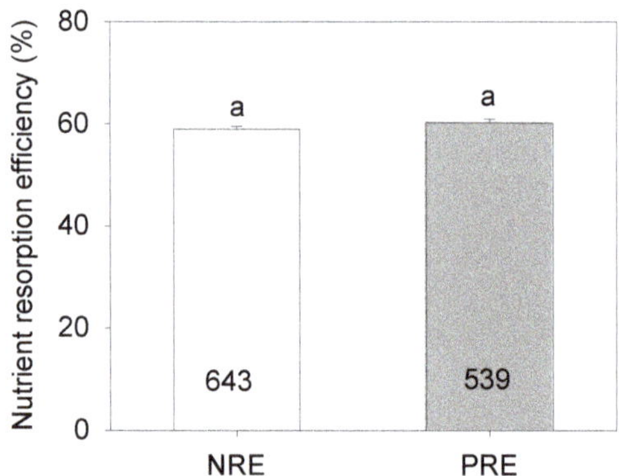

Figure 2. N and P resorption efficiency (%, NRE and PRE) in planted forests (mean ± SE). Values in the bars represent the number of observations. Different lowercase letters represent statistical differences at $p = 0.05$ between NRE and PRE.

Table 1. ANOVA table of the linear mixed-effects models for N and P resorption efficiency (%, NRE and PRE). *: <0.05, ***: <0.001.

Resorption	Fixed Term	df	F, p
NRE	Climate zones	3	3.56 *
	Broadleaf vs. Conifer	1	0.56
	Deciduous vs. Evergreen	1	2.48
	Non-N-fixing tree vs. N-fixing tree	1	15.85 ***
PRE	Climate zones	3	0.79
	Broadleaf vs. Conifer	1	4.74 *
	Deciduous vs. Evergreen	1	12.93 ***
	Non-N-fixing tree vs. N-fixing tree	1	5.06 *

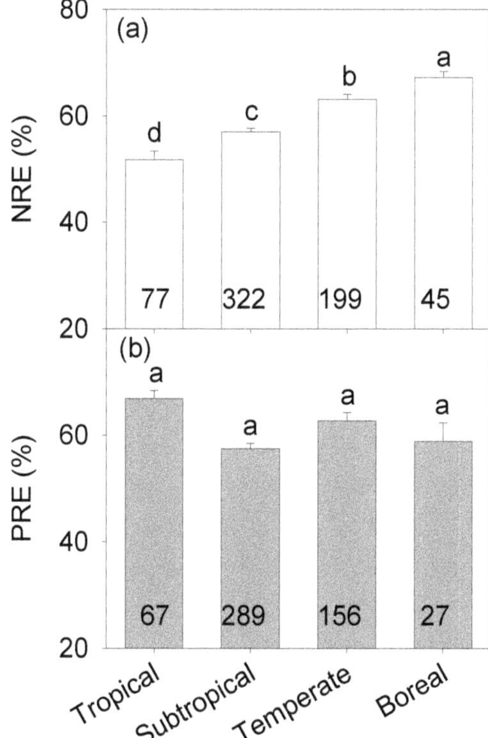

Figure 3. N and P resorption efficiency (%, NRE (**a**) and PRE (**b**)) among different climate zones in planted forests (mean ± SE). Values in the bars represent the number of observations. Different lowercase letters represent statistical differences at $p = 0.05$ among the climate zones.

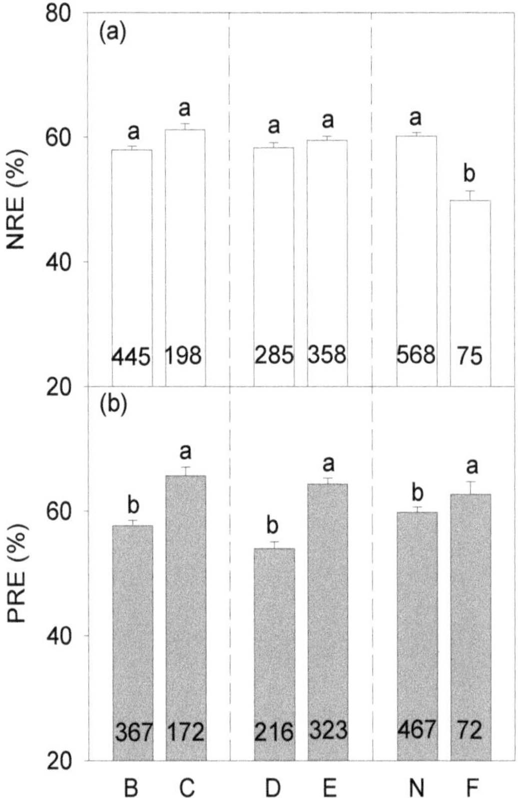

Figure 4. N and P resorption efficiency (%, NRE (**a**) and PRE (**b**)) between broadleaf (B) vs. conifer (C), deciduous (D) vs. evergreen (E), and non-N-fixing (N) vs. N-fixing (F) tree groups in planted forests (mean $\pm$ SE). Values in the bars represent the number of observations. Different lowercase letters represent statistical differences at $p = 0.05$ between the different functional types.

3.2. Controls of Nutrient Resorption in Planted Forests

Variations in NRE and PRE at the global scale were not controlled by the same explanatory variables. Our multivariate regression analyses showed that NRE was regulated negatively by MAP and positively by green leaf P concentration (P_{green}) and latitude (LAT) ($p < 0.001$, Table 2). For PRE, green leaf N concentration (N_{green}) and P_{green} had positively and negatively impacted on it, respectively ($p < 0.001$, Table 2). The stand ages of forests had little impact on both NRE and PRE.

Table 2. Stepwise regressions of N and P resorption efficiency (%, NRE and PRE) with stand age (year, Age), absolute latitude (°, LAT), mean annual precipitation (mm, MAP), and green leaf N and P concentrations (mg g^{-1}, N_{green} and P_{green}, respectively) in planted forests. ***: <0.001.

	Variables	Regression	n	r^2	Excluded Variables
	MAP	$y = 67.48 - 0.01 \times MAP$	476	0.09 ***	Age, N_{green}, P_{green}, LAT
NRE	MAP, P_{green}	$y = 62.65 - 0.01 \times MAP + 2.70 \times P_{green}$	476	0.12 ***	Age, N_{green}, LAT
	MAP, P_{green}, LAT	$y = 54.68 - 0.01 \times MAP + 2.36 \times P_{green} + 0.21 \times LAT$	476	0.14 ***	Age, N_{green}
PRE	P_{green}	$y = 66.11 - 4.12 \times P_{green}$	476	0.05 ***	Age, N_{green}, LAT, MAP
	P_{green}, N_{green}	$y = 60.31 - 5.33 \times P_{green} + 0.41 \times N_{green}$	476	0.07 ***	Age, LAT, MAP

4. Discussion

4.1. Nutrient Resorption Patterns in Planted Forests

Our results showed that the mean values of NRE and RPE were 58.98% $\pm$ 0.53% and 60.21% $\pm$ 0.77%, respectively, for the planted forests worldwide—much higher than the assumed 50% for both NRE and PRE in most models [1]. Consistent with Zhang et al. that the NRE of *Metasequoia glyptostroboides* increased along latitude at the regional scale [27], we found that NRE significantly increased from tropical to boreal zones with increasing latitude. This pattern may result from three main reasons associated with low temperatures at high latitudes. First, decomposition and nutrient mineralization are high at low latitudes, which may subsequently enhance the availability of soil N [43,44]. Second, warmer temperatures and higher water availability could stimulate nutrient movement as well as root N uptake at low latitudes [45]. In high-latitude regions, however, the delivery of mass-flow nutrients from the soil to plant tissues is inhibited, which in turn, may reduce the process of nutrient resorption [20,46]. Third, the metabolic activity of plants may be inhibited at high latitudes [27]. Nutrient resorption is recognized as one of the efficient nutrient conservation mechanisms for plants to adapt to environmental conditions [12]. Plants at high latitudes usually grow quickly in order to finish development in relatively short growing seasons via high nutrient retrieval strategies that can enhance growth rate and lower dependence on the supply of soil nutrients [1,5,47]. Interestingly, PRE did not linearly increase or decrease from tropical to boreal zones, which is in line with the findings of Zhang et al., who observed a curved correlation of PRE with latitude [27]. PRE may largely depend on P availability in soils [13,48]. For example, a negative correlation between PRE and soil P was found in planted forests [27,33,49]. Those results suggest that NRE and PRE are differentially regulated by climate and soil nutrient availability.

The resorption of N and P varied among functional groups. First, the PREs of coniferous trees were significantly higher than those of broadleaf trees. Similar observations were also found in natural forests [18,20]. Coniferous trees are usually planted in nutrient-poor environments. The higher nutrient resorption could thus help them survive via reducing their dependence on soil nutrient supplies [16,50]. High nutrient resorption reflects the adaptation of plants to oligotrophic habitats [10,51]. Interestingly, NRE did not differ between coniferous and broadleaf trees. This may result from the limited number of NRE observations in coniferous planted forests, as also reported by Yuan and Chen [5]. Second, we also found that evergreen tree species presented higher PRE, whereas the NRE of deciduous species did not differ from that of the evergreen species, which is completely in line with the findings in the forest plantations in the Brazilian Amazon [29]. Possible reasons may include that evergreen species (1) are mostly planted in P-limited low latitudes [3,20] and (2) have high nutrient use efficiency and thus low nutrient loss rates associated with prolonged nutrient retention time in leaf biomass and leaf longevity [12,52]. Third, our results showed that N-fixing trees had significantly lower NRE but higher PRE than non-N-fixing trees. N-fixing trees can acquire N from the atmosphere, and they are thus less dependent on the internal N cycling process [53,54]. However, N-fixing trees may contain higher leaf N and may thus enhance PRE in order to maintain N to P stoichiometric homeostasis. These findings indicate that plant functional types exert a strong influence on nutrient resorption.

4.2. Controls of Nutrient Resorption in Planted Forests

In this study, the global-scale NRE and PRE patterns were not mainly regulated by the same ecological factors. Our multivariate regression analyses showed that NRE was primarily controlled by climate variables, supporting the observations of some previous studies [1,20,27,48]. Specially, we found that NRE was regulated negatively by MAP and positively by LAT. Climates (e.g., temperature and precipitation) most commonly affect soil N availability via microbial activities [44]. That is, soil N availability is lower under hostile climates (e.g., dry and cold at high latitude) than under climates that are favored by microbes (e.g., wet and warm at low latitude). Soil nutrient availability usually has a negative impact on nutrient resorption efficiency [41,55], leading to the phenomena that NRE correlated positively with LAT and negatively with MAP. Contrary to what we found here, Achat et al. showed that climates had only minor effects on the NRE of forests in 102 permanent forest sites across France [9]. The gap may result from the rather narrow range of climatic conditions in the study by Achat et al. [9] compared with the global climate scale considered in our study. In addition, NRE was also regulated positively by green leaf P (P_{green}). N and P in plants are usually closely related, which means changes of one nutrient could alter the other [56,57]. For example, plants may elevate NRE to maintain their N and P stoichiometric homeostasis by increasing P_{green}, and vice versa. Generally, our results suggest that the predicted global warming and changed precipitation regimes may profoundly affect N cycling in planted forests.

For the resorption of P, N_{green} had positive impacts on it, which is in accordance with the results of Wang et al. [58] and Yan et al. [4], suggesting that coupled relationships between N and P may exist in planted forests. Furthermore, we found that P_{green} had negative effects on PRE, which is supported by the findings by Kobe et al. [59]. Green leaf nutrient status is supposed to represent soil nutrient availability, whose negative impacts on nutrient resorption efficiency is widely accepted [41,59,60]. Therefore, green leaf nutrient status may negatively relate to nutrient resorption efficiency. We also found that climate variables were not the main explanatory factors for PRE, supporting the idea that soil properties and soil parent materials other than climate had more impact on soil P availability [48]. Green leaf nutrient concentration may be a good indicator for PRE in planted forests.

5. Conclusions

To our best knowledge, this synthesis provides the first comprehensive analysis of NRE and PRE in planted forests worldwide. Generally, NRE significantly increased along climate zones, while no significant differences were observed for PRE among climate zones. These results suggest differential impacts of climates on NRE and PRE. In terms of plant functional groups, the PRE of conifer trees was higher than that of broadleaf trees; evergreen trees had higher PRE than deciduous trees; non-N-fixing trees had higher NRE but lower PRE than N-fixing trees. These findings indicate that plant functional types exert strong impacts on nutrient resorption. Additionally, multivariate regression analyses showed that variations in NRE were mainly regulated by mean annual precipitation (MAP) and latitude, indicating that the predicted global warming and changed precipitation regimes may profoundly affect N cycling in planted forests. Variations in PRE were dominantly controlled by green leaf nutrient concentrations (N and P), suggesting green leaf nutrient concentrations may be good indicators for PRE in planted forests.

Supplementary Materials: The following are available online at http://www.mdpi.com/1999-4907/10/3/201/s1.

Author Contributions: Conceptualization: X.X.; Data curation: D.J.; Formal analysis: D.J.; Funding acquisition: D.J., H.R., and X.X.; Project administration: X.X.; Supervision: X.X.; Writing—original draft: D.J.; Writing—review and editing: Q.G., Q.L., Y.L., J.V., Z.S., H.R., and X.X.

Funding: This study was financially supported by the National Key Research and Development Program of China (2016YFD0600204), the Recruitment Program for Young Professionals (Thousand Youth Talents Plan), the Jiangsu Specially-Appointed Professors Program, the Six Talent Peaks Program of Jiangsu Province (JY-041& TD-XYDXX-006), the Nanjing Forestry University Science Fund for Distinguished Young Scholars (the "5151" Talent Program), the Jiangsu Collegiate Science and Technology Fund for the Excellent Innovative Research Teams, the Priority Academic Program Development of Jiangsu Higher Education Institutions (PAPD), the Doctorate Fellowship Foundation of Nanjing Forestry University, the Research Innovation Program for College Graduates of Jiangsu Province (KYLX16_0833), and the Scientific and Technological Innovation Program for College Students of Nanjing Forestry University (DXSKC-201617).

Acknowledgments: We thank many lab members for their help with the manuscript preparation.

Conflicts of Interest: The authors declare no conflict of interest.

References

1. Vergutz, L.; Manzoni, S.; Porporato, A.; Novais, R.F.; Jackson, R.B. Global resorption efficiencies and concentrations of carbon and nutrients in leaves of terrestrial plants. *Ecol. Monogr.* **2012**, *82*, 205–220. [CrossRef]
2. Sokolov, A.P.; Kicklighter, D.W.; Melillo, J.M.; Felzer, B.S.; Schlosser, C.A.; Cronin, T.W. Consequences of considering carbon-nitrogen interactions on the feedbacks between climate and the terrestrial carbon cycle. *J. Clim.* **2008**, *21*, 3776–3796. [CrossRef]
3. Aerts, R. Nutrient Resorption from Senescing Leaves of Perennials: Are there General Patterns? *J. Ecol.* **1996**, *84*, 597–608. [CrossRef]
4. Yan, T.; Lu, X.T.; Zhu, J.J.; Yang, K.; Yu, L.Z.; Gao, T. Changes in nitrogen and phosphorus cycling suggest a transition to phosphorus limitation with the stand development of larch plantations. *Plant Soil* **2018**, *422*, 385–396. [CrossRef]
5. Yuan, Z.Y.; Chen, H.Y.H. Global-scale patterns of nutrient resorption associated with latitude, temperature and precipitation. *Glob. Ecol. Biogeogr.* **2009**, *18*, 11–18. [CrossRef]
6. Wang, Z.Q.; Fan, Z.X.; Zhao, Q.; Wang, M.C.; Ran, J.Z.; Huang, H.; Niklas, K.J. Global data analysis shows that soil nutrient levels dominate foliar nutrient resorption efficiency in herbaceous species. *Front. Plant Sci.* **2018**, *9*, 1–8. [CrossRef] [PubMed]
7. Killingbeck, K.T. Nutrients in senesced leaves: Keys to the search for potential resorption and resorption proficiency. *Ecology* **1996**, *77*, 1716–1727. [CrossRef]
8. Schmidt, M.; Veldkamp, E.; Corre, M.D. Tree-microbial biomass competition for nutrients in a temperate deciduous forest, central Germany. *Plant Soil* **2016**, *408*, 227–242. [CrossRef]
9. Achat, D.L.; Pousse, N.; Nicolas, M.; Augusto, L. Nutrient remobilization in tree foliage as affected by soil nutrients and leaf life span. *Ecol. Monogr.* **2018**, *88*, 408–428. [CrossRef]
10. Aerts, R.; Chapin, F.S. The mineral nutrition of wild plants revisited: A re-evaluation of processes and patterns. *Adv. Ecol. Res.* **2000**, *30*, 1–67. [CrossRef]
11. Deng, M.F.; Liu, L.L.; Jiang, L.; Liu, W.X.; Wang, X.; Li, S.P.; Yang, S.; Wang, B. Ecosystem scale trade-off in nitrogen acquisition pathways. *Nat. Ecol. Evol.* **2018**, *2*, 1724–1734. [CrossRef] [PubMed]
12. Brant, A.N.; Chen, H.Y.H. Patterns and mechanisms of nutrient resorption in plants. *Crit. Rev. Plant Sci.* **2015**, *34*, 471–486. [CrossRef]
13. Yuan, Z.Y.; Chen, H.Y.H. Negative effects of fertilization on plant nutrient resorption. *Ecology* **2015**, *96*, 373–380. [CrossRef] [PubMed]
14. Sun, Z.Z.; Liu, L.L.; Peng, S.S.; Penuelas, J.; Zeng, H.; Piao, S.L. Age-related modulation of the nitrogen resorption efficiency response to growth requirements and soil nitrogen availability in a temperate Pine plantation. *Ecosystems* **2016**, *19*, 698–709. [CrossRef]
15. Sedjo, R.A. The potential of High-Yield Plantation Forestry for Meeting Timber Needs. In *Planted Forests: Contributions to the Quest for Sustainable Societies*; Boyle, J.R., Winjum, J.K., Kavanagh, K., Jensen, E.C., Eds.; Springer: Dordrecht, The Netherlands, 1999; pp. 339–359.
16. Yan, T.; Lu, X.T.; Yang, K.; Zhu, J.J. Leaf nutrient dynamics and nutrient resorption: A comparison between larch plantations and adjacent secondary forests in Northeast China. *J. Plant Ecol.* **2016**, *9*, 165–173. [CrossRef]
17. Sun, X.; Kang, H.; Chen, H.Y.H.; Bjorn, B.; Samuel, B.F.; Liu, C. Biogeographic patterns of nutrient resorption from *Quercus variabilis* Blume leaves across China. *Plant Biol.* **2016**, *18*, 505–513. [CrossRef] [PubMed]
18. Tang, L.Y.; Han, W.X.; Chen, Y.H.; Fang, J.Y. Resorption proficiency and efficiency of leaf nutrients in woody plants in eastern China. *J. Plant Ecol.* **2013**, *6*, 408–417. [CrossRef]
19. Zhang, H.Y.; Lu, X.T.; Hartmann, H.; Keller, A.; Han, X.G.; Trumbore, S.; Phillips, R.P. Foliar nutrient resorption differs between arbuscular mycorrhizal and ectomycorrhizal trees at local and global scales. *Glob. Ecol. Biogeogr.* **2018**, *27*, 875–885. [CrossRef]
20. Yan, T.; Zhu, J.J.; Yang, K. Leaf nitrogen and phosphorus resorption of woody species in response to climatic conditions and soil nutrients: A meta-analysis. *J. For. Res.* **2018**, *29*, 905–913. [CrossRef]
21. Han, W.X.; Tang, L.Y.; Chen, Y.H.; Fang, J.Y. Relationship between the relative limitation and resorption efficiency of nitrogen vs phosphorus in woody plants. *PLoS ONE* **2013**, *8*, e83366. [CrossRef] [PubMed]

22. Zhang, H.; Wang, J.N.; Wang, J.Y.; Guo, Z.W.; Wang, G.G.; Zeng, D.H.; Wu, T.G. Tree stoichiometry and nutrient resorption along a chronosequence of *Metasequoia glyptostroboides* forests in coastal China. *For. Ecol. Manag.* **2018**, *430*, 445–450. [CrossRef]

23. Maeda, Y.; Tashiro, N.; Enoki, T.; Urakawa, R.; Hishi, T. Effects of species replacement on the relationship between net primary production and soil nitrogen availability along a topographical gradient: Comparison of belowground allocation and nitrogen use efficiency between natural forests and plantations. *For. Ecol. Manag.* **2018**, *422*, 214–222. [CrossRef]

24. Xu, S.; Zhou, G.; Tang, X.; Wang, W.; Wang, G.; Ma, K.; Han, S.; Du, S.; Li, S.; Yan, J.; et al. Different spatial patterns of nitrogen and phosphorus resorption efficiencies in China's forests. *Sci. Rep.* **2017**, *7*, 10584. [CrossRef] [PubMed]

25. Yuan, Z.Y.; Li, L.H. Soil water status influences plant nitrogen use: A case study. *Plant Soil* **2007**, *301*, 303–313. [CrossRef]

26. Austin, A.T.; Yahdjian, L.; Stark, J.M.; Belnap, J.; Porporato, A.; Norton, U.; Ravetta, D.A.; Schaeffer, S.M. Water pulses and biogeochemical cycles in arid and semiarid ecosystems. *Oecologia* **2004**, *141*, 221–235. [CrossRef] [PubMed]

27. Zhang, H.; Guo, W.H.; Yu, M.K.; Wang, G.G.; Wu, T.G. Latitudinal patterns of leaf N, P stoichiometry and nutrient resorption of *Metasequoia glyptostroboides* along the eastern coastline of China. *Sci. Total Environ.* **2018**, *618*, 1–6. [CrossRef] [PubMed]

28. Zhou, L.L.; Addo-Danso, S.D.; Wu, P.F.; Li, S.B.; Zou, X.H.; Zhang, Y.; Ma, X.Q. Leaf resorption efficiency in relation to foliar and soil nutrient concentrations and stoichiometry of *Cunninghamia lanceolata* with stand development in southern China. *J. Soils Sediments* **2016**, *16*, 1448–1459. [CrossRef]

29. Machado, M.R.; Sampaio, P.D.; Ferraz, J.; Camara, R.; Pereira, M.G. Nutrient retranslocation in forest species in the Brazilian Amazon. *Acta Sci.-Agron.* **2016**, *38*, 93–101. [CrossRef]

30. Ye, G.F.; Zhang, S.J.; Zhang, L.H.; Lin, Y.M.; Wei, S.D.; Liao, M.M.; Lin, G.H. Age-related changes in nutrient resorption patterns and tannin concentration of *Casuarina equisetifolia* plantations. *J. Trop. For. Sci.* **2012**, *24*, 546–556. [CrossRef]

31. Lal, C.B.; Annapurna, C.; Raghubanshi, A.S.; Singh, J.S. Foliar demand and resource economy of nutrients in dry tropical forest species. *J. Veg. Sci.* **2001**, *12*, 5–14. [CrossRef]

32. Yuan, Z.Y.; Li, L.H.; Han, X.G.; Huang, J.H.; Jiang, G.M.; Wan, S.Q.; Zhang, W.H.; Chen, Q.S. Nitrogen resorption from senescing leaves in 28 plant species in a semi-arid region of northern China. *J. Arid Environ.* **2005**, *63*, 191–202. [CrossRef]

33. Pang, D.B.; Wang, G.Z.; Li, G.J.; Sun, Y.L.; Liu, Y.G.; Zhou, J.X. Ecological stoichiometric characteristics of two typical plantations in the Karst ecosystem of southwestern China. *Forests* **2018**, *9*, 56. [CrossRef]

34. MacDicken, K.; Jonsson, Ö.; Piña, L.; Marklund, L.; Maulo, S.; Contessa, V.; Adikari, Y.; Garzuglia, M.; Lindquist, E.; Reams, G.; et al. *Global Forest Resources Assessment 2015: How Are the World'S Forests Changing?* Food and Agriculture Organization of the United Nations (FAO): Rome, Italy, 2015; pp. 1–54.

35. Van Dijk, A.I.J.M.; Keenan, R.J. Planted forests and water in perspective. *For. Ecol. Manag.* **2007**, *251*, 1–9. [CrossRef]

36. Fang, J.Y.; Chen, A.P.; Peng, C.H.; Zhao, S.Q.; Ci, L.J. Changes in forest biomass carbon storage in China between 1949 and 1998. *Science* **2001**, *292*, 2320–2322. [CrossRef] [PubMed]

37. Goodale, C.L.; Apps, M.J.; Birdsey, R.A.; Field, C.B.; Heath, L.S.; Houghton, R.A.; Jenkins, J.C.; Kohlmaier, G.H.; Kurz, W.; Liu, S.; et al. Forest carbon sinks in the northern Hemisphere. *Ecol. Appl.* **2002**, *12*, 891–899. [CrossRef]

38. Netzer, F.; Schmid, C.; Herschbach, C.; Rennenberg, H. Phosphorus-nutrition of European beech (*Fagus sylvatica* L.) during annual growth depends on tree age and P-availability in the soil. *Environ. Exp. Bot.* **2017**, *137*, 194–207. [CrossRef]

39. Weatherall, A.; Proe, M.F.; Craig, J.; Cameron, A.D.; Midwood, A.J. Internal cycling of nitrogen, potassium and magnesium in young Sitka spruce. *Tree Physiol.* **2006**, *26*, 673–680. [CrossRef] [PubMed]

40. Chen, F.S.; Niklas, K.J.; Liu, Y.; Fang, X.M.; Wan, S.Z.; Wang, H. Nitrogen and phosphorus additions alter nutrient dynamics but not resorption efficiencies of Chinese fir leaves and twigs differing in age. *Tree Physiol.* **2015**, *35*, 1106–1117. [CrossRef] [PubMed]

41. Reed, S.C.; Townsend, A.R.; Davidson, E.A.; Cleveland, C.C. Stoichiometric patterns in foliar nutrient resorption across multiple scales. *New Phytol.* **2012**, *196*, 173–180. [CrossRef] [PubMed]

42. Van Heerwaarden, L.M.; Toet, S.; Aerts, R. Current measures of nutrient resorption efficiency lead to a substantial underestimation of real resorption efficiency: Facts and solutions. *Oikos* **2003**, *101*, 664–669. [CrossRef]

43. Oleksyn, J.; Reich, P.B.; Zytkowiak, R.; Karolewski, P.; Tjoelker, M.G. Nutrient conservation increases with latitude of origin in European *Pinus sylvestris* populations. *Oecologia* **2003**, *136*, 220–235. [CrossRef] [PubMed]

44. Bengtson, P.; Falkengren-Grerup, U.; Bengtsson, G. Relieving substrate limitation-soil moisture and temperature determine gross N transformation rates. *Oikos* **2005**, *111*, 81–90. [CrossRef]

45. Reich, P.B.; Oleksyn, J. Global patterns of plant leaf N and P in relation to temperature and latitude. *Proc. Natl. Acad. Sci. USA* **2004**, *101*, 11001–11006. [CrossRef] [PubMed]

46. Cramer, M.D.; Hawkins, H.J.; Verboom, G.A. The importance of nutritional regulation of plant water flux. *Oecologia* **2009**, *161*, 15–24. [CrossRef] [PubMed]

47. Wright, I.J.; Westoby, M. Nutrient concentration, resorption and lifespan leaf traits of Australian sclerophyll species. *Funct. Ecol.* **2003**, *17*, 10–19. [CrossRef]

48. Augusto, L.; Achat, D.L.; Jonard, M.; Vidal, D.; Ringeval, B. Soil parent material-A major driver of plant nutrient limitations in terrestrial ecosystems. *Glob. Chang. Biol.* **2017**, *23*, 3808–3824. [CrossRef] [PubMed]

49. Zeng, F.P.; Chen, X.; Huang, B.; Chi, G.Y. Distribution changes of phosphorus in soil-plant systems of larch plantations across the chronosequence. *Forests* **2018**, *9*, 563. [CrossRef]

50. Yuan, Z.Y.; Li, L.H.; Han, X.G.; Huang, J.H.; Wan, S.Q. Foliar nitrogen dynamics and nitrogen resorption of a sandy shrub *Salix gordejevii* in northern China. *Plant Soil* **2005**, *278*, 183–193. [CrossRef]

51. Liu, J.T.; Gu, Z.J.; Shao, H.B.; Zhou, F.; Peng, S.Y. N–P stoichiometry in soil and leaves of *Pinus massoniana* forest at different stand ages in the subtropical soil erosion area of China. *Environ. Earth Sci.* **2016**, *75*, 1091. [CrossRef]

52. Escudero, A.; Del Arco, J.M.; Sanz, I.C.; Ayala, J. Effects of leaf longevity and retranslocation efficiency on the retention time of nutrients in the leaf biomass of different woody species. *Oecologia* **1992**, *90*, 80–87. [CrossRef] [PubMed]

53. Norris, M.D.; Reich, P.B. Modest enhancement of nitrogen conservation via retranslocation in response to gradients in N supply and leaf N status. *Plant Soil* **2009**, *316*, 193–204. [CrossRef]

54. Zhao, G.S.; Shi, P.L.; Wu, J.S.; Xiong, D.P.; Zong, N.; Zhang, X.Z. Foliar nutrient resorption patterns of four functional plants along a precipitation gradient on the Tibetan Changtang Plateau. *Ecol. Evol.* **2017**, *7*, 7201–7212. [CrossRef] [PubMed]

55. Hayes, P.; Turner, B.L.; Lambers, H.; Laliberte, E. Foliar nutrient concentrations and resorption efficiency in plants of contrasting nutrient-acquisition strategies along a 2-million-year dune chronosequence. *J. Ecol.* **2014**, *102*, 396–410. [CrossRef]

56. Deng, M.F.; Liu, L.L.; Sun, Z.Z.; Piao, S.L.; Ma, Y.C.; Chen, Y.W.; Wang, J.; Qiao, C.L.; Wang, X.; Li, P. Increased phosphate uptake but not resorption alleviates phosphorus deficiency induced by nitrogen deposition in temperate *Larix principis-rupprechtii* plantations. *New Phytol.* **2016**, *212*, 1019–1029. [CrossRef] [PubMed]

57. Yu, Q.; Chen, Q.; Elser, J.J.; He, N.; Wu, H.; Zhang, G.; Wu, J.; Bai, Y.; Han, X. Linking stoichiometric homoeostasis with ecosystem structure, functioning and stability. *Ecol. Lett.* **2010**, *13*, 1390–1399. [CrossRef] [PubMed]

58. Wang, Z.N.; Lu, J.Y.; Yang, H.M.; Zhang, X.; Luo, C.L.; Zhao, Y.X. Resorption of nitrogen, phosphorus and potassium from leaves of lucerne stands of different ages. *Plant Soil* **2014**, *383*, 301–312. [CrossRef]

59. Kobe, R.K.; Lepczyk, C.A.; Iyer, M. Resorption efficiency decreases with increasing green leaf nutrients in a global data set. *Ecology* **2005**, *86*, 2780–2792. [CrossRef]

60. Tsujii, Y.; Onoda, Y.; Kitayama, K. Phosphorus and nitrogen resorption from different chemical fractions in senescing leaves of tropical tree species on Mount Kinabalu, Borneo. *Oecologia* **2017**, *185*, 171–180. [CrossRef] [PubMed]

Article

Nitrogen and Phosphorus Resorption in Relation to Nutrition Limitation along the Chronosequence of Black Locust (*Robinia pseudoacacia* L.) Plantation

Jian Deng [1,2,*], **Sha Wang** [2], **Chengjie Ren** [3], **Wei Zhang** [3], **Fazhu Zhao** [4], **Xianfang Li** [1,2], **Dan Zhang** [2], **Xinhui Han** [3] and **Gaihe Yang** [3]

[1] Shaanxi Key Laboratory of Chinese Jujube, Yan'an University, Yan'an 716000, China; lixianfang810702@163.com

[2] College of Life Science, Yan'an University, Yan'an 716000, China; wwang97625@163.com (S.W.); xtmy1008@126.com (D.Z.)

[3] College of Agronomy, Northwest A&F University, Yangling 712100, China; rencj1991@nwsuaf.edu.cn (C.R.); zwgwyd@163.com (W.Z.); hanxinhui@nwsuaf.edu.cn (X.H.); ygh@nwsuaf.edu.cn (G.Y.)

[4] College of Urban and Environmental Science, Northwest University, Xi'an 710069, China; zhaofazhu@nwu.edu.cn

* Correspondence: dengjian@yau.edu.cn; Tel.: +86-0911-2332030

Received: 2 February 2019; Accepted: 12 March 2019; Published: 15 March 2019

Abstract: Plant nitrogen (N) and phosphorus (P) resorption is an important strategy to conserve N and P in the face of nutrient limitation. However, little is known about the variation of N and P resorption efficiency (NRE and PRE) and their correlation with leaves and soil C:N:P stoichiometry in black locust forests (*Robinia pseudoacacia* L.) of different ages. In this study, we measured C, N, and P concentrations in soil, green leaves, and senesced leaves from black locust forests of different ages (i.e, 10-, 20-, 30-, 36-, and 45-year-old), and calculated the NRE, PRE, and C:N:P stoichiometry ratios. The NRE and PRE tended to increase and then decrease with stand age, ranging from 46.8% to 57.4% and from 37.4% to 58.5%, with averages of 52.61 and 51.89, respectively. The PRE:NRE decreased with increased stand ages. The C:P and N:P of soil and green leaves increased with stand ages, indicating the increase of P limitation. In the senesced leaves, C:P and N:P were lower than in green leaves and first increased and then decreased with stand age. The PRE was significantly negatively correlated with the C:P and N:P of soil and green leaves. The NRE was significantly correlated with the C concentration of green leaves, P of the senesced leaves, and C:N. Results suggested that the NRE and PRE responded differently to soil and plant nutrients in black locust forests of different ages. In addition, the black locust plantations would alter the conservation and use strategy of nutrients in the ecosystem through a plant-mediated pathway. Future studies should elucidate the central nutrient utilization strategy of black locust in response to a nutrient-poor environment and determine how it is involved in regulating nutrient resorption.

Keywords: nutrition resorption; ecological stoichiometry; plant-soil feedback; stand age; *Robinia pseudoacacia* L.

---

1. Introduction

Plant nutrient resorption is the process by which a plant transfers nutrients to its growing tissues before the branches and leaves wither, effectively lengthening the retention time of nutrients in plants [1,2]. This process reduces the reliance of plants on soil nutrients, helping to maintain the ecological stoichiometry balance in plants; it also reduces nutrient leaching after litter decomposition, minimizing the ecosystem nutrient loss [2,3]. Nutrient resorption from senescing organs is one of the most important mechanisms for plants to adapt to nutrient limitation and improve nutrient utilization efficiency through conserving and optimizing the use of nutrients. This is also an important strategy to regulate nutrient cycling within plant–soil systems. Nutrient resorption can be defined as nutrient resorption efficiency (NuRE), i.e, the proportion of nutrients withdrawn from green leaves prior to senescence [4]. Therefore, research on the plant NuRE is helpful to reveal the response mechanism of plants to a nutrient-poor environment.

Nitrogen (N) and phosphorus (P), strong limiting factors of natural ecosystems, play an important role in plant growth and metabolism [5,6]. Previous studies have reported that natural forests in low latitudes areas (e.g., tropical forests) are usually limited by N [7] and those in high latitudes are limited by P [6]. In addition, chronosequence studies have shown that young forests are usually vulnerable to N limitation, while ageing forests are usually vulnerable to P limitation, especially in areas with nutrient deficiency or imbalanced N and P input [8,9]. Plant N and P resorption efficiency (NRE and PRE, respectively) attract ample attention because they play an important role in revealing the adaptability of plants to a nutrient limiting environment [10,11]. Vergutz estimated that the mean NRE and PRE for global terrestrial plants were 62.1% and 64.9%, respectively [12]. However, the NRE and PRE varies greatly depending on the species [13] and life-form [14]; NRE and PRE are also easily influenced by environmental factors (e.g., soil nutrient availability, elevation, temperature, precipitation, etc.) [13,15,16] or disturbances [17].

In forest plantations, plant photosynthetic characteristics, soil nutrient supply, and plant nutrition demands usually vary with stand age [18]. These changes may have substantial effects on plant nutrient conservation strategies, thus leading to the change of NuRE with stand age [10,19]. However, previous studies have shown inconsistent trends (e.g., increase [18], decrease [20], and no significant change [21]) of NuRE in response to stand age, which may be related to the tree species and nutrition status [10,21]. The measurement of C:N:P stoichiometry is an effective way to investigate plant and soil nutrient status. For instance, leaf N:P was used in a previous study to indicate the N (N:P < 14) or P (N:P > 16) limitation of the plants [22]. Therefore, studying NRE and PRE characteristics and their possible links with nutrition stoichiometry, together with forest age changes, may help to reveal the adaptability of forests to varying levels of nutrients [18,19]. This is of great significance to the sustainable management of plantations that are subject to nutrient restriction [10].

Due to its nitrogen fixation, fast growth, and tolerance to drought and barren lands, black locust (*Robinia pseudoacacia* L.) has become one of the most widely introduced tree species for ecological restoration and timber production worldwide [23–25]. The Loess Plateau, China, is a very ecologically fragile region that suffers from severe soil erosion and low availability of nutrients [26]. To restore the region, black locust has been introduced as the dominant plantation species with the aim to aid in soil and water conservation [27,28]. Black locust can significantly improve the soil quality by accumulating soil C and N [25,29]. However, black locust plantations grown in the long-term (for approximately 35 years) suffer from withering branches and a decline in productivity, which may indicate forest degradation [30]. Soil nutrient limitation (usually P limitation in the arid areas) [29,31] and water deficits [32,33] also may become serious with stand development. However, few studies have been conducted on how the resorption of N and P vary with the age of the black locust plantation. Therefore, we studied the adaptability of black locust to the changing nutrient status, especially the variation of NuRE with stand age, which may reveal the possible mechanism of degradation of the black locust plantation.

In this study, we examined the changes of nutrients in soil and leaves (both mature leaves and withered leaves) of black locust plantations covering an age range from 10 to 45 years old in the Loess Plateau of China. The objectives of this study were to quantify the variations of N and P resorption across the black locust plantation chronosequence and to find out whether there are any relationships between NuRE and nutrient stoichiometry indexes in leaves and soil. These data were then used to reveal the responses of plant nutrient use strategy to the altered leaf and soil nutrient status in black locust plantations of different ages. Because nutrient resorption is an important nutrient conservation mechanism and the plant nutrient demand may increase with stand age, we hypothesized that the N and P resorption will: (1) increase across the black locust plantation chronosequence, and (2) show significant negative relationships with leaf and soil N and P contents. The results may provide novel insight into the understanding of the nutrient cycling of black locust forests in arid and semi-arid areas.

2. Materials and Methods

2.1. Research Area

The research was conducted in the Loess Plateau at the Wuliwan catchment of Ansai County, Shaanxi Province, China (36°39′–36°52′ N, 109°20′–109°21′ E). The site has a hilly loess landscape and an elevation of 1010–1400 m above sea level. The study area has a typical semiarid climate with a mean annual rainfall of 505 mm, an annual average temperature of 8.8 °C, and a frost-free period of 159 days. Approximately 60% of the precipitation occurs between July and September, and the precipitation varies greatly in different years. The soil in this region is mainly loessal soil, developed from wind-accumulated loess, which has extremely poor resistance to erosion. The proportion of the differently sized soil particles in the 0–20 cm layer was 63.6% sand, 29.2% silt, and 7.2% clay. Artificial forests have been planted at the study site since the 1970s to mitigate the serious soil erosion on steeply sloped crop land [29]; therefore, the large area of plantations have different stand ages. The dominant plantation species is black locust and *Caragana korshinskii* Kom.

2.2. Experimental Design and Sampling

Based on the afforestation records of the local forestry department, black locust forests of five age classes (i.e., 10, 20, 30, 36, and 45 years old during field sampling in 2016, hereafter referred to as RP10, RP20, RP30, RP36, and RP45, respectively) were selected to represent a chronosequence. All of the sample lands were converted from cropland subjected to similar farming practices, namely, mainly planting maize (*Zea mays* L.), foxtail millet (*Setaria italic* L.), and broomcorn millet (*Panicum miliaceum* L.) with watering mainly through the rainfall and a small amount of manure was the main fertilizer. After trees were planted, the sample lands were fenced to prevent grazing and other human disturbance including thinning, wood usage, or other practices. Detailed information of the sampled lands is shown in Table 1. The climatic and edaphic conditions of the forests of five age classes were similar. Three replicate fields were measured for each black locust stand age. The three replicate fields for each stand age were within 1.5 km of one another to ensure consistent climatic variables and other conditions. In each replicate field, we randomly selected three replicated plots (20 m × 20 m) for sampling. All sites were topographically similar (i.e., slope aspect and slope degree) with similar elevation. Finally, we established 45 sampling plots (5 stand ages × 3 replicate fields × 3 sample plots).

Green and senesced leaves were collected in mid-August and mid-October 2016, respectively, when the green leaves were at peak biomass, and senesced leaves were ready to fall. At least five individual trees were selected in each sample plot. At least 20 branches were chosen around each tree from the lower, middle, and upper canopy. Approximately 50 g leaves were sampled from each individual (picked using a long reach pruner), all the leaves from the same sample plots were mixed (about 250–300 g), and then a quantity sufficient for chemical analysis (about 100 g) was taken out for analysis. The leaves with obvious diseases and/or insect pests were excluded. To avoid nutrient decomposition and leaching, the fully senesced leaves were collected from trees rather than from litter in mid-October, when they were completely yellow, dry, without any signs of deterioration, and would fall from the branch with a gentle touch [10,34]. The sample leaves were transfer to the laboratory in paper bags, then oven-dried at 105 °C for 30 min and maintained at 60 °C until a constant mass was reached. Dried sample leaves were ground and sieved with a 0.1 mm sieve.

Table 1. Field site information for *Robinia pseudoacacia* L. plantations with different stand ages.

Stand Ages	10	20	30	36	45
Location	36°52′00″ N 109°21′38″ E	36°52′08″ N 109°21′06″ E	36°51′36″ N 109°21′06″ E	36°51′37″ N 109°21′05″ E	36°52′16″ N 109°20′54″ E
Elevation (m)	1332.31	1245.2	1283.6	1255	1121.34
Slope (°)	37	30	35	30	28
Aspect (°)	ES38	ES20	EN16	EN15	ES10
Bulk density (g cm^{-3})	1.09	1.29	1.05	1.01	0.98
pH	8.56	8.62	8.58	8.51	8.60
Canopy density (%)	48	55	64	63	60
Stand density (trees hm^{-2})	1540	1535	1350	1305	1240
DBH (cm)	6.33	15.32	17.25	18.27	19.10
Degree of herb cover (%)	46.83	59.13	83.87	76.85	73.40
Number of herb species	13	24	21	22	23

ES and EN means East by South and Ease by North; DBH means diameter at breast height.

The soil samples were collected in mid-August 2016. In each sample plot, we selected five sample points along the diagonal. Because the surface soil (0–10 cm) is the most closely related to plant nutrients [10,29], we collected soil samples at each point from the 0–10 cm layer using a soil auger with a 4 cm diameter. The soil samples from the same plot were adequately mixed and reduced by coning and quartering to achieve appropriate quantities (about 400 g). Then, the soil samples were transported to the laboratory, air-dried at room temperature (25–28 °C), and passed through a 0.25 mm sieve in preparation for soil C, N, and P analysis. To avoid the impact of roots and litter, all samples were collected 80 cm from trees after the litter layer was removed and the plant roots, stones and debris were removed before drying through hand picking and sieving. A vegetation survey was also conducted in each plot in August 2016 using five randomly selected 1 m × 1 m small plots.

2.3. Laboratory Analysis and Determination

The TOC analyzer (Total organic carbon analyzer, Shimadzu Corp., Kyoto, Japan) was used to measure the total C from soil and leaf samples (0.1 mol L^{-1} HCl were used to destroy the carbonates in soil). The total nitrogen (TN) content was measured using the semimicro-Kjeldahl method with a Kjeldahl auto-analyzer (KDY-9830, Beijing, China) after digestion in sulfuric acid hydrogen peroxide (for leaves) or sulfuric acid perchloric acid (for soil). Total phosphorus (TP) content was measured using the molybdenum antimony colorimetric method by using an ultraviolet spectrometer subsystem (UV-6100, Shanghai, China) after digestion following the methods as in TN measurement [35].

2.4. Calculations and Statistical Analysis

NuRE was used to quantify the nutrient resorption, which was calculated as follows [19]:

$$\text{NuRE} = (1 - \frac{W_1}{W_2} \times \text{MLCF}) \times 100\% \tag{1}$$

where NuRE is the nutrient resorption efficiency, and W_1 and W_2 are the nutrient concentrations (TN or TP, g kg^{-1}) in senesced and green leaves, respectively. MLCF is the mass loss correction factor to correct the unbiased resorption value, which is 0.784 for broad-leaved deciduous trees as introduced by Vergutz et al. [12].

The soil and leaf C:N:P stoichiometry ratios were calculated as the mass ratio. There were no missing values in our data. All data were distributed normally after the Kolmogorov–Smirnov test ($p > 0.05$ for each null hypothesis). Then, the Bartlett test was used to investigate the homogeneity of variance, and $p > 0.05$ for each null hypothesis. There were no significant differences for each soil and leaf properties between different replicate fields of the same stand ages, or between different sample plots of the same replicate fields ($p > 0.05$). Therefore, one-way analysis of variance (ANOVA) was used to test the differences of NuRE, nutrients and stoichiometric ratios among the plantations of different ages (the significant level was $p < 0.05$). A least significant difference (LSD) test was conducted for post-hoc multiple comparisons. A Pearson correlation analysis was used to estimate the relationship between NuRE and other leaf or soil nutrition properties. The relationships between the stand age and other properties were estimated by linear regression. All of the statistical analyses were conducted using SAS (SAS Institute Inc., Cary, NC, USA), and the figures were plotted in Origin 2016 (OriginLab Corporation, Northampton, MA, USA).

3. Results

3.1. Leaf and Soil C, N, and P Concentrations

The concentrations of the C and N in green leaves significantly increased from RP10 to RP36; however, the concentrations were decreased from RP36 to RP45 ($p < 0.05$). The concentrations of C and N ranged from 380.81–475.38 g kg^{-1} and 22.38–32.49 g kg^{-1} in all stand ages, respectively (Table 2). The P concentration of green leaves showed no significant variation among the different stand ages, except P concentration in RP45 was significantly lower ($p < 0.05$).

Table 2. Leaf and soil C, N, and P concentrations in different stand aged black locust forests.

Field Site	Soil (g kg^{-1})			Green Leaves (g kg^{-1})			Senesced Leaves (g kg^{-1})		
	C	N	P	C	N	P	C	N	P
RP10	4.65 ± 0.27	0.50 ± 0.03	0.55 ± 0.07	380.81 ± 12.89	22.38 ± 0.94	2.07 ± 0.06	352.61 ± 16.54	15.16 ± 0.46	1.18 ± 0.05
RP20	4.97 ± 0.09	0.61 ± 0.05	0.49 ± 0.04	439.85 ± 13.32	25.05 ± 0.54	2.15 ± 0.04	348.77 ± 12.22	13.62 ± 1.09	1.14 ± 0.02
RP30	8.26 ± 0.71	0.87 ± 0.07	0.60 ± 0.04	467.31 ± 8.51	32.02 ± 0.75	2.07 ± 0.05	409.41 ± 12.37	17.90 ± 0.82	1.16 ± 0.06
RP36	8.79 ± 0.35	0.87 ± 0.04	0.53 ± 0.03	475.38 ± 6.11	32.49 ± 0.43	2.17 ± 0.11	425.59 ± 11.48	18.38 ± 0.51	1.25 ± 0.05
RP45	14.41 ± 0.91	1.20 ± 0.13	0.57 ± 0.03	421.53 ± 12.06	28.71 ± 0.80	1.72 ± 0.06	453.97 ± 24.16	19.4 ± 0.52	1.37 ± 0.05

The data are shown as mean ± standard error.

The C and N concentrations of senesced leaves increased from RP20 to RP45 with stand age ($p < 0.05$). The concentrations of C and N ranged from 347.27–453.97 g kg^{-1} and 11.64–19.40 g kg^{-1} in all stand ages (Table 2). The P concentration of senesced leaves showed an increasing trend with stand age, with a range of 1.14–1.37 g kg^{-1}. The C, N, and P concentrations were all higher in green leaves than that in senesced leaves in each stand age. The only exception was the C concentrations in RP45, which was higher in senesced leaves ($p < 0.05$).

The C and N concentrations of soil significantly increased with stand ages ($p < 0.05$), ranging from 4.56–14.41 g kg^{-1} and 0.50–1.20 g kg^{-1}, respectively. The soil P concentrations showed no significant changes with stand ages (Table 2).

3.2. Leaf and Soil C:N:P Stoichiometry Ratios

The C:P and N:P in green leaves increased with stand age ($p < 0.05$), ranging from 184.22–245.60 and 10.83–16.72 with an average of 218.50 and 14.11, respectively (Figure 1). The C:N of green leaves declined as a whole with stand age ($p < 0.05$), ranging from 14.60–17.56 with an average of 15.70. The C:N, C:P, and N:P of senesced leaves all increased and then decreased with stand age ($p < 0.05$), with ranges of 22.62–29.95, 274.35–355.60, and 10.73–15.54 with averages of 23.69, 326.77, and 13.84, respectively (Figure 1). The highest C:N values in senesced leaves were measured in RP20; the highest C:P and N:P in senesced leaves were measured in RP30. Generally, the C:N and C:P of senesced leaves were higher than that of green leaves in all stand ages ($p < 0.05$), and the N:P of senesced leaves were higher from RP10 to RP30 and lower from RP36 to RP45 compared with that of green leaves ($p < 0.05$).

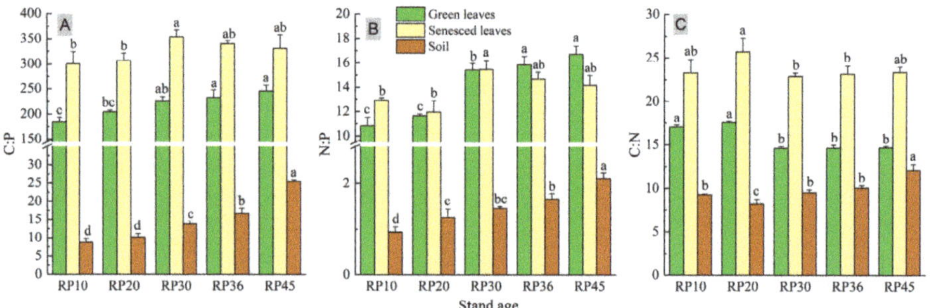

Figure 1. Leaf and soil C:P ratio (**A**), N:P ratio (**B**) and C:N ratio (**C**) in black locust forests with different stand ages. Different lowercase letters represent significant differences among stand ages at $p < 0.05$, and the same letters indicate no significance.

The soil C:P and N:P increased with stand age ($p < 0.05$), ranging from 8.66–25.47 and 0.93–2.11 with averages of 14.98 and 1.48, respectively. The soil C:N decreased by 11.8% from RP10 to RP20, and then increased by 47.8% from RP30 to RP45 ($p < 0.05$), with an average C:N of 9.83 (Figure 1).

3.3. NRE and PRE in Forests of Different Stand Ages and Their Relationship

The NRE and PRE varied among stands of different ages, with the tendency to increase and then decrease with stand age ($p < 0.05$). The NRE and PRE values ranged from 46.8% to 57.4% and from 37.4% to 58.5%, with averages of 52.61 and 51.89, respectively. The highest values for NRE and PRE were obtained in RP20 (Figure 2A). The NRE was higher than the PRE in RP10 and RP20, and lower in other stand ages ($p < 0.05$). A significantly positive correlation was found between NRE and PRE ($R^2 = 0.316$, $p < 0.05$) (Figure 2B). In addition, the PRE:NRE decreased with the age of the stand ($p < 0.05$) (Figure 2C).

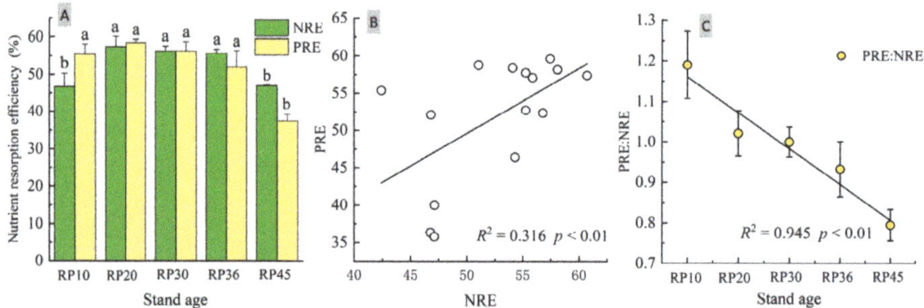

Figure 2. (**A**) Changes of N and P resorption efficiency (NRE and PRE), (**B**) PRE:NRE along the chronosequence of black locust forests, and (**C**) the relationship between NRE and PRE. Data are shown as the mean ± standard deviation. Different lowercase letters represent significant differences among stand ages at $p < 0.05$, and the same letters indication no significant difference.

3.4. Relationships between NuRE, Nutrient Concentrations, and Stoichiometry Ratios in Leaves and Soil

Results from the Pearson correlation analysis showed that the NRE was significantly positively correlated with the C of green leaves and P of senesced leaves ($p < 0.05$); the NRE was significantly negatively correlated with the C:N of senesced leaves among the different stand ages ($p < 0.05$), whereas no significant correlation was found with other indicators (Table 3). The PRE was significantly correlated with most of the nutrient concentration and stoichiometry indicators in leaves and soil ($p < 0.05$). The exception for this was that the soil P; C, N, and C:N of green leaves; and C:N, C:P, and N:P of senesced leaves were not significant. The PRE was positively correlated with the P concentrations of green leaves, whereas it was negatively correlated with other significant related indicators (Table 3).

Table 3. Pearson correlations between nutrient resorption efficiency, nutrient concentrations, and stoichiometry ratios in leaf and soil samples.

		NRE		PRE	
		R	*p*	*R*	*p*
Soil	C	−0.303	0.273	−0.879	<0.001
	N	−0.151	0.592	−0.773	<0.001
	P	−0.116	0.681	-0.258	0.353
	C:N	−0.489	0.064	−0.881	<0.001
	C:P	−0.289	0.296	−0.878	<0.001
	N:P	−0.109	0.700	−0.756	<0.001
Green leaves	C	0.711	<0.001	0.127	0.651
	N	0.432	0.108	−0.188	0.503
	P	0.525	0.044	0.947	<0.001
	C:N	−0.041	0.884	0.489	0.064
	C:P	0.085	0.763	−0.666	<0.001
	N:P	0.058	0.836	−0.645	<0.001
Senesced leaves	C	−0.126	0.656	−0.716	<0.001
	N	−0.292	0.292	−0.675	<0.001
	P	−0.529	0.043	−0.930	<0.001
	C:N	0.504	0.048	0.229	0.413
	C:P	0.345	0.209	−0.045	0.873
	N:P	0.029	0.919	−0.146	0.603

NRE and PRE represent the N and P resorption efficiency, respectively. *R* is the Pearson correlation coefficient.

4. Discussion

4.1. Leaf and Soil Nutrient Concentration and Stoichiometry of Black Locust Forests of Different Ages

The nutrient cycle and transformation process of C, N, and P between soil and plants during vegetation restoration is the basis of material circulation in terrestrial ecosystems [36]. Soil provides the necessary nutrients for plant growth, and plants impact the soil nutrients through litter restitution, root growth, and secretions. As a result of plant growth, this correlation would also change with stand ages [37]. Previous studies have shown that the artificial vegetation in the Loess Plateau may face a nutrient imbalance with the increase of stand age, mainly manifesting as P limitation in restored ecosystems [29]. In this study, the soil C and N concentrations increased by 209.9% and 139.2% ($p < 0.05$) from 10-year-old to 45-year-old black locust forests, respectively (Table 1), whereas the soil P concentrations showed no significant variation among the stand ages. The different variation of soil C, N, and P may have occurred because the accumulation of soil C and N are primarily driven by biological factors (e.g., decomposition of plant litter and dead roots), but P transformations in soil are mainly driven by biochemical mineralization (e.g., phosphate decomposition, which takes a long time [29]. Compared to P, the black locust trees can more effectively import N into the soil with symbiosis of N-fixing microbes [38]. Therefore, a significant increase in soil C:P and N:P ratios was found in the older black locust forests ($p < 0.05$) (Figure 1). Our previous study has found a significant decrease of soil-available P and increase of available N along with the increased stand age [39]; therefore, we can speculate that the soil may be relatively deficient in P in older black locust plantations on the Loess Plateau. This provides further support for conclusions from previous studies [29,31,40].

Nutrient concentrations and stoichiometry ratios in leaves change in different growth stages [40,41]. The results showed that the average concentrations of C and N in leaves from black locust forests of different ages were 436.98 g kg^{-1} and 28.13 g kg^{-1}, respectively (Table 2). The C concentration was similar to the average C content of leaves (438 g kg^{-1}) in the Loess Plateau [42], whereas the N content was much higher than the average level in both the Loess Plateau (21.61 g kg^{-1}) [43] and in China (18.60 g kg^{-1}) [44]. The C and N concentration of green leaves increased and then decreased with the increase of stand ages. The concentrations peaked in the RP36 (Table 2). The overall change of C and N in black locust leaves showed the following trend: middle age forest > older forest > young forest, which was consistent with Ma et al.'s results [40]. This may be due to the accumulation of structural substances with more C in the middle- and older-aged plants compared with younger plants. The average P concentrations in green leaves from black locust forests of different ages was 2.01 g kg^{-1} (Table 2), which was close to or slightly less than the global average P content (2.00 g kg^{-1}) [45] and the average of leaves from the Loess Plateau (2.09 g kg^{-1}) [43]. However, P concentration of the green leaves did not show significant differences with the change of stand ages. The N:P ratios of green leaves increased significantly from 10.83 to 16.72 along the chronosequence of black locust forests ($p < 0.05$) (Figure 1). Our results suggested that the black locust plantations would shift from relative N-limitation to relative P-limitation across the chronosequence based on the criteria proposed by Güsewell [22] (i.e., leaves N:P <14, 14–16, and >16 indicating N limitation, N and P co-limitation, and P limitation, respectively). In other words, the young black locust plantations were relatively more N-limited, and the older plantations were relatively more P-limited. Similar findings have also been reported in *Larix olgensis* A. Henry [9], *Larix kaempferi* (Lamb.) Carr. [10], and other plantations [46], which showed that long-term ecosystem development tended to cause a shift from N- to P-limitation. This may be a result of the relatively easy absorption of soil N (with N-fixing microbes in the rhizosphere), but relatively inadequate P absorption by the black locust [38]. However, our research only tested the soil properties in the top layer, thus further study on nutrient effectiveness and root absorption characteristics of deep soil should also be investigated.

Leaf litter is an important way for plants to return nutrients to the soil. Typically, at least some of the nutrients in senesced leaves will be transferred to the branches before falling, resulting in lower levels of nutrients in the senesced leaves compared to the green leaves [12,14,47]. The data obtained in our study also agreed with this rule. The average concentrations of C, N, and P of senesced leaves in stands of different ages were 398.07 g kg^{-1}, 16.89 g kg^{-1}, and 1.22 g kg^{-1} (Table 2), respectively. Compared with the average level of nutrients of senesced leaves in broad-leaved forests in China (the C, N, and P concentrations was 479.9 g kg^{-1}, 13.2 g kg^{-1}, and 1.06 g kg^{-1}, respectively) [48], our results showed relatively high concentrations of N and P, and low concentrations of C. Previous studies have concluded that the abundance of N and P in litter is conducive to microbial (especially bacterial) activity, thereby promoting rapid decomposition of litter, facilitating nutrient release [49,50]. In this study, the C:N and C:P of senesced leaves was 22.62–29.95 and 274.35–355.60 in black locust forests of different ages, respectively (Figure 1). These ratios are significantly lower than the average value of C:N and C:P in senesced leaves in the broad-leaved forests of China (36.36 and 452.73, respectively), demonstrating that the leaf litter in black locust forests are easy to decompose and are beneficial to nutrient release. In addition, our findings showed increasing N and P contents in senesced leaves in older stands, indicating that more N and P would be returned to the soil through litter decomposition.

4.2. N and P Resorption Correlated with Soil and Leaf Nutrients across the Plantation Chronosequence

Plants absorb nutrients from the soil and allocate them to different organs to accumulate or participate in various life activities; the nutrients will be reabsorbed before the leaves fall, thus, prolonging the retention time of nutrients in plants [14]. In this study, the N and P resorption efficiency of black locust forests of different ages were 46.8%–57.4% and 37.4%–58.5%, respectively (Figure 2A); these efficiencies are significantly lower than that of global terrestrial forests (62.1% and 64.9%, respectively) [12], and this may be due to species-dependent differences. The NRE and

PRE were significant correlated with each other, and both of the NRE and PRE showed significantly increasing and then decreasing trends along the chronosequence of black locust forests (Figure 2), which was consistent with previous research in *Pinus massoniana* Lamb and *Metasequoia glyptostroboides* Hu et Cheng plantations of different ages [19,51]. This variation may be because young plants have rapid biomass production, thus need a higher N and P resorption efficiency to supply an appropriate amount of nutrients [19]. Black locust has greater N absorbing efficiency than P absorbing efficiency with symbiosis of N-fixing microbes around roots. Meanwhile, the rapid growth of young trees requires more P for production of genetic material. These reasons may likely result in a greater PRE than NRE in 10-year-old and 20-year-old forests [40]. A similar trend was also reported in *Eucalyptus urophylla* S.T. Blake × *E. grandis* W. Mill ex Maiden forests [20]. In the older growth stage of black locust forests, the PRE significantly decreased with the increasing stand ages, causing a deficit of P in the soil and trees. This disagrees with our first hypothesis, which may be because the wood biomass became an effective reservoir of P in the later stage of plant growth, thus reducing the P resorption from fallen leaves [52,53]. This indicated that nutrient resorption may not be the main strategy for the preservation of P elements in the older black locust forests. In addition, the NRE decreased with the age of the stand, which is contrary to our first hypothesis. Previous research does not support that tree growth depended more on N resorption to supply its N requirements with the development of the stand [18,37], but it is consistent with the view that decreased NRE was a result of sufficient soil N supply after decomposition [19]. The increased N concentration in soil and senesced leaves with the older stands and relatively low C:N of senesced leaves in our study also supported this view.

The nutrition resorption may be individually or collectively affected by soil nutrient conditions and nutrient contents of green leaves and senesced leaves [12,54]. There is still great uncertainty about the degree of influence and direction of these factors. For example, a study conducted by Aerts reported that there was no significant correlation between NuRE and nutrient concentrations in soil and green leaves, but a negative correlation was found with the nutrient content in senesced leaves [1,55]. However, a study conducted by Vergutz et al. based on data at a global scale concluded that the NuRE depends more on the nutrient state of green leaves than on the senesced leaves [12]. In addition, other studies have also shown that plants subject to N or P limitation should have higher N or P resorption efficiency [56,57]. Killingbeck proposed that nutrient transfer and resorption may be only an inherent feature of species [58], indicating that genetic differences in plants may be the main factor determining nutrient resorption [54,59]. Our results demonstrated that the NRE of black locust forests was significantly correlated with the C concentration in green leaves, P in senesced leaves, and C:N ($p < 0.05$). No significant correlations were found with soil nutrients (Table 3). This was consistent with a previous study that reported no impact of soil nutrients on NRE [11,55], whereas the PRE was significantly correlated with most of the nutrient content and C:N:P stoichiometry ratios of soil and leaves ($p < 0.05$). Intriguingly, the PRE was negatively correlated with P limitation indicators (i.e., N:P and C:P of soil and leaves) (Figure 3), suggesting that the PRE would be lower with greater P limitation. The decreasing PRE:NRE also indicated relatively lower rates of P resorption with increased plantation age. This was partly in accordance with our second hypothesis but in contrast with the "relative resorption hypothesis". The "relative resorption hypothesis" revealed that N and P would show an increase in resorption when plants are growing in N- or P-limited conditions [11,60], indicating that the leaves nutrients resorption may be not be the main strategy of black locust to cope the P-limited conditions. Similar results were also observed in *Metasequoia glyptostroboides* and *Larix gmelinii* Rupr. plantations [10,19]. These studies suggested that more efficient nutrient absorption by roots should also be considered.

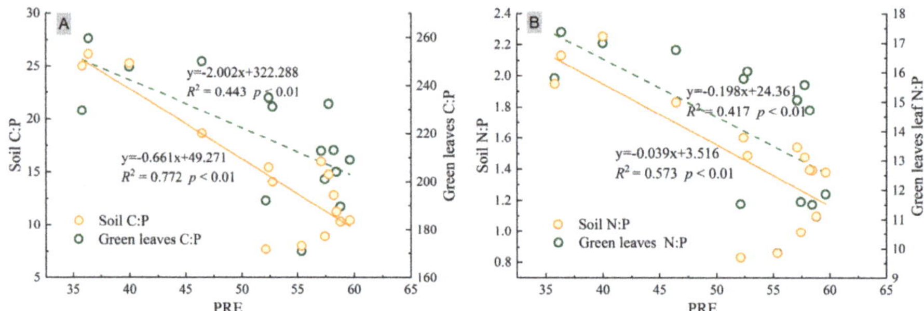

Figure 3. Correlations between PRE and soil and the C:P (**A**) and N:P (**B**) ratios of green leaves in black locust plantations from stands of different ages.

Though the NRE and PRE respond differently to soil and plant nutrients, the results revealed that black locust plantations would alter the conservation and use strategy of nutrients in the ecosystem through a plant-mediated pathway in different stand ages. Based on the present study and previously published studies, we speculate that more nutrients stay in senesced leaves of trees in older forests, possibly prolonging the P cycle and increasing the risk of nutrition loss by leaching in the process of litter decomposition [17,61]. However, litter decomposition may induce increased incorporation of P into the soil microbial biomass, which can promote the recycling speed of nutrients and availability of P in ecosystems [62,63]. Recent studies in P-deficient deciduous forests in Europe have shown that the absorption and return of nutrients also depend on soil P availability [62,64,65]. Our study provided limited information about how the release of litter nutrients and root nutrient absorption affect and react to soil available nutrient change in black locust forests. Since plant nutrient absorption and utilization is a complex process [17,21], future experimental studies considering certain soil nutrient species (e.g., plant available/labile forms, organically-bound forms, etc.), root nutrients absorption, and soil microbes are needed to elucidate the central nutrient utilization strategy of black locust in response to a nutrient-poor environment and how these factors are involved in regulating NuRE in different growth stages.

5. Conclusions

Along with the chronosequence of black locust plantations, the NRE and PRE both increased and then decreased. Imbalanced C, N, and P variation with the increase in stand ages may result in P limitation. The NRE was significantly correlated with the C of green leaves and the P content and C:N of senesced leaves, whereas the PRE was significantly negatively correlated with the C:P and N:P of soil and green leaves. The PRE:NRE decreased with increased stand ages. In summary, our result showed that the NRE and PRE were both changed as the stand aged. The middle-aged black locust trees (approximately 30 years) were more efficient at resorbing P and N from senesced leaves than younger and older stands, but the NRE and PRE responded differently in soil and plant nutrients. The black locust plantations would alter the conservation and use strategy of nutrients in the ecosystem through a plant-mediated pathway with increasing stand age.

Author Contributions: Conceptualization, J.D., X.L., and X.H.; Formal analysis, S.W. and F.Z.; Investigation, J.D., S.W., C.R., and W.Z.; Writing—original draft, J.D. and D.Z.; Writing—review and editing, C.R., F.Z., X.H., and G.Y.

Funding: The study was funded by the Natural Science Basic Research Program of Shaanxi Province, China (2018JQ3041); Young Talent fund of University Association for Science and Technology in Shaanxi; Scientific Research Program Funded by Shaanxi Provincial Education Department (18JK0871), and Innovation and entrepreneurship Training Program of Shaanxi Province (201820086).

References

1. Aerts, R. Nutrient resorption from senescing leaves of perennials: Are there general patterns? *J. Ecol.* **1996**, *84*, 597–608. [CrossRef]
2. Deng, M.; Liu, L.; Jiang, L.; Liu, W.; Wang, X.; Li, S.; Yang, S.; Wang, B. Ecosystem scale trade-off in nitrogen acquisition pathways. *Nat. Ecol. Evol.* **2018**, *2*, 1724–1734. [CrossRef] [PubMed]
3. Lü, X.; Reed, S.C.; Yu, Q.; Han, X. Nutrient resorption helps drive intra-specific coupling of foliar nitrogen and phosphorus under nutrient-enriched conditions. *Plant Soil* **2016**, *398*, 111–120. [CrossRef]
4. Wang, L.; Wang, L.; He, W.; An, L.; Xu, S. Nutrient resorption or accumulation of desert plants with contrasting sodium regulation strategies. *Sci. Rep. UK* **2017**, *7*, 17035. [CrossRef]
5. Vitousek, P.M.; Porder, S.; Houlton, B.Z.; Chadwick, O.A. Terrestrial phosphorus limitation: Mechanisms, implications, and nitrogen and phosphorus interactions. *Ecol. Appl.* **2010**, *20*, 5–15. [CrossRef] [PubMed]
6. Elser, J.J.; Bracken, M.E.; Cleland, E.E.; Gruner, D.S.; Harpole, W.S.; Hillebrand, H.; Ngai, J.T.; Seabloom, E.W.; Shurin, J.B.; Smith, J.E. Global analysis of nitrogen and phosphorus limitation of primary producers in freshwater, marine and terrestrial ecosystems. *Ecol. Lett.* **2007**, *10*, 1135–1142. [CrossRef]
7. Campo, J. Shift from ecosystem P to N limitation at precipitation gradient in tropical dry forests at Yucatan, Mexico. *Environ. Res. Lett.* **2016**, *11*, 95006. [CrossRef]
8. Mediavilla, S.; García-Iglesias, J.; González-Zurdo, P.; Escudero, A. Nitrogen resorption efficiency in mature trees and seedlings of four tree species co-occurring in a Mediterranean environment. *Plant Soil* **2014**, *385*, 205–215. [CrossRef]
9. Chen, L.; Zhang, C.; Duan, W. Temporal variations in phosphorus fractions and phosphatase activities in rhizosphere and bulk soil during the development of *Larix olgensis* plantations. *J. Plant Nutr. Soil Sci.* **2015**, *179*, 67–77. [CrossRef]
10. Yan, T.; Lü, X.; Zhu, J.; Yang, K.; Yu, L.; Gao, T. Changes in nitrogen and phosphorus cycling suggest a transition to phosphorus limitation with the stand development of larch plantations. *Plant Soil* **2018**, *422*, 385–396. [CrossRef]
11. Yuan, Z.Y.; Chen, H.Y.H. Changes in nitrogen resorption of trembling aspen (*Populus tremuloides*) with stand development. *Plant Soil* **2010**, *327*, 121–129. [CrossRef]
12. Vergutz, L.; Manzoni, S.; Porporato, A.; Novais, R.F.; Jackson, R.B. Global resorption efficiencies and concentrations of carbon and nutrients in leaves of terrestrial plants. *Ecol. Monogr.* **2012**, *82*, 205–220. [CrossRef]
13. Yan, T.; Zhu, J.; Yang, K. Leaf nitrogen and phosphorus resorption of woody species in response to climatic conditions and soil nutrients: A meta-analysis. *J. For. Res.* **2018**, *29*, 1–9. [CrossRef]
14. Lu, J.; Duan, B.; Yang, M.; Yang, H.; Yang, H. Research progress in nitrogen and phosphorus resorption from senesced leaves and the influence of ontogenetic and environmental factors. *Acta Pratacult. Sin.* **2018**, *27*, 178–188.
15. Du, B.; Ji, H.; Peng, C.; Liu, X.; Liu, C. Altitudinal patterns of leaf stoichiometry and nutrient resorption in *Quercus variabilis* in the Baotianman Mountains, China. *Plant Soil* **2017**, *413*, 193–202. [CrossRef]
16. Zhang, H.; Guo, W.; Yu, M.; Wang, G.G.; Wu, T. Latitudinal patterns of leaf N, P stoichiometry and nutrient resorption of *Metasequoia glyptostroboides* along the eastern coastline of China. *Sci. Total Environ.* **2018**, *618*, 1–6. [CrossRef] [PubMed]
17. Wang, Z.; Lu, J.; Yang, H.; Zhang, X.; Luo, C.; Zhao, Y. Resorption of nitrogen, phosphorus and potassium from leaves of lucerne stands of different ages. *Plant Soil* **2014**, *383*, 301–312. [CrossRef]
18. Sun, Z.; Liu, L.; Peng, S.; Peñuelas, J.; Zeng, H.; Piao, S. Age-Related Modulation of the Nitrogen Resorption Efficiency Response to Growth Requirements and Soil Nitrogen Availability in a Temperate Pine Plantation. *Ecosystems* **2016**, *19*, 698–709. [CrossRef]
19. Zhang, H.; Wang, J.; Wang, J.; Guo, Z.; Wang, G.G.; Zeng, D.; Wu, T. Tree stoichiometry and nutrient resorption along a chronosequence of *Metasequoia glyptostroboides* forests in coastal China. *For. Ecol. Manag.* **2018**, *430*, 445–450. [CrossRef]
20. Qiu, L.; Hu, H.; Lin, Y.; Ge, L.; Wang, K.; He, Z.; Dong, Q. Nutrient resorption effciency and C:N:P stoichiometry of *Eucalyptus urophylla* × *E. grandis* of different ages in a sandy coastal plain area. *Chin. J. Appl. Environ. Biol.* **2017**, *23*, 739–744.

21. Zhou, L.; Addo-Danso, S.D.; Wu, P.; Li, S.; Zou, X.; Zhang, Y.; Ma, X. Leaf resorption efficiency in relation to foliar and soil nutrient concentrations and stoichiometry of *Cunninghamia lanceolata* with stand development in southern China. *J. Soil Sediment* **2016**, *16*, 1–12. [CrossRef]

22. Güsewell, S. N:P ratios in terrestrial plants: Variation and functional significance. *New Phytol.* **2004**, *164*, 243–266. [CrossRef]

23. Meng, F.; Luo, Q.; Wang, Q.; Zhang, X.; Qi, Z.; Xu, F.; Lei, X.; Cao, Y.; Chow, W.S.; Sun, G. Physiological and proteomic responses to salt stress in chloroplasts of diploid and tetraploid black locust (*Robinia pseudoacacia* L.). *Sci. Rep. UK* **2016**, *6*, 23098. [CrossRef]

24. Sitzia, T.; Campagnaro, T.; Dainese, M.; Cierjacks, A. Plant species diversity in alien black locust stands: A paired comparison with native stands across a north-Mediterranean range expansion. *For. Ecol. Manag.* **2012**, *285*, 85–91. [CrossRef]

25. Zhang, G.; Zhang, P.; Cao, Y. Ecosystem carbon and nitrogen storage following farmland afforestation with black locust (*Robinia pseudoacacia*) on the Loess Plateau, China. *J. For. Res.* **2018**, *29*, 761–771. [CrossRef]

26. Hui, S.; Shao, M. Soil and water loss from the Loess Plateau in China. *J. Arid Environ.* **2000**, *45*, 9–20.

27. Ma, C.; Luo, Y.; Shao, M.; Li, X.; Sun, L.; Jia, X. Environmental controls on sap flow in black locust forest in Loess Plateau, China. *Sci. Rep. UK* **2017**, *7*, 13160. [CrossRef]

28. Liu, G.; Shang, G.; Yao, W.; Yang, Q.; Zhao, M.; Dang, X.; Guo, M.; Wang, G.; Wang, B. Ecological Effects of Soil Conservation in Loess Plateau. *Bull. Chin. Acad. Sci.* **2017**, *32*, 11–19.

29. Deng, J.; Sun, P.; Zhao, F.; Han, X.; Yang, G.; Feng, Y.; Ren, G. Soil C, N, P and Its Stratification Ratio Affected by Artificial Vegetation in Subsoil, Loess Plateau China. *PLoS ONE* **2016**, *11*, e151446. [CrossRef] [PubMed]

30. Wei, J.; Li, Z.; Feng, X.; Zhang, Y.; Chen, W.; Wu, X.; Jiao, L.; Wamg, X. Ecological and physiological mechanisms of growth decline of *Robinia pseudoacacia* plantations in the Loess Plateau of China: A review. *Chin. J. Appl. Ecol.* **2018**, *29*, 2433–2444.

31. Cao, Y.; Chen, Y. Coupling of plant and soil C:N:P stoichiometry in black locust (*Robinia pseudoacacia*) plantations on the Loess Plateau, China. *Trees* **2017**, *31*, 1–12. [CrossRef]

32. Jia, X.; Shao, M.; Zhu, Y.; Luo, Y. Soil moisture decline due to afforestation across the Loess Plateau, China. *J. Hydrol.* **2017**, *546*, 113–122. [CrossRef]

33. Shangguan, Z.; Zheng, S. Ecological properties of soil water and effects on forest vegetation in the Loess Plateau. *Int. J. Sustain. Dev. World* **2006**, *13*, 307–314. [CrossRef]

34. Yan, T.; Lü, X.; Yang, K.; Zhu, J. Leaf nutrient dynamics and nutrient resorption: A comparison between larch plantations and adjacent secondary forests in Northeast China. *J. Plant Ecol.* **2015**, *9*, 165–173. [CrossRef]

35. Bao, S. *Analysis of Soil Agrochemical*; Chinese Agricultural Press: Beijing, China, 2000.

36. He, J.; Lu, Y.; Fu, B. *Frontier of Soil Biology*; Science Press: Beijing, China, 2015.

37. Liu, J.; Gu, Z.; Shao, H.; Zhou, F.; Peng, S. N–P stoichiometry in soil and leaves of *Pinus massoniana* forest at different stand ages in the subtropical soil erosion area of China. *Environ. Earth Sci.* **2016**, *75*, 1091. [CrossRef]

38. Rice, S.K.; Westerman, B.; Federici, R. Impacts of the exotic, nitrogen-fixing black locust (*Robinia pseudoacacia*) on nitrogen-cycling in a pine-oak ecosystem. *Plant Ecol.* **2004**, *174*, 97–107. [CrossRef]

39. Zhang, W.; Liu, W.; Xu, M.; Deng, J.; Han, X.; Yang, G.; Feng, Y.; Ren, G. Response of forest growth to C:N:P stoichiometry in plants and soils during *Robinia pseudoacacia* afforestation on the Loess Plateau, China. *Geoderma* **2019**, *337*, 280–289. [CrossRef]

40. Ma, R.; An, S.; Huang, Y. C, N and P stoichiometry characteristics of different-aged *Robinia pseudoacacia* plantations on the Loess Plateau, China. *Chin. J. Appl. Ecol.* **2017**, *28*, 2787–2793.

41. Schreeg, L.A.; Santiago, L.S.; Wright, S.J.; Turner, B.L. Stem, root, and older leaf N:P ratios are more responsive indicators of soil nutrient availability than new foliage. *Ecology* **2016**, *95*, 2062–2068. [CrossRef]

42. Zheng, S.; Shangguan, Z. Spatial patterns of leaf nutrient traits of the plants in the Loess Plateau of China. *Trees* **2007**, *21*, 357–370. [CrossRef]

43. Ma, L.; Chen, Y.; Zhang, X.; Yang, J.; An, S. Characteristics of Leaf Ecological Stoichiometry of *Robinia pseudoacacia* in Loess Plateau. *Res. Soil Water Conserv.* **2014**, *21*, 57–61.

44. Han, W.; Fang, J.; Guo, D.; Zhang, Y. Leaf nitrogen and phosphorus stoichiometry across 753 terrestrial plant species in China. *New Phytol.* **2005**, *168*, 377–385. [CrossRef] [PubMed]

45. Elser, J.J.; Sterner, R.W.; Gorokhova, E.; Fagan, W.F.; Markow, T.A.; Cotner, J.B.; Harrison, J.F.; Hobbie, S.E.; Odell, G.M.; Weider, L.W. Biological stoichiometry from genes to ecosystems. *Ecol. Lett.* **2008**, *3*, 540–550. [CrossRef]

46. Hayes, P.; Turner, B.L.; Lambers, H.; Laliberté, E.; Bellingham, P. Foliar nutrient concentrations and resorption efficiency in plants of contrasting nutrient-acquisition strategies along a 2-million-year dune chronosequence. *J. Ecol.* **2014**, *102*, 396–410. [CrossRef]
47. Chapin, F.S.; Moilanen, L. Nutritional Controls Over Nitrogen and Phosphorus Resorption from Alaskan Birch Leaves. *Ecology* **1991**, *72*, 709–715. [CrossRef]
48. Xie, Y.; Liang, Y.; Xiao, H.; Zhu, R.; Luo, L.; Guo, W.; Cao, Y.; Zhang, Z.; Pan, Y.; Zheng, N.; et al. Pattern and controlling factors of terrestrial leaf litters decomposition in China. *J. East China Univ. Technol. (Nat. Sci.)* **2018**, *41*, 271–276.
49. William, P.; Silver, W.L.; Burke, I.C.; Leo, G.; Harmon, M.E.; Currie, W.S.; King, J.Y.; Adair, E.C.; Brandt, L.A.; Hart, S.C. Global-scale similarities in nitrogen release patterns during long-term decomposition. *Science* **2007**, *315*, 361–364.
50. Gergócs, V.; Hufnagel, L. The effect of microarthropods on litter decomposition depends on litter quality. *Eur. J. Soil Biol.* **2016**, *75*, 24–30. [CrossRef]
51. Li, R.; Wang, S.L.; Wang, Q.K. Nutrient contents and resorption characteristics in needles of different age *Pinus massoniana* (Lamb.) before and after withering. *Chin. J. Appl. Ecol.* **2008**, *19*, 1443.
52. Sardans, J.; Peñuelas, J. Trees increase their P:N ratio with size. *Glob. Ecol. Biogeogr.* **2015**, *24*, 147–156. [CrossRef]
53. Sardans, J.; Peñuelas, J. Tree growth changes with climate and forest type are associated with relative allocation of nutrients, especially phosphorus, to leaves and wood. *Glob. Ecol. Biogeogr.* **2012**, *22*, 494–507. [CrossRef]
54. Luyssaert, S.; Staelens, J.; De Schrijver, A. Does the commonly used estimator of nutrient resorption in tree foliage actually measure what it claims to? *Oecologia* **2005**, *144*, 177–186. [CrossRef]
55. Aerts, R.; Chapin, F.S. The Mineral Nutrition of Wild Plants Revisited: A Re-evaluation of Processes and Patterns. *Adv. Ecol. Res.* **2000**, *30*, 1–67.
56. Güsewell, S. Nutrient Resorption of Wetland Graminoids Is Related to the Type of Nutrient Limitation. *Funct. Ecol.* **2010**, *19*, 344–354. [CrossRef]
57. Agüero, M.L.; Puntieri, J.; Mazzarino, M.J.; Grosfeld, J.; Barroetaveña, C. Seedling response of Nothofagus species to N and P: Linking plant architecture to N/P ratio and resorption proficiency. *Trees* **2014**, *28*, 1185–1195. [CrossRef]
58. Killingbeck, K.T. Nutrients in Senesced Leaves: Keys to the Search for Potential Resorption and Resorption Proficiency. *Ecology* **1996**, *77*, 1716–1727. [CrossRef]
59. De Marco, A.; Arena, C.; Giordano, M.; De Santo, A.V. Impact of the invasive tree black locust on soil properties of Mediterranean stone pine-holm oak forests. *Plant Soil* **2013**, *372*, 473–486. [CrossRef]
60. Han, W.; Tang, L.; Chen, Y.; Fang, J. Relationship between the Relative Limitation and Resorption Efficiency of Nitrogen vs. Phosphorus in Woody Plants. *PLoS ONE* **2013**, *8*, e83366. [CrossRef] [PubMed]
61. Lado-Monserrat, L.; Lidón, A.; Bautista, I. Litterfall, litter decomposition and associated nutrient fluxes in Pinus halepensis: Influence of tree removal intensity in a Mediterranean forest. *Eur. J. For. Res.* **2015**, *134*, 833–844. [CrossRef]
62. Spohn, M.; Zavišić, A.; Nassal, P.; Bergkemper, F.; Schulz, S.; Marhan, S.; Schloter, M.; Kandeler, E.; Polle, A. Temporal variations of phosphorus uptake by soil microbial biomass and young beech trees in two forest soils with contrasting phosphorus stocks. *Soil Biol. Biochem.* **2018**, *117*, 191–202. [CrossRef]
63. Zavišić, A.; Nassal, P.; Yang, N.; Heuck, C.; Spohn, M.; Marhan, S.; Pena, R.; Kandeler, E.; Polle, A. Phosphorus availabilities in beech (*Fagus sylvatica* L.) forests impose habitat filtering on ectomycorrhizal communities and impact tree nutrition. *Soil Biol. Biochem.* **2016**, *98*, 127–137.

64. Netzer, F.; Schmid, C.; Herschbach, C.; Rennenberg, H. Phosphorus-nutrition of European beech (*Fagus sylvatica* L.) during annual growth depends on tree age and P-availability in the soil. *Environ. Exp. Bot.* **2017**, *137*, 194–207. [CrossRef]

65. Pistocchi, C.; Mészáros, É.; Tamburini, F.; Frossard, E.; Bünemann, E.K. Biological processes dominate phosphorus dynamics under low phosphorus availability in organic horizons of temperate forest soils. *Soil Biol. Biochem.* **2018**, *126*, 64–75. [CrossRef]

Article

Decreased Temperature with Increasing Elevation Decreases the End-Season Leaf-to-Wood Reallocation of Resources in Deciduous *Betula ermanii* Cham. Trees

Yu Cong [1], Mai-He Li [1,2], Kai Liu [1], Yong-Cai Dang [1], Hu-Dong Han [1] and Hong S. He [1,3,*]

[1] School of Geographical Sciences, Northeast Normal University, Changchun 130024, China; congy345@nenu.edu.cn (Y.C.); maihe.li@wsl.ch (M.-H.L.); liuk368@nenu.edu.cn (K.L.); dangyc818@nenu.edu.cn (Y.-C.D.); hanhd254@nenu.edu.cn (H.-D.H.)

[2] Swiss Federal Institute for Forest, Snow and Landscape Research WSL, Zuercherstrasse 111, CH-8903 Birmensdorf, Switzerland

[3] School of Natural Resources, University of Missouri, Columbia, MO 65211, USA

* Correspondence: HeH@missouri.edu; Tel.: +86-1351-421-4004

Received: 27 January 2019; Accepted: 13 February 2019; Published: 15 February 2019

Abstract: Global air temperature has increased and continues to increase, especially in high latitude and high altitude areas, which may affect plant resource physiology and thus plant growth and productivity. The resource remobilization efficiency of plants in response to global warming is, however, still poorly understood. We thus assessed end-season resource remobilization from leaves to woody tissues in deciduous *Betula ermanii* Cham. trees grown along an elevational gradient ranging from 1700 m to 2187 m a.s.l. on Changbai Mountain, northeastern China. We hypothesized that end-season resource remobilization efficiency from leaves to storage tissues increases with increasing elevation or decreasing temperature. To test this hypothesis, concentrations of non-structural carbohydrates (NSCs), nitrogen (N), phosphorus (P), and potassium (K) during peak shoot growth (July) were compared with those at the end of growing season (September on Changbai Mt.) for each tissue type. To avoid leaf phenological effects on parameters, fallen leaves were collected at the end-season. Except for July-shoot NSC and July-leaf K, tissue concentrations of NSC, N, P, and K did not decrease with increasing elevation for both July and September. We found that the end-season leaf-to-wood reallocation efficiency decreased with increasing elevation. This lower reallocation efficiency may result in resource limitation in high-elevation trees. Future warming may promote leaf-to-wood resource reallocation, leading to upward shift of forests to higher elevations. The NSC, N, P, and K accumulated in stems and roots but not in shoots, especially in trees grown close to or at their upper limit, indicating that stems and roots of deciduous trees are the most important storage tissues over winter. Our results contribute to better understand the resource-related ecophysiological mechanisms for treeline formation, and vice versa, to better predict forest dynamics at high elevations in response to global warming. Our study provides resource-related ecophysiological knowledge for developing management strategies for high elevation forests in a rapidly warming world.

Keywords: Alpine treeline; Nitrogen; Non-structural carbohydrates; Phosphorus; Potassium; Remobilization; Storage; Upper limits

---

1. Introduction

Global air temperature has increased and continues to increase, especially in high latitude and high altitude areas. Climate change such as global warming will inevitably influence tree growth rate [1] via changing resource availability and use efficiency [2], and probably also resource allocation and reallocation. Resources can be recycled through remobilization from senescing leaves to storage tissues in deciduous trees which enables plants to reuse these resources for regrowth after dormancy [3,4], and for defense and reproduction [5]. Plants remobilize carbon components and nutrients through internal resources cycling from senescing tissues to maximize the resource use efficiency and to minimize the costs [5]. Thus, remobilization is an internal conservation process which can contribute a substantial annual nutrient supply to increase the resource use efficiency [6]. Owing to the importance of resource resorption for plants, differences in patterns of resource remobilization have been exhaustively studied at the intra- and interspecific levels [3,7,8]. The main recycled carbon components in trees are non-structural carbohydrates (NSCs), which are mainly stored in ligneous tissues, particularly in stems and roots [5,8]. Carbohydrate storage is particularly important for deciduous trees that lack photosynthesizing tissues in early spring [9,10]. As a result of greater asynchrony of supply and demand undergone by deciduous species, more carbon compounds are stored in deciduous than in evergreen species [5]. Woody roots or the stems seem to play a crucial role in resource storage of deciduous species [11,12]. Compared to evergreen species that can directly retain resources in over-wintering leaves, rather than translocating them to stems and roots [4], deciduous species recycle resources back to storage tissues before leaf abscission [5]. The resources withdrawn from senescent tissues can be depleted during spring for new growth [13].

Environmental factors have strong impacts on resource remobilization during leaf senescence, because leaf senescence relies most on photoperiod, temperature, precipitation and nutrient availability. For example, resource remobilization efficiency is higher when leaves senesce prematurely as a consequence of low nutrient availability [14–16], indicating an adaptation to infertile habitats [17]. Similarly, recent studies suggested that climatic warming would increase the proficiency of nutrient resorption due to delayed leaf senescence [18]. However, González-Zurdo et al. (2015) studied leaf nitrogen resorption of three evergreen tree species, and concluded that N resorption efficiency was lower when winter temperature was low [19]. Nutrient remobilization efficiency can be lower under water stress, because drought advances leaf abscission [18,20,21]. Current rapid global warming, for example, may result in less water availability in water-limited areas, and warming, in combination with warming-induced changes in water availability, will influence plant ecophysiology such as resource remobilization. Using elevational gradients as a proxy of decreasing temperature, Kutbay and Ok (2003) found that foliar N and P resorption efficiencies did not change in evergreen *Juniperus oxycedrus* L. subsp. *macrocarpa* (Sibth. & Sm.) Ball trees [22]. However, it is unclear how resource remobilization efficiency in deciduous trees responds to low temperature at the alpine treeline and, contrastingly, to future global warming.

We, therefore, studied the effects of elevation on end-season remobilization of carbon and nutrients in deciduous *Betula ermanii* Cham. (Erman's birch) trees grown along an elevational gradient from lower elevation (1700 m a.s.l.) up to the upper limit (2200 m a.s.l.) on Changbai Mountain. We measured tissue NSC, N, P and K concentrations both at the peak shoot growth period and at the end of growing season, and calculated the resource remobilization efficiency of trees along the elevational gradient, to test our hypothesis that the end-season resource remobilization efficiency from leaves to storage tissues in deciduous *B. ermanii* trees increases with increasing elevations, to increase resource use efficiency of trees in a harsh environment.

2. Materials and Methods

2.1. Study Area

The study was conducted in the Natural Reserve, Changbai Mountain (41°59′ N, 127°59′–128° E) in northeastern China. The area was not disturbed because of its remoteness and high elevation. The area has a temperate continental climate [23] with mean growing season (late May to late September) temperature of 5.87 °C (ranging from 3.37 to 8.82 °C) [24]. The annual precipitation ranges from 700 to 1400 mm (data from Tian-Chi Meteorological Station located at 2623 m a.s.l.) [25] and the annual frost-free period is about 65 to 70 days [26]. Soils on the Changbai Mountain are classified as mountain soddy forest soil [25]. The Changbai Mountain has vertical spectra with four vegetation belts [27]: mixed coniferous broad leaved forests distributed from 740 to 1100 m a.s.l., coniferous forests from 1100 to 1700 m a.s.l., birches (*Betula ermanii*) distributed from 1700 to 2200 m a.s.l., with the birch alpine treeline at 2030 m a.s.l., and tundra above 2000 m a.s.l. [22,26].

2.2. Field Sampling

Birch (*B. ermanii*) forests are distributed between 1700 and 2200 m a.s.l. (upper limit) on the Changbai Mt. The *Betula* elevational gradient can be divided by the alpine treeline (2030 m a.s.l.) into two parts, i.e., below the alpine treeline with trees of > 2 m in height and above the alpine treeline with trees of < 2 m in height. We used this elevational gradient as a proxy of environmental change, mainly air warming [1,28]. To compare patterns of remobilization efficiency across environmental gradients from low elevations to high elevations, samples were taken at the peak-growing season and at the end of growing season. We selected end-July (23 July 2017) to collect samples, since for that time the difference in leaf phenology along the elevational gradient was negligible [29] and the physiological processes such as photosynthesis and growth are most active [30]. The end-season samples were collected from the higher elevations downslope to the lower elevations between September 5 and 9, when the nighttime temperature at an elevation dropped to below 0 °C according to a linear model using climate data from Tian-Chi Meteorological Station located at 2623 m a.s.l. and from the Meteorological Station (738 m a.s.l.) of Changbai Mountain Forest Ecosystem Research facility. Five healthy *B. ermanii* trees (n = 5) with similar age, height and diameter at breast height (DBH) or basal diameter (2107 and 2187 m a.s.l.) were selected at each of the six elevations, i.e., 1700, 1800, 1900, 2027, 2107 and 2187 m a.s.l. (Table 1). Using a temperature lapse rate of 0.6 °C decrease with 100 m increase in elevation, the annual mean air temperature is 3 °C lower at 2187 m than at 1700 m a.s.l. One-year-old shoots (with bark), main stem sapwood (with bark) and fine roots (< 0.5 cm in diameter, with bark) were collected from each selected tree. For leaf samples, mature green leaves were collected on 23 July (mid-season samples), whereas only freshly fallen leaves were collected from the forest floor close to each sample tree (end-season samples), to minimize difference in leaf phenology among elevations. Also, to minimize the effects of diurnal temperature variation and sunlight on samples, all samples were taken around noon and immediately stored in a cool box [31]. The samples were killed in a microwave oven at 600 W for 40 s to minimize the physiological activities, dried to a constant weight at 65 °C and then ground to fine powder to pass through a 1-mm sieve for analysis. We measured the height and DBH of each tree at the first sampling date (July) (Table 1). For juvenile trees with DBH of < 5 cm, basal diameter of trees (about 1 cm above ground surface) was measured in July. As limitations have been imposed on tree-harvesting within the Changbai Mountain Nature Reserve, we, therefore, only measured the biomass of leaves, shoots, stems, and roots of 5 average trees at the two highest elevations (2107 and 2187 m a.s.l.) in July 2017.

Table 1. Characteristics of the plots and the sampling trees *Betula ermanii* Cham. (mean ± 1 standard deviation; n = 5 trees) on the Changbai Mountain (Jilin, NE China).

Site No.	Elevation (m a.s.l.)	Average DBH (cm)	Average Height (m)	Slope Exposure
1	2187	1.0 ± 0.1 [a]	0.9 ± 0.1	West
2	2107	1.1 ± 0.1 [a]	1.5 ± 0.1	West
3	2027	4.7 ± 0.4	4.2 ± 0.1	West
4	1900	9.4 ± 1.0	9.3 ± 0.4	West
5	1800	9.2 ± 0.7	9.2 ± 0.5	West
6	1700	10.9 ± 0.7	11.2 ± 0.8	West

[a] basal diameter (about 1 cm above ground surface).

2.3. Analysis of NSC

Plant tissue (30 mg) was placed into a 10-mL centrifuge tube and mixed with 5 mL of 80% ethanol. The mixture was incubated at 80 °C in a water shaker (SHA-C, Jintan Jingda Instrument Manufacturing Co., Ltd., Jintan, China) for 30 min, cooled to ambient temperature, and thereafter centrifuged at 4000 rpm for 10 min. The sediments were re-extracted twice with 80% ethanol (Sinopharm Chemical Reagent Co., Ltd., Shanghai, China) to extract the soluble sugars [29]. The ethanol-insoluble pellets were used for starch extraction, and the combined supernatants were retained for soluble sugars analysis by the anthrone method. Glucose was used as a standard. Starch was extracted from the ethanol-insoluble pellets after placing in water at 80 °C to remove the ethanol by evaporation. The ethanol-insoluble residues were boiled with 2 mL of distilled water for 15 min. After cooling to room temperature, 2 mL of 9.2M $HClO_4$ (Sinopharm Chemical Reagent Co., Ltd., Shanghai, China) was added to hydrolyze the starch for 15 min, 4 mL distilled water was added and mixed thereafter, and then the mixture was centrifuged at 4000 rpm for 10 min. Subsequently, the solid residues were added with 2 mL of 4.6M $HClO_4$ for one more extraction. Soluble sugars and starch concentrations were both measured at 620 nm using a spectrophotometer (TU-1810, Beijing Purkinje General Instrument Co., Ltd., Beijing, China) [32]. The NSC, soluble sugars and starch concentrations were expressed on a dry matter basis (% d.m.).

2.4. Analysis of Nitrogen, Phosphorus and Potassium

Plant material (0.1 g) was digested in concentrated sulphuric acid and hydrogen peroxide [33]. The digested solutions were used for the determination of nitrogen, phosphorus and potassium. Nitrogen (N) concentration was determined by the idophenol blue colorimetric method on an automatic chemical analyzer (SmartChem 140, AMS-Alliance Instruments, Rome, Italy). Phosphorus (P) was colorimetrically determined by the ammonium molybdate-ascorbic acid method [34] on a spectrophotometer (TU-1810, Beijing Purkinje General Instrument Co., Ltd., Beijing, China). Potassium (K) concentration was measured with a flame photometer (FP6410, Shanghai Precise Scientific Instrument Co., Ltd., Shanghai, China).

2.5. Methods for Evaluating Resource Remobilization

Remobilization efficiency (R) was calculated based on differences in concentrations of mobile carbohydrates and nutrients in tissues collected at the peak shoot growth period and at the end season, using the Equation (1):

$$R\% = 100\% \times (C_s - C_j)/C_j = (C_s/C_j - 1) \times 100\% \tag{1}$$

where C_s (September) and C_j (July) represent concentrations of NSC, N, P or K within each tissue type at the end-season (September) and at the peak shoot growth period (July), respectively. A negative R-value ($C_s < C_j$) indicates resource reallocation from that tissue, whereas a positive R-value ($C_s > C_j$) reflects resource accumulation in that tissue at the end-season. A more negative R-value indicates higher remobilization efficiency, whereas a more positive R-value suggests higher accumulation efficiency.

2.6. Data Analysis

Data (NSC, N, P and K) were confirmed for normality (Kolmogorov-Smirnov test) before statistical analysis. Two-way analyses of variance (ANOVAs) were used to test the effects of elevation, time, and their interactions on the concentrations of NSC, N, P and K in each tissue. Two-way ANOVAs were performed with elevation and tissue type as factors, with random selection of trees as random factor to identify the trends in the remobilization efficiency of NSC, N, P, and K. Differences in the parameters studied among elevations were tested for significance using Tukey's HSD test at $\alpha = 0.05$. Because we had equal sample size, Duncan's test at $\alpha = 0.05$ was used to analyze the difference for remobilization means next to each adjacent tree among the elevations.

3. Results

3.1. NSC Concentration

Elevation did not affect leaf NSC but significantly affected the NSC concentrations in shoots, stem sapwood, and roots (Table 2). Tissue NSC concentrations varied significantly with time (season), and time interacted with elevation to influence the NSC concentrations in aboveground tissues (leaves, shoots, and stem sapwood) (Table 1). Leaf NSC concentration was higher in July than in September, whereas NSC concentrations in sink tissues (shoots, stem, and roots) were higher in September than in July (Figure 1a–h). Root NSC concentrations seemed to be stable across elevations in both July and September (Figure 1d,h). Apart from this, clear elevational trends in NSC were found only in July shoots where the NSC concentration significantly decreased with increasing elevation (Figure 1b), and in September stem sapwood where the NSC concentrations were significantly higher at higher elevations (Figure 1g).

Table 2. Results of two-way nested ANOVAs with elevation and time as fixed factors. F and p values are given. NSC = Non-structural carbohydrates.

	NSC		Nitrogen		Phosphorus		Potassium	
	F	p	F	p	F	p	F	p
Leaves								
Elevations (E)	0.42	0.829	20.15	<0.001	25.57	<0.001	7.21	<0.001
Time (T)	27.80	<0.001	478.52	<0.001	554.96	<0.001	125.46	<0.001
E × T	14.82	<0.001	7.19	<0.001	11.61	<0.001	2.74	0.043
1-year-old-shoots								
Elevations (E)	8.25	<0.001	2.11	0.098	3.96	0.009	8.42	<0.001
Time (T)	253.11	<0.001	301.44	<0.001	222.21	<0.001	465.96	<0.001
E × T	5.77	0.001	5.77	0.001	3.53	0.016	9.37	<0.001
Stem sapwood								
Elevations (E)	23.45	<0.001	206.63	<0.001	215.03	<0.001	22.26	<0.001
Time (T)	36.69	<0.001	306.85	<0.001	113.94	<0.001	28.97	<0.001
E × T	11.69	<0.001	101.70	<0.001	49.27	<0.001	28.55	<0.001
Fine roots								
Elevations (E)	3.63	0.014	7.13	<0.001	10.60	<0.001	5.88	0.001
Time (T)	232.16	<0.001	84.80	<0.001	72.13	<0.001	37.25	<0.001
E × T	1.50	0.228	2.28	0.079	1.21	0.333	0.62	0.688

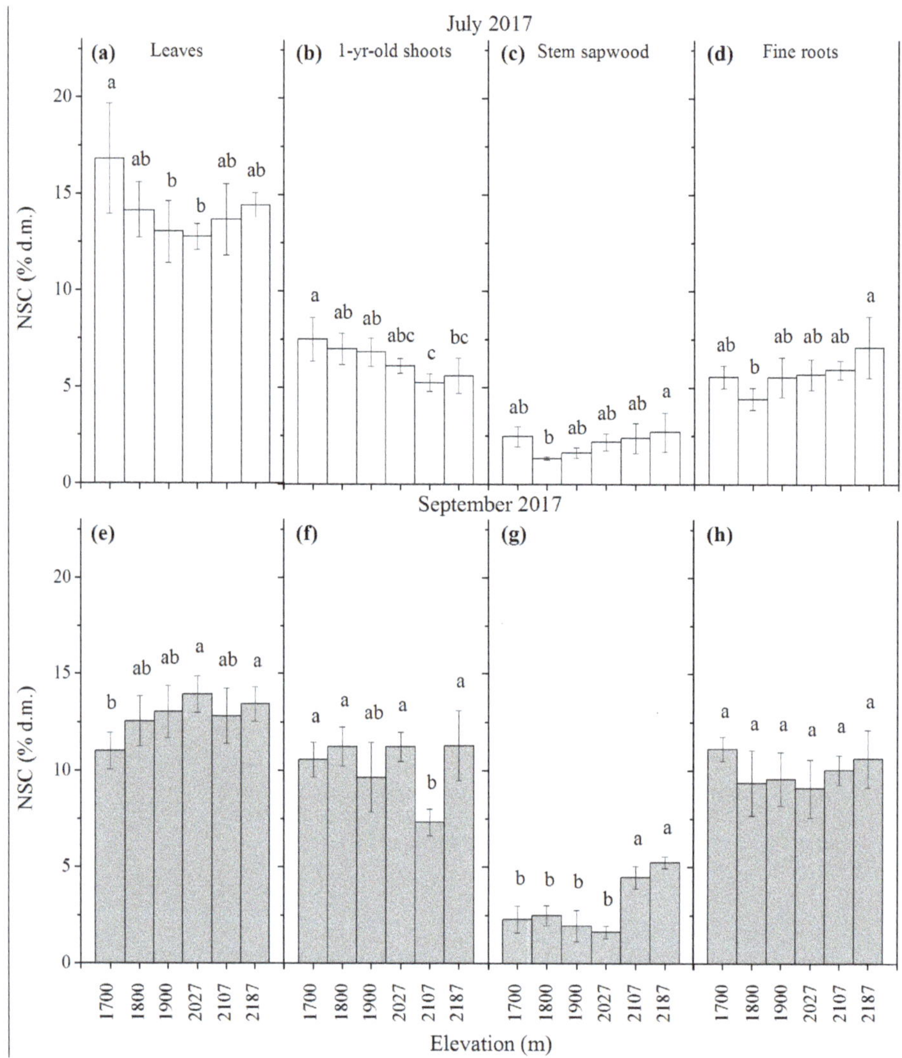

Figure 1. Seasonal tissues concentration (Mean ± 1SD; % of dry matter) changes in non-structural carbohydrate (NSC) compounds of *Betula ermanii* Cham. trees along elevational gradients in Changbai Mountain (*n* = 5 for each elevational site and tissue type). Different letters display significant differences at the 0.05 level among elevations as determined by Tukey's HSD test.

3.2. N Concentration

Elevation significantly affected the N concentrations in leaves, stem sapwood, and roots but not in shoots (Table 2). Tissue N concentrations changed significantly with time (season), and the interaction between time and elevation was significant for N concentrations in aboveground tissues (leaves, shoots, and stem sapwood) (Table 2). Stem N concentration was higher in September than in July, whereas N concentrations in other tissues (leaves, shoots, and roots) were higher in July than in September (Figure 2a–h). In both July and September, trees at higher elevations had higher N concentrations in tissues, except the N concentrations in shoots and roots did not change with elevation in July (Figure 2a–h).

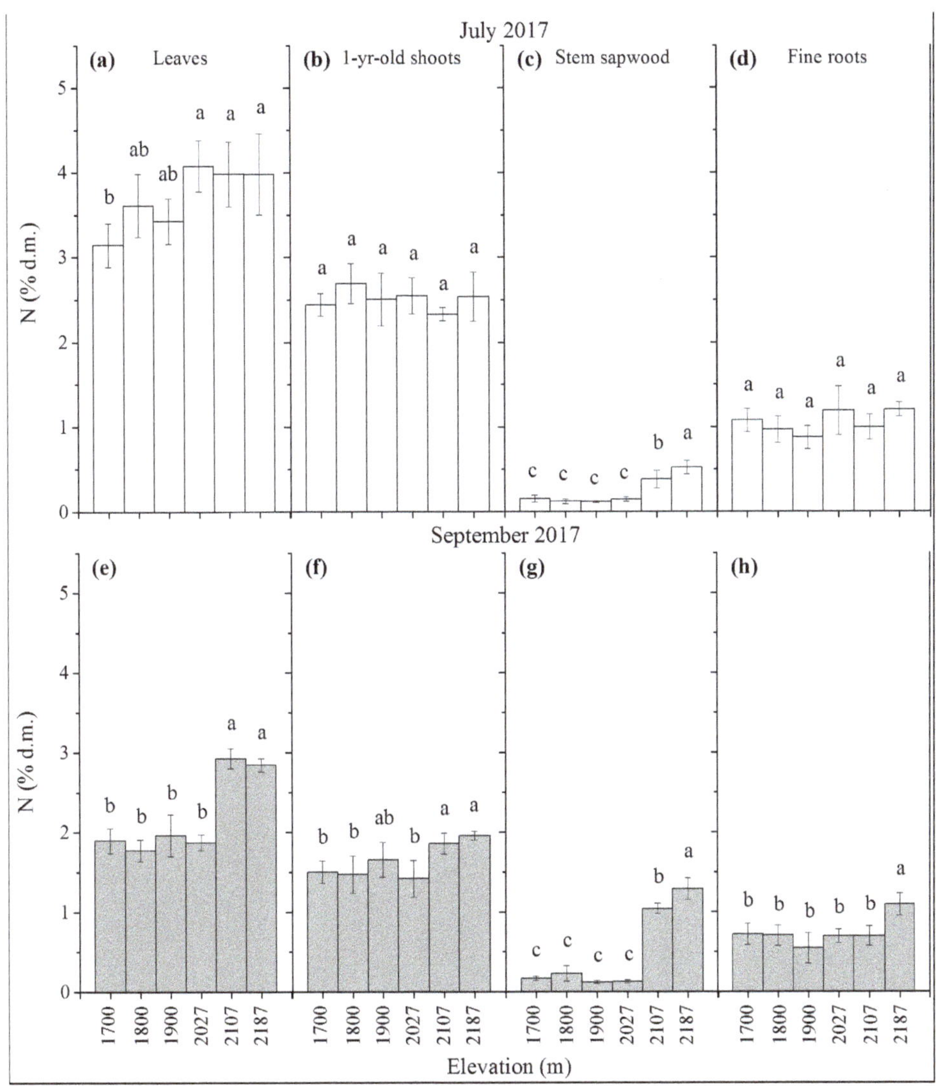

Figure 2. Seasonal tissues concentration (Mean ± 1SD; % of dry matter) changes in total nitrogen (N) of *Betula ermanii* trees along elevational gradients in Changbai Mountain (n = 5 for each elevational site and tissue type). Different letters display significant differences at the 0.05 level among elevations as determined by Tukey's HSD test.

3.3. P Concentration

Both elevation and time (season) significantly affected the tissue P concentrations, and the elevation × time interaction was significant for P concentrations in aboveground tissues (leaves, shoots, and stem sapwood) (Table 2). Leaf and shoot P concentrations were higher in July than in September, whereas stem and root P concentrations were higher in September than in July (Figure 3a–h). Tissue P concentrations were higher in trees at higher elevations in both July and September, except for P in leaves and shoots in July (Figure 3a–h).

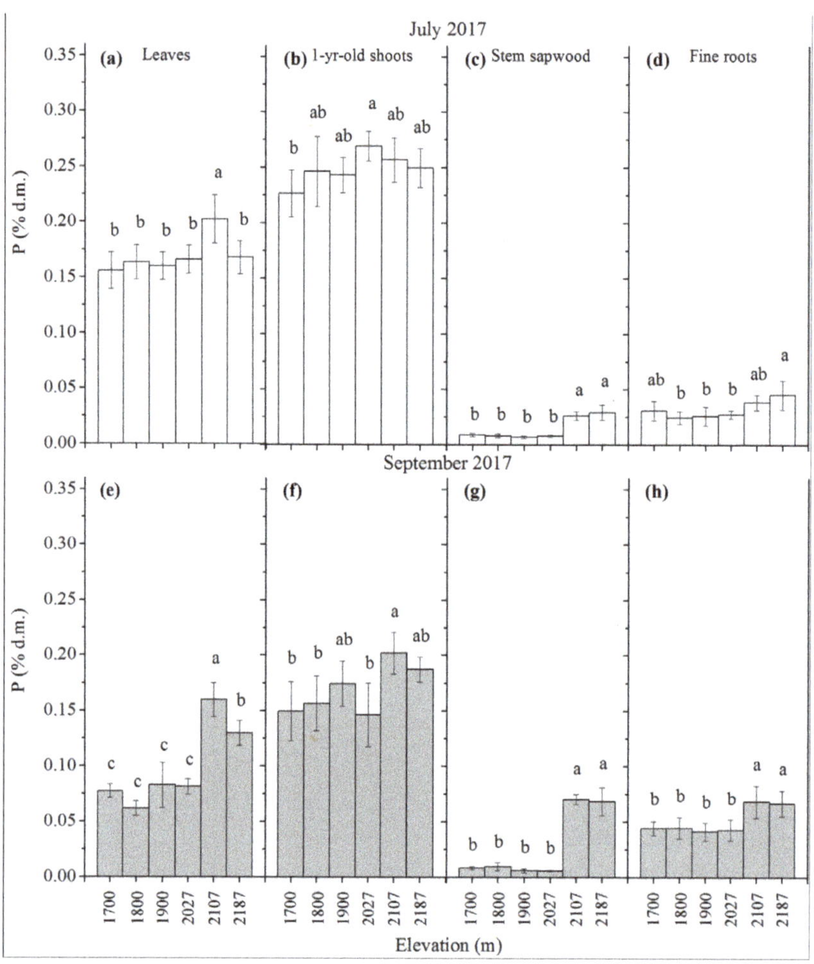

Figure 3. Seasonal tissues concentration (Mean ± 1SD; % of dry matter) changes in total phosphorus (P) of *Betula ermanii* trees along elevational gradients in Changbai Mountain (*n* = 5 for each elevational site and tissue type). Different letters display significant differences at the 0.05 level among elevations as determined by Tukey's HSD test.

3.4. K Concentration

Tissue K concentrations were significantly affected by both elevation and time (season) (Table 2). The interaction between elevation and time was significant for K concentrations in leaves, shoots and stem sapwood (Table 2). Leaf and shoot K concentrations were higher in July than in September, whereas stem and root K concentrations were higher in September than in July (Figure 4a–h). Stem and root K concentrations in July, as well as leaf, shoot and root K in September remained stable among trees at different elevations (Figure 4c,d–f,h). Apart from those, leaf K concentrations significantly decreased but shoot K significantly increased with increasing elevation in July (Figure 4a,b), and stem sapwood had higher K concentrations in trees close to their upper limit in September (Figure 4g).

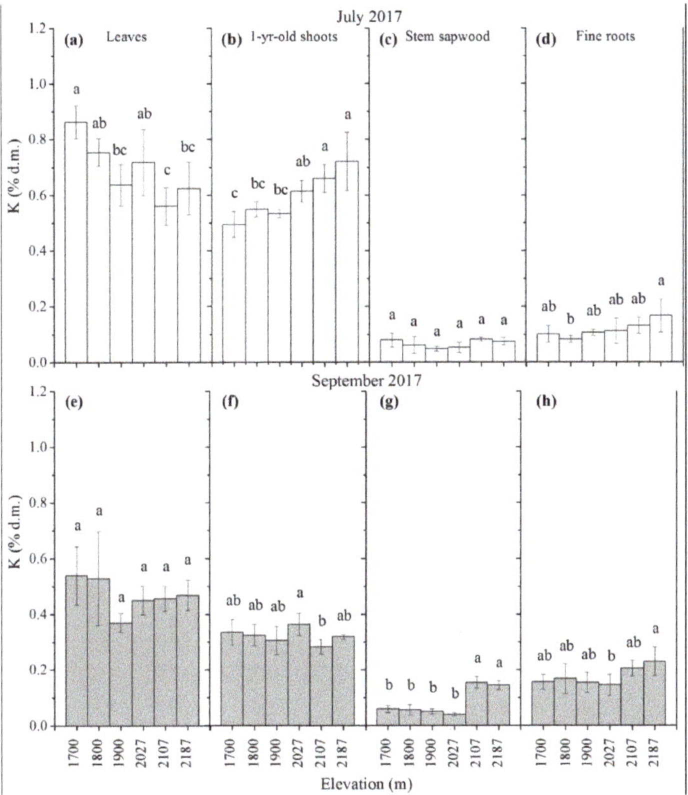

Figure 4. Seasonal tissues concentration (Mean ± 1SD; % of dry matter) changes in total potassium (K) of *Betula ermanii* trees along elevational gradients in Changbai Mountain (*n* = 5 for each elevational site and tissue type). Different letters display significant differences at the 0.05 level among elevations as determined by Tukey's HSD test.

3.5. Elevational Effects on Resource Remobilization

Both elevation and tissue type, and their interaction had significant effects on the remobilization efficiency of NSC, N, P and K (Table 3). At the end of the growing season, leaves reallocated NSC, N, P, and K to other tissues (Figure 5a,e,i,m), and the remobilization efficiency tended to decrease with increasing elevation, and even for an NSC accumulation at 1900 and 2027 m a.s.l. (Figure 5a). Shoots accumulated NSC but remobilized N, P, and K (Figure 5b,f,j,n), and the NSC accumulation (Figure 5b) and K reallocation (Figure 5n) tended to increase but N reallocation (Figure 5f) seemed to decrease with increasing elevation. Stems in trees close to the upper limit accumulated NSC, N, P and K (Figure 5c,g,k,o). Roots accumulated NSC, P and K, but reallocated N to other tissues (Figure 5d,h,l,p). The root NSC accumulation efficiency decreased with increasing elevation (Figure 5d). At the two highest elevations (i.e., 2107 and 2187 m a.s.l.), for example, both the amount of resource reallocation (less negative values at 2187 m compared to 2107 m) and accumulation (positive values) decreased with increasing elevation (Table 4). The mean reallocation efficiency of resources from leaves to other tissues was significantly lower in trees above the alpine treeline than in trees below the treeline (Table 5), and the mean efficiency across the entire transect was 40% (N), 42% (P), 31% (K), and 8% (NSC) (Table 5).

Table 3. Results of two-way nested ANOVAs with elevation and tissue types as fixed factors. The *F* and *p* values are given. R refers to remobilization rate.

	R NSC		R Nitrogen		R Phosphorus		R Potassium	
	F	*p*	*F*	*p*	*F*	*p*	*F*	*p*
Elevations (E)	8.67	**<0.001**	17.20	**<0.001**	10.72	**<0.001**	3.49	**0.016**
Tissue types (T)	287.57	**<0.001**	42.36	**<0.001**	79.04	**<0.001**	84.44	**<0.001**
E × T	5.73	**<0.001**	2.50	**0.005**	3.22	**<0.001**	2.50	**0.005**

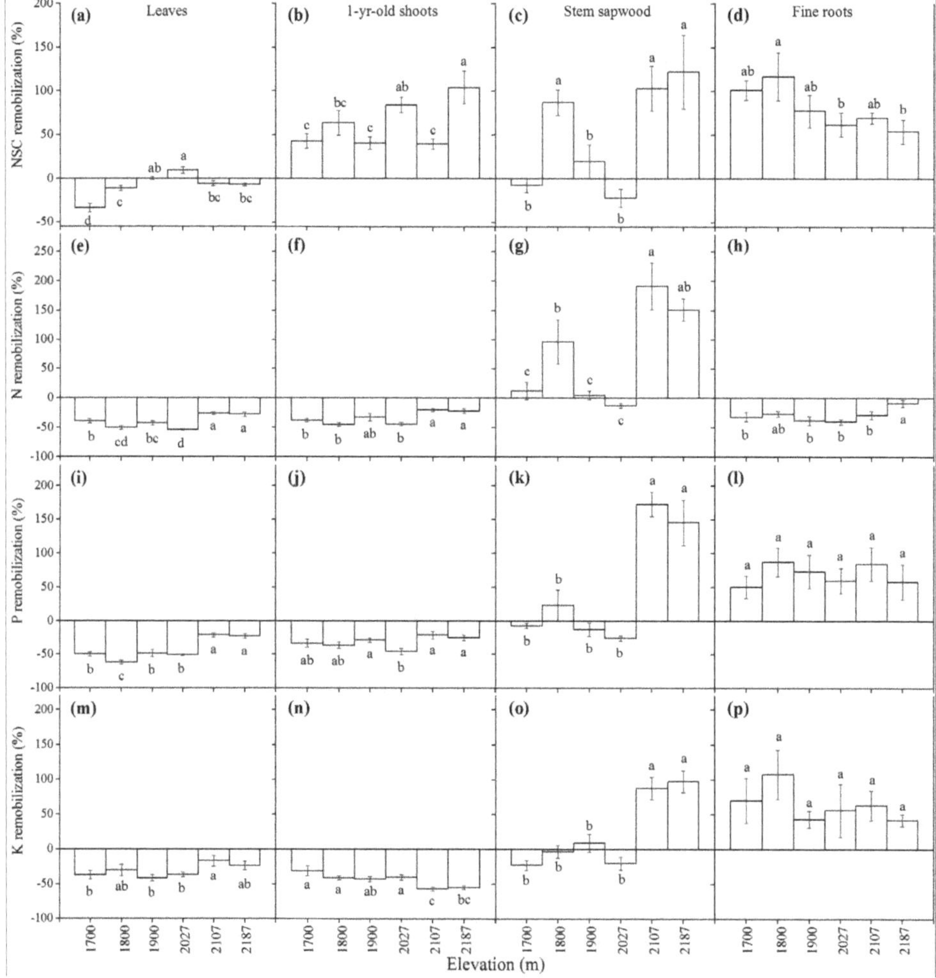

Figure 5. Carbohydrates, N, P and K remobilization efficiency (Mean ± 1SE) (R%) in *Betula ermanii* trees along the altitudinal gradients in Changbai Mountain (*n* = 5 for each elevational site and tissue type). Different letters display significant differences at the 0.05 level among elevations as determined by Duncan's test (A negative *R*-value indicates resource translocation, whereas a positive *R*-value reflects resource accumulation).

Table 4. End-season resource reallocation (mean value $\pm$ 1 standard deviation; n = 5 trees; calculated using biomass $\times$ reallocation rate) for *Betula ermanii* trees at 2187 m (upper limit of distribution) and 2107 m a.s.l. on Changbai Mt. Different letters display significant differences at the 0.05 level among elevations as determined by Tukey's HSD test. Negative values (dark-highlighted) indicate resource "loss" due to resource reallocation, and positive values indicate resource "gain" because of accumulation.

Elevation (m a.s.l.)	NSC (g)	N (g)	P (g)	K (g)
	Leaves			
2187	-1.4 ± 0.4^a	-5.8 ± 1.5^b	-4.3 ± 0.8^b	-5.3 ± 1.9^a
2107	-3.0 ± 1.3^a	-12.9 ± 2.7^a	-10.5 ± 2.8^a	-4.7 ± 3.6^a
	1-year-old shoots			
2187	31.1 ± 8.3^a	-7.0 ± 2.3^a	-7.6 ± 2.2^a	-17.4 ± 4.0^a
2107	26.6 ± 9.9^a	-17.1 ± 7.0^a	-18.2 ± 8.1^a	-43.7 ± 16.3^a
	Stem sapwood			
2187	63.0 ± 21.2^b	77.5 ± 10.8^b	76.7 ± 20.1^b	49.8 ± 7.6^b
2107	240.4 ± 72.6^a	391.5 ± 65.1^a	369.6 ± 50.5^a	193.4 ± 43.3^a
	Fine roots			
2187	22.0 ± 6.7^b	-4.1 ± 2.2^b	22.3 ± 8.8^a	17.0 ± 4.7^a
2107	85.1 ± 15.1^a	-36.7 ± 13.1^a	102.7 ± 34.7^a	59.3 ± 26.5^a

Table 5. Mean end-season reallocation rate (%) of resources (NSC, N, P, K) from leaves to other tissues in *Betula ermanii* trees grown at higher elevations (2107 and 2187 m a.s.l.) and lower elevations (1700, 1800, 1900, and 2027 m), separated by the alpine treeline located at 2030 m a.s.l. on the Changbai Mt., northeastern China.

	Trees across the Entire Transect	Trees below the Alpine Treeline	Trees above the Alpine Treeline
Non-structural carbohydrates (NSC)	8	9	6
Nitrogen (N)	40	47	27
Phosphorus (P)	42	52	22
Potassium (K)	31	36	20

4. Discussion

4.1. Tissue- and Resource-Dependent Reallocation or Accumulation

At the end of season, NSC reallocated from leaves to woody tissues (Figure 5a–d) and N accumulated only in stems (Figure 5g), whereas P and K seemed to store in stems and roots (Figure 5k,l,o,p), indicating tissue- and resource-dependent reallocation or accumulation (Table 4). *B. ermanii* leaves reallocated NSC to other tissues (Figure 5a; Table 4), leading to decreases in leaf NSC (Figure 1e vs. Figure 1a) but increases in NSC concentrations in shoots (Figure 1f vs. Figure 1b), stem sapwood (Figure 1g vs. Figure 1c), and roots (Figure 1h vs. Figure 1d), and thus an NSC accumulation in those storage tissues (Figure 5b–d). In line with previous studies, our results confirmed that leaf carbon compounds in deciduous trees were reallocated and stored in ligneous tissues, particularly in stems and roots [5,8]. Leaf carbon components recycled through remobilization from leaves to storage tissues before leaf senescence are particularly important for regrowth of new leaves in a leafless state of deciduous trees in early spring [5,9,10]. We found that the NSC remobilization efficiency of *B. ermanii* leaves ranged from 6% at high elevation to 33% at low elevation (Figure 5a). Chapin et al. (1990) proposed a carbon remobilization efficiency of ~10% based on an assumption that the loss of leaf weight was a direct result of carbon remobilization [5]. Similarly, Eckstein et al. (1998) proposed a carbon remobilization efficiency of 6%–13% for deciduous species [35].

Nitrogen (N) is the most important stored compound [36], paralleling to carbohydrates in supporting growth and reproduction [37]. Our results indicated that all tissue types decreased in N concentration during leaf senescence except stems (Figure 5e–h), suggesting that stems participated in the N accumulation rather than N translocation. Our results agreed with previous studies, showing that N remobilized from senescing tissues (e.g., leaves and shoots) for storage prior to leaf abscission [3,37,38]. As previously reported for *Betula pendula* Roth., the deciduous broadleaf tree tended to store N in roots and stem [11,12]. Our findings differ from coniferous evergreen species, which store N in the youngest age class of needle [11,13]. N stored in specific tissues with an individual species, dependent on leaf habit [37]. In addition, the mean leaf N remobilization efficiency of 40% (Table 5) was lower than that value reported by Aerts (1996) for deciduous shrubs and trees (mean 54%) [3]. In another study, Vergutz et al. (2012) showed that, based on a leaf mass loss correction, average nutrient resorption was 62% for N during senescence [39].

Apart from mobile carbohydrates and nitrogen (N), plants also store and remobilize macronutrients such as phosphorus (P) and potassium (K) [38,40]. By the process of leaf senescence, P and K is reallocated from leaves and shoots to and stored in roots and stems (Figure 5i–p). The mean efficiency of P remobilized from senescing leaves was 42% (Table 5), which is lower than that reported by Aerts (1996) for deciduous shrubs and trees (50% for N) [3]. Vergutz et al. (2012) even suggested that, in general, leaf average P resorption value was 65%, after correction for leaf mass loss [39]. Similar results were found for K. As K occurs in ionic form but not in macromolecule form [18], K is thus highly mobile and has been transferred to storage tissues (i.e., stems and roots) before leaf abscission. However, our remobilization efficiency from senescing leaves of 31% (Table 5) was much lower than published resorption efficiency. Chapin et al. (1990) and Vergutz et al. (2012) proposed 50 and 70%, respectively [5,39].

As previously reported for deciduous species, carbohydrate remobilization took place in the leaves, indicating that carbon components recycled from senescing leaves (Figure 5a) are used as a carbon source for new growth [5,41]. But unlike carbohydrates, nitrogen was recycled from senescing leaves (Figure 5e), shoots (Figure 5f) and roots (Figure 5h) to support new leaf and shoot growth, showing that senescing tissues play an important role as potential supplier of nitrogen. Besides, our findings were in agreement with other studies that demonstrate N storage in roots and stems in broad-leaved deciduous species [11,12]. P and K were reallocated from leaves and shoots to stems and roots at the end-season (Figure 5i–p), although these patterns differed from those of NSC and N mentioned above (Table 4). In *B. ermanii* trees, 6%–33% of the leaf mobile carbohydrates was recycled at the end-season. However, about half the N and P was remobilized from senescing leaves before leaves were shed, and 36% of the leaf K was resorbed during leaf senescence. Our results support the fact that resource remobilization is composed of a large storage of nutrients but a small storage of carbohydrates [5].

4.2. Elevational Effects on Resource Remobilization

In the present study, we calculated the remobilization efficiency using the resource concentration within the same tissue type measured at two time points (end-season vs. peak growth season), which may provide insights into resource remobilization processes during leaf senescence. This calculation revealed that the end-season remobilization efficiency of NSC, N, P, and K from leaves to storage tissues tended to decrease with increasing elevation (Figure 5a,e,i,m), which does not support our hypothesis that expected an increase in remobilization efficiency at higher elevations, especially at the upper limit. In our previous study [25], we calculated the remobilization efficiency using the end-season difference in the resource concentrations between leaves (source) and shoots (sink), and found that the remobilization efficiency from leaves to shoots increased with increasing elevation. Using that method [25], we re-calculated the end-season remobilization efficiency from leaves to shoots for the present study and found that NSC remobilization efficiency had a bimodal curve of R% with a bigger remobilization efficiency (smaller R% value) at the upper distribution limit (Figure 6a). Leaf to shoot K reallocation efficiency did not show any response to elevational gradients (Figure 6d), and leaf to shoot N and P reallocation efficiency tended to be lower above the treeline compared to below the treeline (Figure 6b,c). This comparison suggests that a standard method for calculating the remobilization efficiency and thus for understanding the reallocation processes of resources is urgently needed. Previous studies proposed that resource remobilization efficiency can be higher at low nutrient availability or in low temperature condition [14,16], representing an adaptation to stress environment [17]. However, Kutbay and Ok (2003) have demonstrated that the absolute and proportional N and P resorption efficiency did not change significantly along elevational gradients [22,29,31,42]. Many reports have revealed the importance of temperature on regulating leaf senescence and fall [18]. For example, leaf senescence and fall were delayed, in response to warmer temperature [43–45]. The timing or phenology of leaf abscission is considered as one of the reasons leading to altered resorption patterns of nutrients. To minimize such leaf phenological effects on resource remobilization, we did not collect end-season leaves from trees, instead, we collected the fallen leaves as end-season samples in the present study. However, *B. ermanii* in low temperature at the upper limit may have less time to complete nutrient resorption from leaves, which may result in a lower resorption of resources [18], as discussed in our previous study [25].

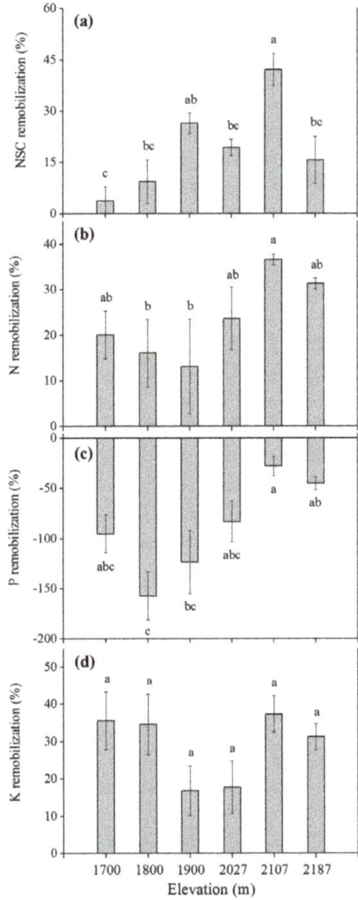

Figure 6. Carbohydrates, N, P and K remobilization efficiency (Mean $\pm$ 1SE) (R%) in *Betula ermanii* trees along the altitudinal gradients in Changbai Mountain (n = 5 for each elevational site). R% was calculated using $R\% = ((C_l - C_s)/C_l) \times 100\% = (1 - C_s/C_l) \times 100\%$, where C_l and C_s represent concentrations in leaves and shoots at the end of growing season, respectively. Different letters display significant differences at the 0.05 level among elevations as determined by Tukey's HSD test (The larger the R-value, the lower the remobilization efficiency is).

Stem and shoot NSC accumulation increased but root NSC accumulation decreased with increasing elevation (Figure 5b–d). Given that the NSC accumulation increased in stems and shoots up to the upper limit, this process may act to resist the harsh cold alpine environment for survival in winter [46], because soluble sugars participate in cell osmotic regulation and prevent intracellular ice formation by decreasing the freezing point of the cytoplasm [47]. Moreover, accumulation of soluble sugars would prime stems for recovery after the stress is alleviated, and stem is an important organ not only for long-distance carbon transport, but also for the regulation of the tree's carbon balance [48].

Similarly, our results showed that N, P, and K were also accumulated in stems, especially in trees grown close to or at their upper limit (Figure 5g,k,o). Oleksyn et al. (1998) and Körner (1999) have reported that nutrient accumulation and conservation are adaptive responses that enhance metabolic activity and growth rates in harsh cold environment [49,50]. In such situations, the tree would internally store more recycling nutrients to recover from harsh environment because of losses in respiration. The reserve storage is a risk-aversion function to minimize risk of catastrophic loss [5]. The greater the risk (high frequency or large losses), the more the tree should invest in internally stored reserves.

5. Conclusions

Many studies have stated that the end-season resource remobilization is strategic since it supports growth at the beginning of bud break in deciduous species [18,49,50]. Except for July-shoot NSC and July-leaf K, tissue concentrations of NSC, N, P, and K did not decrease with increasing elevation for both July and September. Our results partly supported our hypothesis that resources (i.e., mobile carbohydrates, N, P, and K) were reallocated from leaves to storage tissues at the end of growing season in deciduous *B. ermanii* trees. In an economic view, shoots should be an important storage tissue for resources to reduce the transport costs at both the end-season and early growing season. However, our study indicates that stems and roots of deciduous *B. ermanii* trees at high elevations are the most important storage tissues for resources over winter. Inconsistent with our initial hypothesis, we found trees had lower end-season reallocation efficiency of NSC, N, P, and K from leaves to woody tissues at the higher elevations above the alpine treeline compared to lower elevations (Table 5; Figure 5a,e,i,m), and this may result in a resource limitation in high-elevation trees and further limit tree growth at high elevations. Our results contribute to better understand the resource-related ecophysiological mechanisms for treeline formation, and vice versa, to better predict treeline dynamics in response to global warming. Our study provides resource-related ecophysiological knowledge for developing management strategies for high elevation forests in a rapidly warming world. To more closely explore carbon and nutrient remobilization mechanisms, further studies are needed to estimate the pool size of resources remobilized (concentration × biomass, see Table 4) [51] and their stoichiometry [52]. Furthermore, stable isotope labeling experiments would be taken into account (e.g., ^{13}C, ^{15}N) to accurately evaluate carbon and N remobilization from senescing tissues, and to determine which tissues contribute to the most internal resource remobilization [13].

Author Contributions: Conceptualization, Y.C., M.-H.L. and H.-S.H.; methodology, M.-H.L. and H.-S.H.; software, Y.C.; investigation, K.L., Y.-C.D., H.-D.H.; data analysis, Y.C. and M.-H.L.; writing—original draft preparation, Y.C. and M.-H.L.; writing—review and editing, M.-H.L. and H.-S.H.; funding acquisition, M.-H.L. and H.-S.H.

Funding: This research was funded by the National Key Research and Development Program of China (No. 2017YFA0604403, 2016YFA0602301), the National Natural Science Foundation of China (No. 41371076, 41601052, 41501089).

Acknowledgments: We sincerely thank Danae Yu and two anonymous referees for valuable comments on our manuscript, Xinhua Zhou for her assistance in the laboratory and Samuel Allen for English proofreading to improve our manuscript.

Conflicts of Interest: The authors declare no conflict of interest.

References

1. Li, M.; Yang, J. Effects of elevation and microsite on growth of *Pinus cembra* in the subalpine zone of the Austrian Alps. *Annu. For. Sci.* **2004**, *61*, 319–325. [CrossRef]
2. Bhattacharya, A. *Changing Climate and Resource use Efficiency in Plants*; Academic Press: London, UK, 2018.
3. Aerts, R. Nutrient resorption from senescing leaves of perennials: Are there general patterns? *J. Ecol.* **1996**, *84*, 597–608. [CrossRef]
4. Cherbuy, B.; Joffre, R.; Gillon, D.; Rambal, S. Internal remobilization of carbohydrates, lipids, nitrogen and phosphorus in the Mediterranean evergreen oak *Quercus ilex*. *Tree Physiol.* **2001**, *21*, 9–17. [CrossRef] [PubMed]
5. Chapin, F.; Schulze, E.; Mooney, H. The ecology and economics of storage in plants. *Annu. Rev. Ecol. Syst.* **1990**, *21*, 423–447. [CrossRef]
6. Li, Y.; Lan, G.; Xia, Y. Rubber Trees demonstrate a clear retranslocation under seasonal drought and cold stresses. *Front. Plant Sci.* **2016**, *7*. [CrossRef] [PubMed]
7. Chapin, F.; Kedrowski, R.A. Seasonal changes in nitrogen and phosphorus fractions and autumn retranslocation in evergreen and deciduous taiga trees. *Ecology* **1983**, *64*, 376–391. [CrossRef]
8. Millard, P. Ecophysiology of the internal cycling of nitrogen for tree growth. *Zeitschrift für Pflanzenernährung Bodenkunde* **1996**, *159*, 1–10. [CrossRef]
9. Marchi, S.; Sebastiani, L.; Gucci, R.; Tognetti, R. Changes in sink-source relationships during shoot development in olive. *J. Am. Soc. Hortic. Sci.* **2005**, *130*, 631–637. [CrossRef]
10. Marchi, S.; Sebastiani, L.; Gucci, R.; Tognetti, R. Sink-source transition in peach leaves during shoot development. *J. Am. Soc. Hortic. Sci.* **2005**, *130*, 928–935. [CrossRef]
11. Millard, P.; Hester, A.; Wendler, R.; Baillie, G. Interspecific defoliation responses of trees depend on sites of winter nitrogen storage. *Funct. Ecol.* **2001**, *15*, 535–543. [CrossRef]
12. Millard, P.; Wendler, R.; Hepburn, A.; Smith, A. Variations in the amino acid composition of xylem sap of *Betula pendula* Roth. trees due to remobilization of stored N in the spring. *Plant Cell Environ.* **1998**, *21*, 715–722. [CrossRef]
13. Millard, P.; Proe, M.F. Storage and internal cycling of nitrogen in relation to seasonal growth of *Sitka spruce*. *Tree Physiol.* **1992**, *10*, 33–43. [CrossRef] [PubMed]
14. Bridgham, S.D.; Pastor, J.; McClaugherty, C.A.; Richardson, C.J. Nutrient-use efficiency: A litterfall index, a model, and a test along a nutrient-availability gradient in North Carolina peatlands. *Am. Nat.* **1995**, *145*, 1–21. [CrossRef]
15. Maillard, A.; Diquélou, S.; Billard, V.; Laîné, P.; Garnica, M.; Prudent, M.; Garcia-Mina, J.-M.; Yvin, J.-C.; Ourry, A. Leaf mineral nutrient remobilization during leaf senescence and modulation by nutrient deficiency. *Front. Plant Sci.* **2015**, *6*, 317. [CrossRef] [PubMed]
16. Pugnaire, F.I.; Chapin, F.S. Environmental and physiological factors governing nutrient resorption efficiency in barley. *Oecologia* **1992**, *90*, 120–126. [CrossRef] [PubMed]
17. Dissanayaka, D.M.S.B.; Maruyama, H.; Nishida, S.; Tawaraya, K.; Wasaki, J. Landrace of japonica rice, Akamai exhibits enhanced root growth and efficient leaf phosphorus remobilization in response to limited phosphorus availability. *Plant Soil* **2017**, *414*, 327–338. [CrossRef]
18. Estiarte, M.; Peñuelas, J. Alteration of the phenology of leaf senescence and fall in winter deciduous species by climate change: Effects on nutrient proficiency. *Glob. Chang. Biol.* **2015**, *21*, 1005–1017. [CrossRef]
19. González-Zurdo, P.; Escudero, A.; Mediavilla, S. N resorption efficiency and proficiency in response to winter cold in three evergreen species. *Plant Soil* **2015**, *394*, 87–98. [CrossRef]
20. Killingbeck, K.T. Nutrient Resorption. In *Plant Cell Death Processes*; Academic Press: San Diego, CA, USA, 2004; pp. 215–226.
21. Etienne, P.; Diquelou, S.; Prudent, M.; Salon, C.; Maillard, A.; Ourry, A. Macro and micronutrient storage in plants and their remobilization when facing scarcity: the case of drought. *Agriculture* **2018**, *8*, 14. [CrossRef]
22. Kutbay, H.; Ok, T. Foliar N and P resorption and nutrient levels along an elevational gradient in *Juniperus oxycedrus* L. subsp. *macrocarpa* (Sibth. & Sm.) Ball. *Ann. For. Sci.* **2003**, *60*, 449–454.
23. Du, H.; Liu, J.; Li, M.-H.; Büntgen, U.; Yang, Y.; Wang, L.; Wu, Z.; He, H.S. Warming-induced upward migration of the alpine treeline in the Changbai Mountains, northeast China. *Glob. Chang. Biol.* **2018**, *24*, 1256–1266. [CrossRef] [PubMed]

24. Zong, S.; He, H.; Liu, K.; Du, H.; Wu, Z.; Zhao, Y.; Jin, H. Typhoon diverged forest succession from natural trajectory in the treeline ecotone of the Changbai Mountains, Northeast China. *For. Ecol. Manag.* **2018**, *407*, 75–83. [CrossRef]

25. Cong, Y.; Wang, A.; He, H.; Yu, F.; Tognetti, R.; Cherubini, P.; Wang, X.; Li, M.H. Evergreen *Quercus aquifolioides* remobilizes more soluble carbon components but less N and P from leaves to shoots than deciduous *Betula ermanii* at the end-season. *iForest* **2018**, *11*, 517–525. [CrossRef]

26. Yu, D.; Wang, Q.; Liu, J.; Zhou, W.; Qi, L.; Wang, X.; Zhou, L.; Dai, L. Formation mechanisms of the alpine Erman's birch (*Betula ermanii*) treeline on Changbai Mountain in Northeast China. *Trees* **2014**, *28*, 935–947. [CrossRef]

27. Liu, Q.-J.; Li, X.-R.; Ma, Z.-Q.; Takeuchi, N. Monitoring forest dynamics using satellite imagery—A case study in the natural reserve of Changbai Mountain in China. *For. Ecol. Manag.* **2005**, *210*, 25–37. [CrossRef]

28. Li, M.; Yang, J.; Kräuchi, N. Growth responses of *Picea abies* and *Larix decidua* to elevation in subalpine areas of Tyrol, Austria. *Can. J. For. Res.* **2003**, *33*, 653–662. [CrossRef]

29. Li, M.; Jiang, Y.; Wang, A.; Li, X.; Zhu, W.; Yan, C.-F.; Du, Z.; Shi, Z.; Lei, J.; Schönbeck, L.; et al. Active summer carbon storage for winter persistence in trees at the cold alpine treeline. *Tree Physiol.* **2018**, *38*, 1345–1355. [CrossRef]

30. Yamaguchi, D.P.; Nakaji, T.; Hiura, T.; Hikosaka, K. Effect of seasonal change and experimental warming on the temperature dependence of photosynthesis in the canopy leaves of *Quercus serrata*. *Tree Physiol.* **2016**, *36*, 1283–1295. [CrossRef]

31. Li, M.; Xiao, W.-F.; Shi, P.; Wang, S.-G.; Zhong, Y.-D.; Liu, X.-L.; Wang, X.-D.; Cai, X.-H.; Shi, Z.-M. Nitrogen and carbon source–sink relationships in trees at the Himalayan treelines compared with lower elevations. *Plant Cell Environ.* **2008**, *31*, 1377–1387. [CrossRef]

32. Wang, X.; Xu, Z.; Yan, C.; Luo, W.; Wang, R.; Han, X.; Jiang, Y.; Li, M.-H. Responses and sensitivity of N, P and mobile carbohydrates of dominant species to increased water, N and P availability in semi-arid grasslands in northern China. *J. Plant Ecol.* **2017**, *10*, 486–496. [CrossRef]

33. Parkinson, J.A.; Allen, S.E. A wet oxidation procedure suitable for the determination of nitrogen and mineral nutrients in biological material. *Commun. Soil Sci. Plant Anal.* **1975**, *6*, 1–11. [CrossRef]

34. Murphy, J.; Riley, J.P. A modified single solution method for the determination of phosphate in natural waters. *Anal. Chim. Acta* **1962**, *27*, 31–36. [CrossRef]

35. Eckstein, R.L.; Karlsson, P.S.; Weih, W. The significance of resorption of leaf resources for shoot growth in evergreen and deciduous woody plants from a subarctic environment. *Oikos* **1998**, *81*, 567–575. [CrossRef]

36. Millard, P.; Sommerkorn, M.; Grelet, G.A. Environmental change and carbon limitation in trees: a biochemical, ecophysiological and ecosystem appraisal. *New Phytol.* **2007**, *175*, 11–28. [CrossRef] [PubMed]

37. Millard, P.; Grelet, G.-A. Nitrogen storage and remobilization by trees: ecophysiological relevance in a changing world. *Tree Physiol.* **2010**, *30*, 1083–1095. [CrossRef] [PubMed]

38. Nambiar, E.K.S.; Fife, D.N. Nutrient retranslocation in temperate conifers. *Tree Physiol.* **1991**, *9*, 185–207. [CrossRef]

39. Vergutz, L.; Manzoni, S.; Porporato, A.; Novais, R.F.; Jackson, R.B. Global resorption efficiencies and concentrations of carbon and nutrients in leaves of terrestrial plants. *Ecol. Monogr.* **2012**, *82*, 205–220. [CrossRef]

40. Villar-Salvador, P.; Uscola, M.; Jacobs, D.F. The role of stored carbohydrates and nitrogen in the growth and stress tolerance of planted forest trees. *New For.* **2015**, *46*, 813–839. [CrossRef]

41. Gessler, A.; Treydte, K. The fate and age of carbon—Insights into the storage and remobilization dynamics in trees. *New Phytol.* **2016**, *209*, 1338–1340. [CrossRef]

42. Körner, C. The nutritional status of plants from high altitudes. A worldwide comparison. *Oecologia* **1989**, *81*, 379–391. [CrossRef]

43. Chung, H.; Muraoka, H.; Nakamura, M.; Han, S.; Muller, O.; Son, Y. Experimental warming studies on tree species and forest ecosystems: A literature review. *J. Plant Res.* **2013**, *126*, 447–460. [CrossRef] [PubMed]

44. Gunderson, C.A.; Edwards, N.T.; Walker, A.V.; O'Hara, K.H.; Campion, C.M.; Hanson, P.J. Forest phenology and a warmer climate—growing season extension in relation to climatic provenance. *Glob. Chang. Biol.* **2012**, *18*, 2008–2025. [CrossRef]

45. Xu, Z.; Hu, T.; Zhang, Y. Effects of experimental warming on phenology, growth and gas exchange of treeline birch (*Betula utilis*) saplings, Eastern Tibetan Plateau, China. *Eur. J. For. Res.* **2012**, *131*, 811–819. [CrossRef]

46. Sheen, J.; Zhou, L.; Jang, J.-C. Sugars as signaling molecules. *Curr. Opin. Plant Biol.* **1999**, *2*, 410–418. [CrossRef]
47. Morin, X.; Améglio, T.; Ahas, R.; Kurz-Besson, C.; Lanta, V.; Lebourgeois, F.; Miglietta, F.; Chuine, I. Variation in cold hardiness and carbohydrate concentration from dormancy induction to bud burst among provenances of three European oak species. *Tree Physiol.* **2007**, *27*, 817–825. [CrossRef] [PubMed]
48. Furze, M.E.; Trumbore, S.; Hartmann, H. Detours on the phloem sugar highway: Stem carbon storage and remobilization. *Curr. Opin. Plant Biol.* **2018**, *43*, 89–95. [CrossRef] [PubMed]
49. Oleksyn, J.; Modrzýnski, J.; Tjoelker, M.G.; Z·ytkowiak, R.; Reich, P.B.; Karolewski, P. Growth and physiology of *Picea abies* populations from elevational transects: common garden evidence for altitudinal ecotypes and cold adaptation. *Funct. Ecol.* **1998**, *12*, 573–590. [CrossRef]
50. Körner, C. *Alpine Plant Life: Functional Plant Ecology of High Mountain Ecosystems*; Springer: Berlin, Genmary, 1999.
51. Li, M.; Kräuchi, N.; Dobbertin, M. Biomass distribution of different-aged needles in young and old Pinus cembra trees at highland and lowland sites. *Trees* **2006**, *20*, 611–618. [CrossRef]
52. Wang, A.; Wang, X.; Tognetti, R.; Lei, J.-P.; Pan, H.-L.; Liu, X.-L.; Jiang, Y.; Wang, X.-Y.; He, P.; Yu, F.-H.; et al. Elevation alters carbon and nutrient concentrations and stoichiometry in *Quercus aquifolioides* in southwestern China. *Sci. Total Environ.* **2018**, *622*, 1463–1475. [CrossRef] [PubMed]

Article

Importance of the Local Environment on Nutrient Cycling and Litter Decomposition in a Tall Eucalypt Forest

Jessie C. Buettel *, Elise M. Ringwaldt, Mark J. Hovenden and Barry W. Brook

Discipline of Biological Sciences, School of Natural Sciences, University of Tasmania, Hobart 7000, Australia;
Elise.Ringwaldt@utas.edu.au (E.M.R.); Mark.Hovenden@utas.edu.au (M.J.H.); Barry.Brook@utas.edu.au (B.W.B.)
* Correspondence: jessie.buettel@utas.edu.au; Tel.: +614-57-666-016

Received: 28 February 2019; Accepted: 3 April 2019; Published: 16 April 2019

Abstract: The relative abundance of nitrogen-fixing species has been hypothesised to influence tree biomass, decomposition, and nitrogen availability in eucalypt forests. This prediction has been demonstrated in experimental settings (two-species mixtures) but is yet to be observed in the field with more realistically complex communities. We used a combination of (a) field measurements of tree-community composition, (b) sampling of soil from a subset of these sites (i.e., the local environment), and (c) a decomposition experiment of forest litter to examine whether there is a local-scale effect of the nitrogen-fixing *Acacia dealbata* Link (presence and abundance) on nitrogen availability, and whether increases in this essential nutrient led to greater biomass of the canopy tree species, *Eucalyptus obliqua* L'Hér. Average *A. dealbata* tree size was a significant predictor of forest basal area in 24 plots (12% deviance explained) and, when combined with average distance between trees, explained 29.1% variance in *E. obliqua* biomass. However, static patterns of local nitrogen concentration were unrelated to the presence or size of *A. dealbata*, despite our experiments showing that *A. dealbata* leaf litter controls decomposition rates in the soil (due to three times higher N). Such results are important for forest management in the context of understanding the timing and turnover of shorter-lived species like acacias, where higher N (through either litter or soil) might be better detected early in community establishment (when growth is faster and intraspecific competition more intense) but with that early signal subsequently dissipated.

Keywords: forests; nutrients; disturbance; management; diversity; biomass

---

1. Introduction

Understanding the mechanisms that control the biomass and density of trees in forests is important for ecological, economic, and conservation motivations [1]. For instance, growing forests sequester substantial amounts of carbon [2], and as global atmospheric carbon dioxide (CO_2) continues to rise, understanding tree growth and the mechanisms that drive carbon accumulation and storage has become a pressing issue. Abiotic factors like climate, soil type, and disturbance frequency are known to be among the major determinants of the productivity of plant communities [3]. However, in recent decades, interest has shifted towards how community properties (e.g., number and relative abundance of species), and functional traits (e.g., [4]), influence productivity [5]. This is particularly important in studies of community structure, because these features are amenable to experimentation and as a management tool. Additionally, since different plant species vary in their nutrient requirements [6], it is possible that nutrient cycling is a fundamental feature contributing to the observed link between diversity and productivity.

The availability of nitrogen (N) limits productivity in many terrestrial ecosystems, and its increase typically promotes plant-community productivity [7,8]. This effect is further facilitated through complementarity, because differences in N-use efficiency lead to positive feedbacks in N-cycling rates [9]. Nitrogen cycling is heavily dependent on the input and output of various forms of N [10,11], which is often supplied in various quantities by different plant species (and their litter). The plant litter can comprise a mixture of fine detritus (e.g., leaf litter, bark, twigs) and coarse material (branches and whole stems of fallen trees), which are then decomposed and transformed into available nutrients, such as nitrogen [12], via ammonification, nitrification, and N mineralisation. Nitrification is the dominant process in most soils, where nitrifying organisms are often constrained by, and are dependent on, the rate of ammonification [13]. Because of the scarcity of N in soils, soil-N transformations have a strong influence on the potential productivity of an ecosystem [14,15].

It is well known that decomposition rates can vary due to differences in leaf quality (C:N ratio) among plant species, and often increase when litter of different species is combined [16,17] due to complementary stimulation among microbial groups [18]. For example, sclerophyllous eucalypts produce litter that has low nutrient concentrations and low rates of decomposition [19] whereas, by contrast, the litter of N-fixing acacias has comparatively high N concentrations and decomposition rates [20,21]. This has led to speculation that Australian forest communities containing species with more readily decomposable and nutrient-rich litter (such as *Acacia* spp.) might have higher litter decomposition rates than those with *Eucalyptus* litter alone [22,23]. However, this idea has yet to be explored in natural forest systems or linked to overall tree size and biomass.

Here we use field and laboratory data to determine the local-scale effect of an N-fixing tree species (*Acacia dealbata* Link) on the size and total biomass of the dominant eucalypt species (*Eucalyptus obliqua* L'Hér) in a tall hardwood forest community in Tasmania, southern Australia. We used a combination of:

1. Field measurements on the standing pattern (size structure, density, and species composition) of mature trees, used as a proxy for productivity;
2. Sampling and analysis of the soil-nutrient composition in the local environment (i.e., across field sites) and assessment of its relationship to site biomass; and
3. Evaluation of the association between forest-litter quality (different species mixes) and quantity, and the rate of decomposition, as a measure of turnover.

2. Materials and Methods

2.1. Study Site

The field sites were located in a tall eucalypt forest at an altitudinal range of 310–440 m above sea level, on the foothills of Mount Wellington, in south-eastern Tasmania (42.895° S–147.268° E). Vegetation in the region is dominated by *E. obliqua*, a widespread and abundant tall tree found across south-eastern Australia, with the occasional occurrence of *Eucalyptus regnans* F.Muell and *Eucalyptus delegatensis* R.T.Baker. The community also contains two acacia species as sub-dominants: *A. dealbata* and *Acacia melanoxylon* R.Br. The understorey species included *Bedfordia salicina* D.C., *Coprosma quadrifida* (Labill.) B.L.Rob., *Pittosporum bicolor* Hook., and *Pomaderris apetala* Labill. The site is predominately covered by grey-brown podzolic soils on mudstone, classified as Kurosol under the Australian Soil Classification system [24]. Kurosol soil is known as being acidic and nutrient poor [25] with vegetation structure strongly dependent on rainfall [24]. The climate has warm, dry summers and cool, moist winters, with a mean annual precipitation of 720 mm and a mean annual temperature of 17.5 °C. We used the basal area of trees per unit area as a proxy for long-term integrated 'productivity' (>four decades of canopy-tree-biomass accumulation). The forest stand is even-aged, being a natural regeneration following a 1967 stand-replacing fire. The few trees that had obviously lived through the fire (e.g., diameter at breast height >100 cm were avoided during field measurements or excluded from analysis). For this study, we were interested in local heterogeneity across space, rather than temporal comparisons across different site histories. Landscape features were homogeneous across sites (e.g., little variation in topography, canopy, understory, and rock cover).

2.2. Site Selection and Survey Technique

To collect data on community structure and to examine the relationship between tree size and total basal area we surveyed 24 sites using a modified point-quarter method [26], where each site was chosen using a random bearing and distance greater than 50 m from a randomly selected starting point located within the forest (Table S2). The tree closest in proximity to this distance and bearing was designated the starting tree. Each starting tree was identified to species level, and diameter at breast height (DBH in cm) was measured. From the starting tree, the nearest neighbour in each of four compass-bearing quadrants (northeast, northwest, southeast and southwest) was selected. These four trees were also identified to species level and their DBH was also measured, with distance from the starting tree determined using a laser meter. This nearest-neighbour selection method was repeated for each of the four initial trees, giving a total of 17 trees (including the starting tree) at each of the 24 sites.

2.3. Soil Sampling and Processing

Soil samples were collected during early autumn from within the study community. Nine of the 24 sites were selected to represent variation in the proportion of *Acacia* (nitrogen fixers) present, relative to total basal area of all trees (Table S1). At each of these *Acacia* sites, five sampling locations were selected using a random bearing (in NW, SE, SW, NE directions) and a random distance from the starting tree. Five soil cores (2×15 cm) were randomly placed in a 2×2 m quadrant within each location for a total of 25 soil core samples per site. A site's soil samples were then mixed and homogenised by passing through a 2 mm sieve, which removed large roots, litter, and rocks. To determine relative soil water content, a 10 g subsample from each site was dried in an oven at 110 °C for 48 h. A 5 g soil subsample from each site was also ground in a ball mill (MM200 Mixer Mill, Retsch GmbH, Haan, Germany) for total C and N determination via a controlled combustion process in a PerkinElmer 2400 Series II Elemental Analyser (PerkinElmer, Australia [27]).

2.4. Assessing Field Nutrient Availability

To determine nutrient availability in the forest community, a 10 g subsample of each sieved soil sample was extracted with 40 mL of 2 M KCl to determine available NH_4^+ and NO_3^- concentrations (Table S3). The tubes containing the soil and solution were shaken for 1 h, then centrifuged for 3 min at 4000 rpm. A total of 10 mL of supernatant was removed with a syringe and filtered through a Whatman No. 42, 2 μm syringe filter. The filtered extracts were analysed for NH_4^+ and NO_3^- concentrations in a SmartChem Discrete Analyser (Westco Scientific Instruments, Brookfield, CT, USA). Additionally, a portion of the soil was also assessed to determine phosphorus (P) concentrations in the soil at each site using the extraction methods described in [28]: 2 g of soil was mixed with 40 mL 0.005 M $NaCl_2$ solution and shaken for 17 h, then the mixed solution was extracted and analysed using the syringe method described above, with available P concentration determined using a SmartChem Discrete Analyser.

2.5. Determining Nitrogen Mineralisation Rate

The N mineralisation rate of the soil from each of the nine subsampled sites was determined using published protocols [29]. Briefly, 20 g of fresh soil from each composite sample was incubated in a sealed 500 mL glass jar placed in a darkened incubator at 25 °C for 34 days. Jars were ventilated regularly to prevent soils becoming anoxic. Net N mineralisation, nitrification, and ammonification were calculated as the difference between initial total available N, NO_3^-, and NH_4^+ concentrations and those measured following the incubation period (Table S4). The rates of nitrogen mineralisation ($mg\ N\ g^{-1}$ soil) in relation to the proportion of *A. dealbata* in the community was analysed from incubation measurements from soil taken from nine sites, as the sum of NH_4^+ and NO_3^-. Each of the nine sites had a different proportion of *A. dealbata*, and this natural cross-site variation was used to determine whether: (a) There were local-scale differences between nitrogen mineralisation rates and (b) there was a relationship between the proportion of *A. dealbata* and mineralisation rates of nitrogen.

Results were analysed with a linear regression, using the resampling technique described below in Section 2.6.

2.6. Statistical Methods

All statistical analyses were done in Program R v3.3.2 (R Core Development Team 2016, Vienna, Austria). A likelihood-based generalised linear modelling framework was used to assess experimental predictors and dependencies. For each data set, an *a priori* model set was created, and strength of support was evaluated by using Akaike's information criterion, adjusted for small sample sizes (AIC_c). Model weights were calculated using the MuMIn package [30]. Akaike weights (w_i) were used to assess model support; these represent the relative likelihood of a given model compared to others in the evaluation set (which can be of varying complexity and non-nested), by adjusting the likelihood for bias associated with fitting the parameters [31,32]. The relative support from the data for competing models was evaluated using the evidence ratio (ER), the ratio of the respective model's w_i. The percentage deviance (%DE) explained by each model relative to the null (intercept only) was used to assess structural goodness of fit.

In cases where multiple measurements of the dependent variable were taken at each level of a predictor (see Section 3.4), resampling was used (with one outcome selected randomly for any given predictor level and repeated 10,000 times) to avoid pseudoreplication, with the models fitted repeatedly to each subsample. The median prediction from the set of models fitted to the resampled data was used to determine the slope of the predictor and 95% confidence intervals. As a check, these were also compared to models fitted to the medians of the raw data (i.e., ignoring within-sample heterogeneity).

The 24 plots were analysed to assess the influence of *Acacia* sp. on *Eucalyptus* sp., and total site productivity. In this case, the *a priori* models represented additive combinations of the following predictors (of varying complexity): Proportion of *A. dealbata* biomass, the average size of *A. dealbata*, the number of species, or the variance of the pairwise distances between trees. Dependent variables were total basal area of a site or, alternatively, the average size of the eucalypts (DBH cm).

To test whether the availability of soil nutrients (C and N) had an influence on the average size of eucalypts or the total basal area (see Section 3.3), fitted linear models were restricted to either one or two terms due to sample-size limitations (nine total sites). Average size of *A. dealbata* was determined to be the most important factor influencing site productivity during the 24-plot analysis (Section 3.1 and see Results), and therefore was included as a control term in all *a priori* models. The other predictor variable assessed was either: C, N, or pre-total mineralisation of N. Phosphorus was also considered but there were negligible differences across sites.

2.7. Laboratory Litter Experiment

The impact of litter quality and quantity on decomposition rate was determined experimentally by manipulating the relative contribution of litter from the three co-occurring species in the field sites (*E. obliqua*, *A. dealbata*, and *P. apetala*). Newly senesced leaves of the three tree species were collected from each of the 24 sites. The leaf litter was combined to give a representative sample of each of the three species across the entire field site and subsequently dried in an oven at 110 °C for 24 h. Leaves were shredded in an adapted coffee grinder and then ground to powder in a ball mill (MM200 Mixer Mill, Retsch, Haan Germany).

We developed seven treatments ranging from 100% contribution of a single species to equal proportions (i.e., an even mixture; Table 1). The total weight of litter was controlled at 2 g, with 10% of the total weight as sieved pasteurised sand (20 g), which eliminated the effect of existing soil organic matter interfering with the effect of litter amendment on the incubations. Three replicates were prepared for each treatment, each with a different dominant species contributing the largest percentage of litter to the mix, yielding a total of 63 replicates, with three 'no litter' controls. Litter was apportioned into appropriate weights based on the percentage contribution of the species (from 0% to 100% contribution: Table 1) and placed in a 60 mL specimen container in a 500 mL glass jar containing the pasteurised sand. A 2 mL soil slurry solution was added to each jar, being prepared by mixing 50 g (fresh weight) of soil taken from the study site (Section 2.1), with 100 mL of deionised water. The soil was added to the incubation containers with a minimum of additional soil organic matter such that there was a consistency across each treatment. Additionally, samples were supplemented with 2.6 mL of deionised water to bring them to 60% field capacity, which is within the range of soil water content (50%–70%) that has been shown to encourage optimal microbial activity [33]. Field capacity was determined by flooding a 3 g subsample of soil and allowing it to stand for 30 min before draining the free water on a porous ceramic plate and then weighing to obtain water content at field capacity [34]. The subsample was then oven-dried for 24 h at 105 °C to obtain soil dry weight. To control for the respiration rate of the organic matter present in the soil slurry, inoculum blanks consisting of sand and the soil slurry (but no added litter) were also prepared. The sealed jars were incubated in the dark at 25 °C. Gas samples were obtained after 1, 2, 4, 7, 10, 14, 18, 23, 30, and 35 days of incubation.

Table 1. Relative proportion of litter across the seven treatments (ranging from an even mixture of 1, to one dominated by a single species (0). The three species (*Eucalyptus obliqua* L'Hér, *Acacia dealbata* Link, and *Bedfordia salicina* D.C.) are the most common found across the site.

Evenness	*E. obliqua*	*A. dealbata*	*B. salicina*
	100%	0%	0%
0	0%	100%	0%
	0%	0%	100%
	95%	2.5%	2.5%
0.2	2.5%	95%	2.5%
	2.5%	2.5%	95%
	88%	6%	6%
0.4	6%	88%	6%
	6%	6%	88%
	79%	10.5%	10.5%
0.6	10.5%	79%	10.5%
	10.5%	10.5%	79%
	65%	17.5%	17.5%
0.8	17.5%	65%	17.5%
	17.5%	17.5%	65%
	55%	22.5%	22.5%
0.9	22.5%	55%	22.5%
	22.5%	22.5%	55%
	33.3%	33.3%	33.3%
1	33.3%	33.3%	33.3%
	33.3%	33.3%	33.3%

Daily C mineralisation results for the litter experiment were analysed to determine the amount of potentially mineralisable C using a nonlinear differential model (two-pool model with one pool constant), as follows:

$$\frac{dC}{dt} = C_a.k_a e^{(-k_a t)} + r, \tag{1}$$

where $\frac{dC}{dt}$ is the cumulative amount of C mineralised at time t (per day in this case), C_a is the amount of C in the fast or labile C pool, $k_a t$ is the respiration rate constant of CO_2 evolution from the active pool, and r is the non-labile soil C respiration rate, which is effectively a constant over timescales less than a year (e.g., [33]). We assumed the labile C pool to have come directly from recent plant inputs (e.g., labile C compounds in litter [35]). Decomposition rates were corrected for CO_2 release from the inoculum by subtracting the daily mean CO_2 release of inoculum blanks from each experimental unit.

We then explored the differences in the rate of decomposition of the labile carbon per day and the proportion of *Acacia* in the litter over time. A linear regression was fitted to the resampled nitrogen composition taken at each proportion of *A. dealbata* (see next section). To determine the differences in carbon mineralisation of litter over time, each of three species: *E. obliqua*, *A. dealbata*, and *Bedfordia* sp. were measured for cumulative CO_2 evolution during litter incubation over a 40 day period and analysed using the Michaelis–Menten equation (a three-term nonlinear saturating model). Although the Michaelis–Menten equation (MME) is commonly used as a simple representation of enzyme kinetics, it can also serve as a useful two-parameter characterisation of any system that exhibits some type of reaction velocity with a non-linear saturation (via estimation of a maximal rate and half-saturation parameter, with a start at the origin). We agree that soils, being complex mixtures, are unlikely to follow any single, simple reaction dynamics. We instead assume that the overall response will be MME-like but fit the model as a phenomenological rather than overtly mechanistic descriptor of the aggregate system response. The fact that it fits extremely well (See results, R^2 ranging from 90%–94%) gives confidence in this approach.

3. Results

3.1. Relationship between Size of Trees and Total Basal Area across All 24 Plots

The average size of *A. dealbata* (biomass) was the best predictor for total basal area across the 24 plots ($df = 3$, $wAIC_c = 0.32$), explaining 12% of the total deviance. There was little information-theoretic support for the saturated model (containing all three variables), nor for the two predictor variables (i) variance of distances between trees ($df = 3$, $wAIC_c = 0.13$) and (ii) proportion of *A. dealbata* (biomass) ($df = 3$, $wAIC_c = 0.10$); these were no better supported than the null expectation. Average tree size of *A. dealbata* (biomass) in additive combination with the variance of distances between trees was the best model for predicting average size of eucalypts ($df = 4$, $wAIC_c = 0.46$, %DE = 29.1), followed by the saturated model ($df = 5$, $wAIC_c = 0.35$, %DE = 36.5). A post hoc analysis suggested that of the two best predictors for average size of eucalypts, variance of distances between trees was 1.6 times better supported than the average size of *A. dealbata* (biomass). See Supplementary Materials for all raw data associated with the linear models and associated analyses.

3.2. Local Soil Nutrient Availability

There was a positive relationship between soil mineral N availability and soil C content (%) ($R^2 = 0.51$, slope = 1.680, 95% CI = 0.4–2.90; Figure 1a), indicating that nitrogen availability was dependent upon the amount of organic matter in the soil. However, neither soil total N (%) content nor available mineral nitrogen concentrations were related to the basal area of *A. dealbata* across the nine sites sampled ($R^2 = 0.03$, slope = −0.01, 95% CI = −0.03–0.02 and $R^2 = 0.04$, slope = −0.29, 95% CI = −1.40–0.81, respectively; Figure 1). Conversely, *A. dealbata* average size did show a strong negative relationship to soil mineral N availability ($R^2 = 0.47$, slope = −0.61, 95% CI = −1.10–0.12), but not to soil N content (%) ($R^2 = 0.19$, slope = −0.01, 95% CI = −0.03–0.01). This result shows that none of the predictors for nutrient availability in the soil were predictive of total basal area of *A. dealbata*.

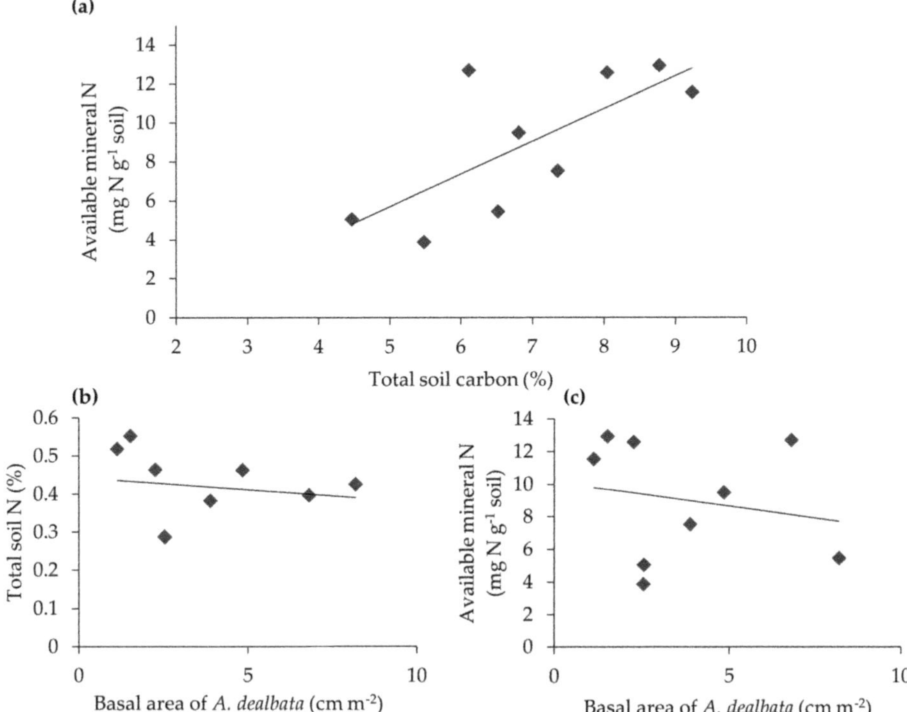

Figure 1. Relationship between total basal area of *A. dealbata* and the amount of nitrogen at each soil site in two different forms: (**a**) Relationship between total soil carbon (%) and available mineral nitrogen (N). Concentrations are determined from soil collected from each of the nine sites in a eucalypt forest. (**b**) Total soil organic matter quantity in terms of total soil nitrogen (%) and the total basal area at each soil site and (**c**) Available mineral N present at each site and the relationship to total basal area of *A. dealbata*.

3.3. Local C and N Dynamics

Since C mineralisation is an enzymatic process, its rate is dependent upon the availability of substrate; as expected, there was a clear relationship in our analysis between C mineralisation rate and soil organic matter content, whether expressed as total soil C or N (R^2 = 0.47, slope = 1371, CI % = 275–2467 and R^2 = 0.65, slope = 96.0, CI % = 42.4–149.6, Figure 2). Thus, the rate of soil organic matter decomposition was dependent upon the quantity of organic matter present.

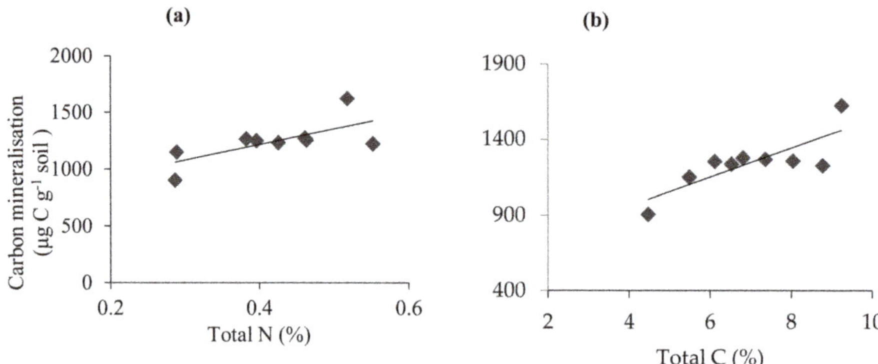

Figure 2. Relationship between total carbon mineralised (μg C g^{-1} soil) and organic matter quality demonstrated as two different forms. (**a**) Total soil nitrogen (N%) $R^2 = 0.47$ and (**b**) total soil carbon (C%) $R^2 = 0.65$. Data were determined from soil collected at nine different sites in a eucalypt forest.

There was a strong positive correlation between field-mineral-N availability and measured nitrification rate in the laboratory ($r = 0.79$). Therefore, the measurement of soil nitrogen availability of field samples corresponded well with measured rates of nitrogen transformation. There was also a strong relationship between rates of N transformation, ammonification, and nitrification ($r = 0.64$), demonstrating that the rate of nitrification was dependent on the rate of net ammonification, and vice versa.

However, there was only a weak correlation between nitrification and total N mineralisation ($r = 0.33$). Indeed, nitrification rates were relatively stable across the range of sites in comparison to the wide range of ammonification rates (Figure 3). There was by contrast, a strong positive relationship between net ammonification and net N mineralisation rates ($r = 0.94$, Figure 3), indicating that the supply of mineral N in this system was related to the conversion of organic N to ammonium.

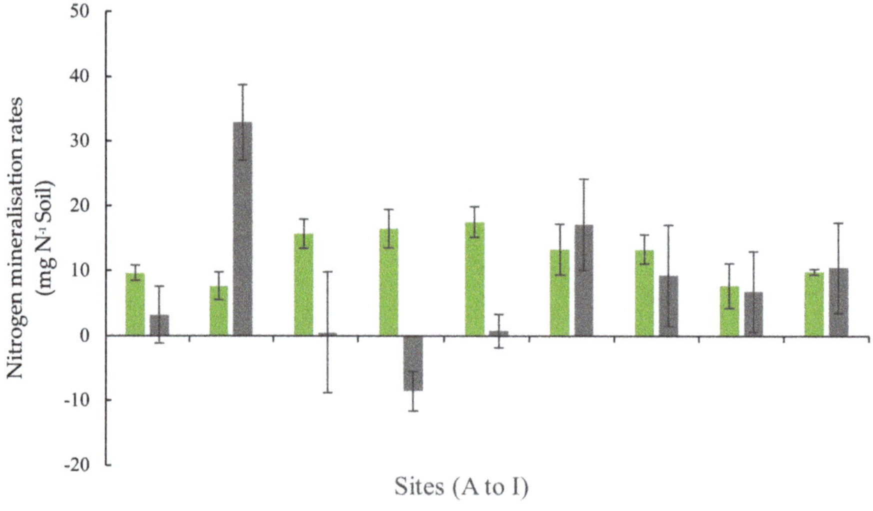

(**a**)

Figure 3. *Cont.*

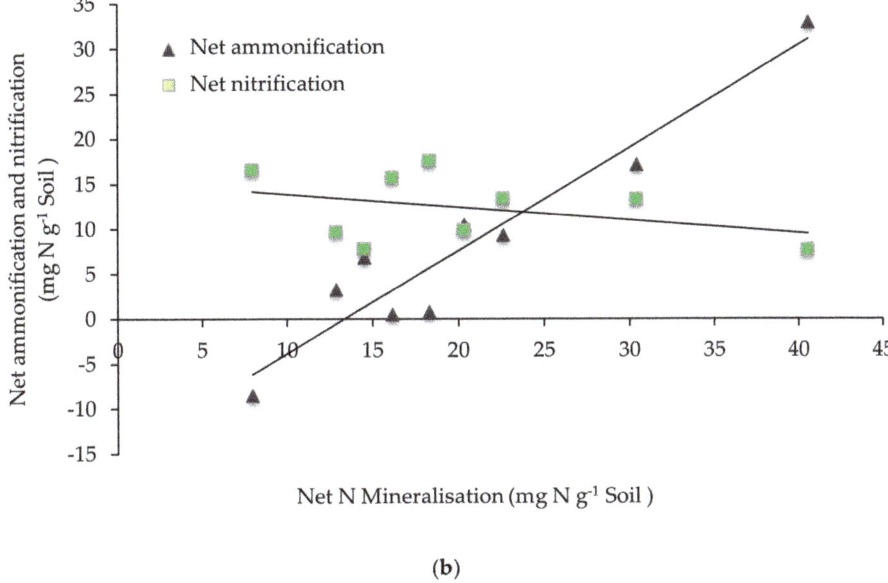

(**b**)

Figure 3. Rates of nitrogen (N) transformation obtained from incubation measurements from soil taken from nine sites. (**a**) Rates of ammonification (mg N g^{-1} soil); as the conversion of C and N (%) into NH_4^+; and the rates of nitrification (mg N g^{-1} soil); as the conversion of NH_4^+ into NO_3^-. One standard error (+ and −) is shown for each site measurement. (**b**) Relationship between net N mineralisation rates and net ammonification and net nitrification rates from soil collected from the nine sites.

3.4. Relationship between Litter Quality, Quantity, and Litter Decomposition

Field samples from communities with *A. dealbata* had almost three times more nitrogen in the litter (2.57% ± 0.11%) compared to *E. obliqua* (0.53% ± 0.03%) and *Bedfordia* sp. (0.80% ± 0.05%), and thus a lower C:N ratio. A higher proportion of *A. dealbata* in the litter (%) subsequently led to a higher release of labile carbon during the litter-decomposition experiment (Table 2; Figure 4). Both approaches to fitting this relationship statistically: (i) The median of the resampled points and (ii) the initial linear model to the median points, yielded nearly identical R^2 (indicating a negligible effect of pseudoreplication).

Table 2. Litter decomposition table showing the intercept, slope, and corresponding 95% confidence intervals for each, the initial model (fitted to the raw data) and the median of model fits to resampled data points; goodness of fit is indicated by the R^2.

Model (Median)	Intercept (α)	95% CI	Slope (β)	95% CI	Sigma	R^2%
Initial	2838	1818–3859	177	155–199	2829	81.8
Resampled	2877	290–5509	169	36–267	2192	81.7

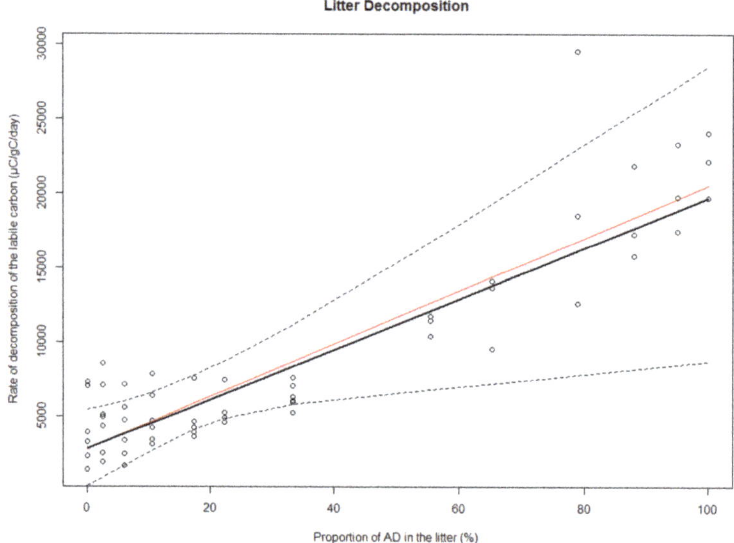

Figure 4. The relationship between the rate of labile carbon decomposition and of the nitrogen fixed *A. dealbata* in the litter. The red line represents the initial linear regression model to the median points (R^2: 81.8%), and the black line shows the median of the resampled points (R^2: 81.7%). Dotted lines represent the confidence bounds for the resampled data.

The resampled fits of the Michaelis–Menten saturating equation (MM) for the cumulative CO_2 evolved from the decomposition of the litter was similar for each of the three species (Table 3; Figure 5). Although both median and resampled fits yielded a similar R^2, the resampled model did reveal subtle deviations in the data, but had little to no influence on the R^2, with *A. dealbata* showing the largest variation of R^2 between models (1.9% difference). *A. dealbata* had the lowest half-saturation point for the cumulative carbon mineralisation, reaching it faster than *E. obliqua* and *Bedfordia*. However, all species released a comparable total amount of CO_2 per gram of litter, as indicated by the similar asymptotes of the MM equation (Table 3).

Table 3. The model fit (R^2) for each species for the cumulative CO_2 evolved from the decomposition of the litter: Initial Michaelis–Menten equation fit to the median data points, and the resampled data (median model prediction). Shown is the asymptote (α), half saturation point (β), and corresponding 95% confidence intervals, for each species.

Model (Median)	Asymptote (α)	95% CI	Half Saturation Point (β)	95% CI	Sigma	R^2 (%)
			Eucalyptus obliqua			
Initial	92,189	84,612–101,407	9.9	7.8–12.7	2210	93.3
Resampled	90,526	72,756–116,217	9.8	5.7–18.0	8355	93.7
			Acacia dealbata			
Initial	89,561	81,141–100,055	5.6	3.9–8.0	3769	90.9
Resampled	89,563	74,410–105,699	5.7	3.5–9.1	6729	92.8
			Bedfordia sp.			
Initial	92,476	76,899–117,482	10.1	5.8–18.0	4637	89.1
Resampled	84,947	67,487–113,376	7.9	4.4–7.9	8859	90.2

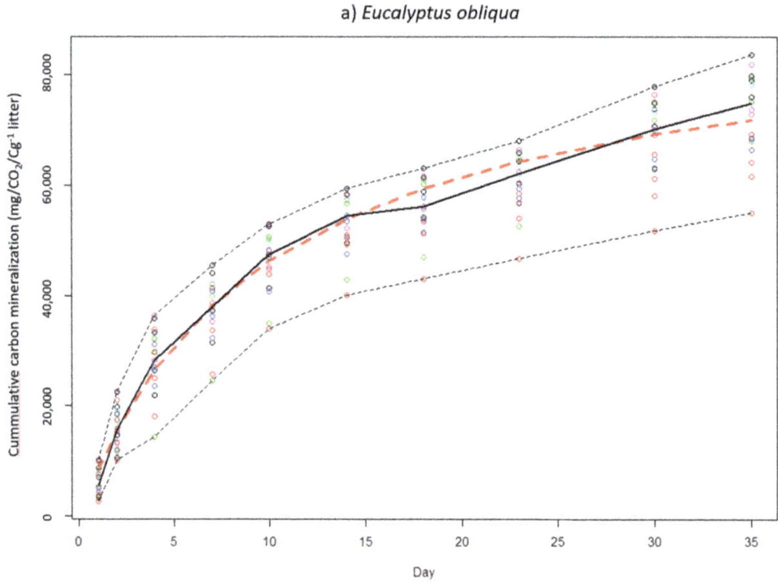

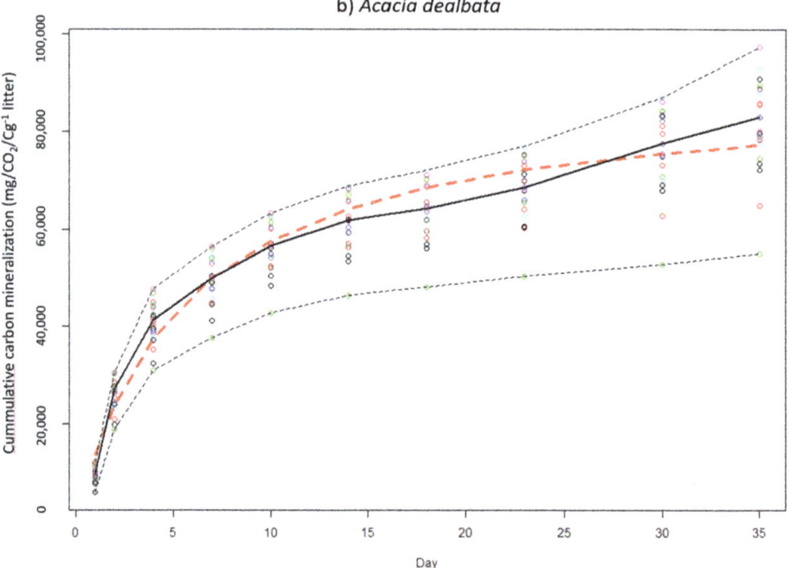

Figure 5. *Cont.*

c) Bedfordia salicina

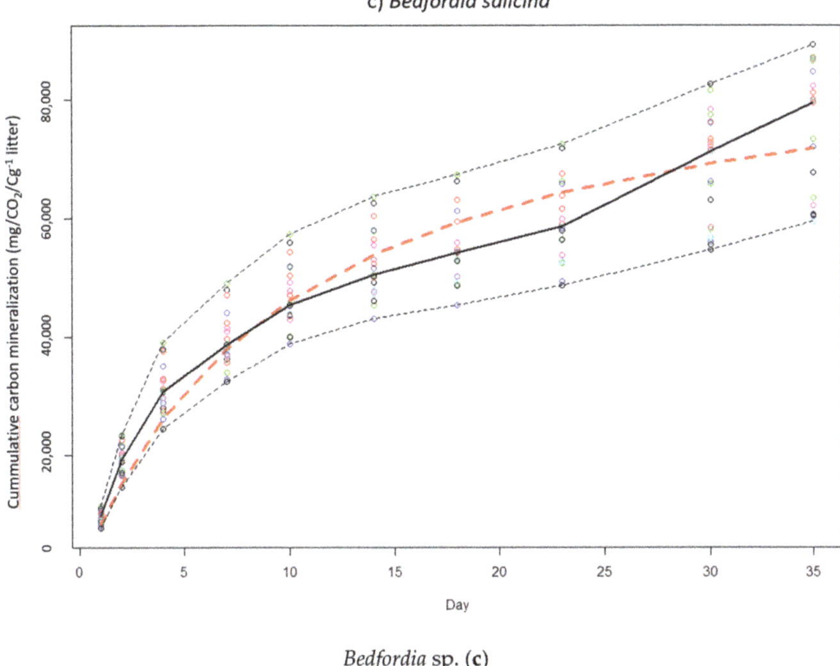

Bedfordia sp. (c)

Figure 5. Cumulative carbon mineralisation rate over time (40 days) against all data points for each species. The red dotted line represents the Michaelis–Menten equation (MM) fit to the median data points. The black line is the median of the resampled data fits using the MM equation, with corresponding black dotted 95% confidence bounds: (**a**) *E. obliqua* initial MM fit (R^2: 93.3) and resampled median (R^2: 93.7%); (**b**) *A. dealbata* initial MM fit (R^2: 90.9%) and resampled median (R^2: 92.8%); (**c**) *Bedfordia* sp. initial MM fit (R^2: 89.1%) and resampled median (R^2: 90.2%).

4. Discussion

We have shown that experimental mixtures of plant species from Australian eucalypt forests that contain a higher proportion of nitrogen-fixing *A. dealbata* litter have higher decomposition rates (Figures 4 and 5). This was driven primarily by the three-fold increase in nitrogen (N) content in *A. dealbata* litter compared to, for example, *B. salicina* (another understorey tree) and the canopy-forming *E. obliqua*. This result is supported mechanistically by previous studies (e.g., [36,37]) showing that decomposition rates of litter mixtures are driven predominantly by substrate quality, particularly the amount of C and N in the litter mixtures. Plant functional traits, such as nitrogen fixation, are key drivers of higher-quality substrates in forest ecosystems, with the potential to mediate and control plant abundance, biomass, and composition via inherent differences between species in quality (C:N, lignin:N) and quantity (%C, %N) of plant litter [38]. Indeed, the quantity of carbon (C) in the soil at our Tasmanian forest sites was the best predictor of average eucalypt tree size, followed by the average size of *A. dealbata* (Figure 1). Studies that have examined this phenomenon in other forests have demonstrated that the identity and functional traits of litter-creating species can surpass species diversity as the main driver of decomposition [39–42].

Given that *A. dealbata* presence and abundance indirectly controls litter decomposition via nitrogen availability and was consequently a useful predictor of average *E. obliqua* tree size, this research supports the hypothesis that N availability and cycling would be higher in sites with a greater *A. dealbata* biomass. Our results align with those of [23], who demonstrated that the presence of *Acacia mearnsii* (another N-fixer) influenced the quantity and rates of N and P cycling in *Eucalyptus globulus* mixed forests compared to monocultures, in an 11-year experimental trial. Similar relationships have also been found in boreal forests, grasslands, and tropical forests, through field N fertilisation trials [42–44] and in laboratory experiments [45,46]. However, in our study, there was no observable relationship between *A. dealbata* biomass and the total amount of organic matter (C and N) or the concentrations of NH_4^+, NO_3^- or with total mineral N and P in the soil, despite high heterogeneity among sites in N availability and cycling (Figure 3). Further, we found no correlation between total available soil-mineral N and average *A. dealbata* tree size. As *A. dealbata* was the sub-dominant canopy species in our forest system, and there was 30% more biomass of *A. dealbata* relative to other sites, the potential for N-rich soil from N-fixers with high N litter should have been realised. As such, this null result demands an explanation.

There are several plausible reasons as to why a higher N content was not observed at sites with high *A. dealbata* biomass (Figure 1), with these inferences potentially useful for forest management during early succession and development. Two leading candidates are: (1) *A. dealbata* is a nitrogen fixer and therefore is not reliant on high soil N to grow, whereas this might be limiting for other species (such as *E. obliqua*); and (2) due to the relatively short lifespan of *A. dealbata* (~40 years) compared to *E. obliqua* (>100 years) [47], a higher N litter might have been present during the early stages of forest succession, when growth was faster and intraspecific competition more intense, but that early signal has since dissipated (e.g., [48,49]). The development of a forest is characterised by both changes in structure and succession, with concomitant consequences for productivity of standing forests and regrowth [50]. For example, restoration of forest ecosystems with multiple forest species helps rebuild multiple functions, such as species structure, litter composition, and soil nutrients of an area [9,51]. Selective logging of natural forests is a form of disturbance, and management plans require some older trees to be left in a logged area for animal habitat and seedling regeneration. Our results suggest that complimentary tree species, such as *A. dealbata* and *E. obliqua*, should be left in such areas, to maintain the overall functionality, especially during early successional development [52]. However, our experimental results only show that complementarity and nitrogen facilitation can occur, whereas determining this relationship explicitly within a natural setting (such as [53,54] is challenging to demonstrate, especially in the face of inevitable inter-site heterogeneity).

In terms of confounding effects, there are also many factors that can influence environmental and topographic heterogeneity within eucalypt forests, which might act to mask any relationship and thus cause the field results to deviate from the expectations of the experimental incubations. For example, the realised distribution of *A. dealbata* litter at the field sites might be influenced by localised erosion or mineral dilution of litter and topographic variation, causing the build-up of litter in soil depressions or exclusions from mounds, but which were not reflected in controlled soil collections. The few studies that have examined patterns of litter fall dispersal and collection in field conditions support the notion of high heterogeneity in litter-sample collection, finding large disparities between the location of the litter source (local tree canopy) and the actual litterfall sample that was collected [55]. Similarly, [56] found that under-representation of certain components of vegetation, such as litter, can have important implications for the estimation of forest productivity and nutrient cycling, leading to a scale mismatch between sample points and the character of the soil across the forest.

While this study focussed in detail on N transformations and cycling within a Tasmanian eucalypt forest system, N losses could not be determined explicitly or assessed in the field. Leaching of N from the soil at field sites might contribute to the low N availability; this suggests that future studies should focus on determining how much N is below the 15 cm of topsoil that was collected in our sampling protocol. Stable isotope tracing is one method that could be used to track N pathways through the soil, providing information of N availability and movement at various depth stratifications. If leaching does occur, then it is likely that N will accumulate further down the soil profile. This effect might have been exacerbated by the overwhelming dominance of the deep-rooted *E. obliqua* (numerically and in terms of biomass), whose leaves, due to variations in litter characteristics between the two species such as litter size, mass, decomposability and chemical composition [57,58], dominate the surface litter composition, and yet it is able to access historically deposited nutrients that are now only found deep in the soil profile.

5. Conclusions

In conclusion, we have presented an experimental and field-observational snapshot of nutrient dynamics within a typical eucalypt forest community from southern Australia. We have demonstrated the importance of understanding litter decomposition, but also the need to quantify how different sources of N can impact N availability and cycling. Repeated sampling across a larger range of seasons in Tasmanian eucalypt forests (including wet and dry periods) would be important to provide a temporal component to a spatially complex process. Our results also highlight the difficulty in defining ecosystem processes and function using any single mechanism in forests—even one so apparently clear cut as N availability—because of the many processes that ultimately interact to determine the size of trees and relative abundance of species.

Supplementary Materials: The following are available online at http://www.mdpi.com/1999-4907/10/4/340/s1, Figure S1: Phosphorous content in the soil from the nine sites surveyed for soil samples, Table S1: Proportion of basal area (relative to the total basal area of each site) for each species present in the field 24 sites, Table S2: Summary statistics of key characteristics within each of the 24 field sites surveyed, Table S3: Data on soil nutrient availability from the soil samples taken from each of the subset of nine sites that were subject to soil profiling, Table S4: Data on soil nutrient mineralisation from the soil samples taken from each of the nine soil-sampling sites. Abbreviations given in the footnote.

Author Contributions: J.C.B. and M.J.H. conceived and designed the experiments; J.C.B. performed the experiments; B.W.B., J.C.B, and E.M.R. analysed the data; and all authors contributed to the manuscript drafts and writing and gave final approval for publication.

Funding: This work was funded by Australian Research Council grant FL160100101.

Conflicts of Interest: The authors declare no conflict of interest.

References

1. Becker, K.; Wulfmeyer, V.; Berger, T.; Gebel, J.; Münch, W. Carbon farming in hot, dry coastal areas: An option for climate change mitigation. *Earth Syst. Dyn.* **2013**, *4*, 237–251. [CrossRef]
2. Law, B.E.; Hudiburg, T.W.; Berner, L.T.; Kent, J.J.; Buotte, P.C.; Harmon, M.E. Land use strategies to mitigate climate change in carbon dense temperate forests. *Proc. Natl. Acad. Sci. USA* **2018**, *115*, 3663–3668. [CrossRef] [PubMed]
3. Johnson, K.H.; Vogt, K.A.; Clark, H.J.; Schmitz, O.J.; Vogt, D.J. Biodiversity and the productivity and stability of ecosystems. *Trends Ecol. Evol.* **1996**, *11*, 372–377. [CrossRef]
4. Wright, I.J.; Westoby, M. Differences in seedling growth behaviour among species: Trait correlations across species, and trait shifts along nutrient compared to rainfall gradients. *J. Ecol.* **1999**, *87*, 85–97. [CrossRef]
5. Stirling, G.; Wilsey, B. Empirical relationships between species richness, evenness, and proportional diversity. *Am. Nat.* **2001**, *158*, 286–299. [CrossRef] [PubMed]
6. Krebs, C.J. *Ecology*, 5th ed.; Benjamin Cummings: Boston, MA, USA, 2001.
7. Vitousek, P.M.; Howarth, R.W. Nitrogen limitation on land and in the sea: How can it occur? *Biogeochemistry* **1991**, *13*, 87–115. [CrossRef]
8. Wardle, D.A.; Bardgett, R.D.; Klironomos, J.N.; Setälä, H.; Van Der Putten, W.H.; Wall, D.H. Ecological linkages between aboveground and belowground biota. *Science* **2004**, *304*, 1629–1633. [CrossRef]
9. Forrester, D.I.; Bauhus, J. A Review of Processes Behind Diversity—productivity relationships in forests. *Curr. For. Rep.* **2016**, *2*, 45–61. [CrossRef]
10. Harrison, K.A.; Bol, R.; Bardgett, R.D. Preferences for different nitrogen forms by coexisting plant species and soil microbes. *Ecology* **2007**, *88*, 989–999. [CrossRef]
11. Parton, W.; Silver, W.L.; Burke, I.C.; Grassens, L.; Harmon, M.E.; Currie, W.S.; King, J.Y.; Adair, E.C.; Brandt, L.A.; Hart, S.C. Global-scale similarities in nitrogen release patterns during long-term decomposition. *Science* **2007**, *315*, 361–364. [CrossRef]
12. Sharma, J.; Sharma, Y. Nutrient cycling in forest ecosystems—A review. *Agric. Rev.* **2004**, *25*, 157–172.

13. Rennenberg, H.; Dannenmann, M.; Gessler, A.; Kreuzwieser, J.; Simon, J.; Papen, H. Nitrogen balance in forest soils: Nutritional limitation of plants under climate change stresses. *Plant Biol.* **2009**, *11*, 4–23. [CrossRef]

14. Reich, P.B.; Grigal, D.F.; Aber, J.D.; Gower, S.T. Nitrogen mineralization and productivity in 50 hardwood and conifer stands on diverse soils. *Ecology* **1997**, *78*, 335–347. [CrossRef]

15. McGuire, A.D.; Melillo, J.M.; Joyce, L.; Kicklighter, D.W.; Grace, A.; Moore, B., III; Vorosmarty, C.J. Interactions between carbon and nitrogen dynamics in estimating net primary productivity for potential vegetation in North America. *Glob. Biogeochem. Cycles* **1992**, *6*, 101–124. [CrossRef]

16. Conn, C.; Dighton, J. Litter quality influences on decomposition, ectomycorrhizal community structure and mycorrhizal root surface acid phosphatase activity. *Soil Biol. Biochem.* **2000**, *32*, 489–496. [CrossRef]

17. King, R.F.; Dromph, K.M.; Bardgett, R.D. Changes in species evenness of litter have no effect on decomposition processes. *Soil Biol. Biochem.* **2002**, *34*, 1959–1963. [CrossRef]

18. Chapman, S.K.; Newman, G.S. Biodiversity at the plant–soil interface: Microbial abundance and community structure respond to litter mixing. *Oecologia* **2010**, *162*, 763–769. [CrossRef] [PubMed]

19. Forrester, D.I.; Bauhus, J.; Cowie, A.L. Nutrient cycling in a mixed-species plantation of *Eucalyptus globulus* and *Acacia mearnsii*. *Can. J. For. Res.* **2005**, *35*, 2942–2950. [CrossRef]

20. Parrotta, J.A. Productivity, nutrient cycling, and succession in single-and mixed-species plantations of *Casuarina equisetifolia*, *Eucalyptus robusta*, and *Leucaena leucocephala* in Puerto Rico. *For. Ecol. Manag.* **1999**, *124*, 45–77. [CrossRef]

21. Attiwill, P.M.; Leeper, G.W. *Forest Soils and Nutrient Cycles*; Melbourne University Press: Carlton, VIC, Australia, 1987.

22. Adams, M.; Attiwill, P. Nutrient cycling and nitrogen mineralization in eucalypt forests of south-eastern Australia. *Plant Soil* **1986**, *92*, 341–362. [CrossRef]

23. Forrester, D.I.; Bauhus, J.; Cowie, A.L.; Vanclay, J.K. Mixed-species plantations of Eucalyptus with nitrogen-fixing trees: A review. *For. Ecol. Manag.* **2006**, *233*, 211–230. [CrossRef]

24. Isbell, R. *The Australian Soil Classification*; CSIRO Publishing: Collingwood, Australia, 2016.

25. Warren, R.J.; Giladi, I.; Bradford, M.A. Competition as a mechanism structuring mutualisms. *J. Ecol.* **2014**, *102*, 486–495. [CrossRef]

26. Mitchell, K. Quantitative analysis by the point-centered quarter method. *arXiv* **2010**, arXiv:1010.3303.

27. Jimenez, R.R.; Ladha, J.K. Automated elemental analysis: A rapid and reliable but expensive measurement of total carbon and nitrogen in plant and soil samples. *Commun. Soil Sci. Plant Anal.* **1993**, *24*, 1897–1924. [CrossRef]

28. Mendham, D.S.; Smethurst, P.J.; Holz, G.K.; Menary, R.C.; Grove, T.S.; Weston, C.; Baker, T. Soil analyses as indicators of phosphorus response in young eucalypt plantations. *Soil Sci. Soc. Am. J.* **2002**, *66*, 959–968. [CrossRef]

29. Osanai, Y.; Bougoure, D.S.; Hayden, H.L.; Hovenden, M.J. Co-occurring grass species differ in their associated microbial community composition in a temperate native grassland. *Plant Soil* **2013**, *368*, 419–431. [CrossRef]

30. Barton, K.; Barton, M.K. Mu-MIn: Multi-model inference. R Package Version 1.43.6. 2019. Available online: https://cran.r-project.org/web/packages/MuMIn/MuMIn.pdf (accessed on 18 January 2019).

31. Anderson, D.R. *Model Based Inference in the Life Sciences: A Primer on Evidence*; Springer Science & Business Media: New York, NY, USA, 2007.

32. Burnham, K.P.; Anderson, D.R. *Model Selection and Multimodel Inference: A Practical Information—Theoretic Approach*; Springer Science & Business Media: New York, NY, USA, 2003.

33. Pendall, E.; Osanai, Y.U.I.; Williams, A.L.; Hovenden, M.J. Soil carbon storage under simulated climate change is mediated by plant functional type. *Glob. Chang. Biol.* **2011**, *17*, 505–514. [CrossRef]

34. Osanai, Y.; Flittner, A.; Janes, J.K.; Theobald, P.; Pendall, E.; Newton, P.C.; Hovenden, M.J. Decomposition and nitrogen transformation rates in a temperate grassland vary among co-occurring plant species. *Plant Soil* **2012**, *350*, 365–378. [CrossRef]

35. Dijkstra, F.A.; Hobbie, S.E.; Reich, P.B. Soil processes affected by sixteen grassland species grown under different environmental conditions. *Soil Sci. Soc. Am. J.* **2006**, *70*, 770–777. [CrossRef]

36. Freschet, G.T.; Cornwell, W.K.; Wardle, D.A.; Elumeeva, T.G.; Liu, W.; Jackson, B.G.; Onipchenko, V.G.; Soudzilovskaia, N.A.; Tao, J.; Cornelissen, J.H. Linking litter decomposition of above-and below-ground organs to plant–soil feedbacks worldwide. *J. Ecol.* **2013**, *101*, 943–952. [CrossRef]

37. García-Palacios, P.; Maestre, F.T.; Kattge, J.; Wall, D.H. Climate and litter quality differently modulate the effects of soil fauna on litter decomposition across biomes. *Ecol. Lett.* **2013**, *16*, 1045–1053. [CrossRef] [PubMed]

38. Makkonen, M.; Berg, M.P.; Handa, I.T.; Hättenschwiler, S.; Van Ruijven, J.; Van Bodegom, P.M.; Aerts, R. Highly consistent effects of plant litter identity and functional traits on decomposition across a latitudinal gradient. *Ecol. Lett.* **2012**, *15*, 1033–1041. [CrossRef] [PubMed]

39. Hobbie, S.E. Effects of plant species on nutrient cycling. *Trends Ecol. Evol.* **1992**, *7*, 336–339. [CrossRef]

40. Wardle, D.; Bonner, K.; Nicholson, K. Biodiversity and plant litter: Experimental evidence which does not support the view that enhanced species richness improves ecosystem function. *Oikos* **1997**, *79*, 247–258. [CrossRef]

41. Jiang, L.; Wan, S.; Li, L. Species diversity and productivity: Why do results of diversity-manipulation experiments differ from natural patterns? *J. Ecol.* **2009**, *97*, 603–608. [CrossRef]

42. Hyvönen, R.; Ågren, G.I.; Linder, S.; Persson, T.; Cotrufo, M.F.; Ekblad, A.; Freeman, M.; Grelle, A.; Janssens, I.A.; Jarvis, P.G.; et al. The likely impact of elevated [CO_2], nitrogen deposition, increased temperature and management on carbon sequestration in temperate and boreal forest ecosystems: A literature review. *New Phytol.* **2007**, *173*, 463–480. [CrossRef] [PubMed]

43. Nordin, A.; Näsholm, T.; Ericson, L. Effects of simulated N deposition on understorey vegetation of a boreal coniferous forest. *Funct. Ecol.* **1998**, *12*, 691–699. [CrossRef]

44. Matson, P.; Lohse, K.A.; Hall, S.J. The globalization of nitrogen deposition: Consequences for terrestrial ecosystems. *Ambio* **2002**, *31*, 113–119. [CrossRef] [PubMed]

45. Wedin, D.A.; Tilman, D. Species effects on nitrogen cycling: A test with perennial grasses. *Oecologia* **1990**, *84*, 433–441. [CrossRef] [PubMed]

46. Temperton, V.M.; Mwangi, P.N.; Scherer-Lorenzen, M.; Schmid, B.; Buchmann, N. Positive interactions between nitrogen-fixing legumes and four different neighbouring species in a biodiversity experiment. *Oecologia* **2007**, *151*, 190–205. [CrossRef] [PubMed]

47. Koch, A.J.; Driscoll, D.A.; Kirkpatrick, J. Estimating the accuracy of tree ageing methods in mature *Eucalyptus obliqua* forest, Tasmania. *Aust. For.* **2008**, *71*, 147–159. [CrossRef]

48. Bouillet, J.-P.; Laclau, J.-P.; Gonçalves, J.L.d.M.; Voigtlaender, M.; Gava, J.L.; Leite, F.P.; Hakamada, R.; Mareschal, L.; Mabiala, A.; Tardy, F.; et al. Eucalyptus and Acacia tree growth over entire rotation in single- and mixed-species plantations across five sites in Brazil and Congo. *For. Ecol. Manag.* **2013**, *301*, 89–101. [CrossRef]

49. Epron, D.; Nouvellon, Y.; Mareschal, L.; Moreira, R.M.e.; Koutika, L.-S.; Geneste, B.; Delgado-Rojas, J.S.; Laclau, J.-P.; Sola, G.; Gonçalves, J.L.d.M.; et al. Partitioning of net primary production in Eucalyptus and Acacia stands and in mixed-species plantations: Two case-studies in contrasting tropical environments. *For. Ecol. Manag.* **2013**, *301*, 102–111. [CrossRef]

50. Ryan, M.; Binkley, D.; Fownes, J.H. Age-related decline in forest productivity: Pattern and process. In *Advances in Ecological Research*; Elsevier: Amsterdam, The Netherlands, 1997; Volume 27, pp. 213–262.

51. Aerts, R.; Honnay, O. Forest restoration, biodiversity and ecosystem functioning. *BMC Ecol.* **2011**, *11*, 29. [CrossRef] [PubMed]

52. Villela, D.M.; Nascimento, M.T.; De Aragao, L.E.O.; Da Gama, D.M. Effect of selective logging on forest structure and nutrient cycling in a seasonally dry Brazilian Atlantic forest. *J. Biogeogr.* **2006**, *33*, 506–516. [CrossRef]

53. Tilman, D.; Reich, P.; Knops, J.; Wedin, D.; Mielke, T.; Lehman, C. Diversity and productivity in a long-term grassland experiment. *Science* **2001**, *294*, 843–845. [CrossRef]

54. Cardinale, B.J.; Wright, J.P.; Cadotte, M.W.; Carroll, I.T.; Hector, A.; Srivastava, D.S.; Loreau, M.; Weis, J.J. Impacts of plant diversity on biomass production increase through time because of species complementarity. *Proc. Natl. Acad. Sci. USA* **2007**, *104*, 18123–18128. [CrossRef]

55. Lowman, M.D. Litterfall and leaf decay in three Australian rainforest formations. *J. Ecol.* **1988**, *76*, 451–465. [CrossRef]

56. Clark, D.A.; Brown, S.; Kicklighter, D.W.; Chambers, J.Q.; Thomlinson, J.R.; Ni, J.; Holland, E.A. Net primary production in tropical forests: An evaluation and synthesis of existing field data. *Ecol. Appl.* **2001**, *11*, 371–384. [CrossRef]

57. Bini, D.; Dos Santos, C.A.; Bouillet, J.-P.; De Morais Goncalves, J.L.; Cardoso, E.J.B.N. Eucalyptus grandis and Acacia mangium in monoculture and intercropped plantations: Evolution of soil and litter microbial and chemical attributes during early stages of plant development. *Appl. Soil Ecol.* **2013**, *63*, 57–66. [CrossRef]
58. Forrester, D.; Pares, A.; O'hara, C.; Khanna, P.; Bauhus, J. Soil organic carbon is increased in mixed-species plantations of Eucalyptus and nitrogen-fixing Acacia. *Ecosystems* **2013**, *16*, 123–132. [CrossRef]

Article

Storage and Climatic Controlling Factors of Litter Standing Crop Carbon in the Shrublands of the Tibetan Plateau

Xiuqing Nie [1,2,3], Dong Wang [1,3], Lucun Yang [2,4] and Guoying Zhou [2,4,*]

1   Key Laboratory of Tree Breeding and Cultivation of the State Forestry Administration, Research Institute of Forestry, Chinese Academy of Forestry, Beijing 100091, China; niexiuqing123@163.com (X.N.); dwang@caf.ac.cn (D.W.)
2   Key Laboratory of Tibetan Medicine Research, Northwest Institute of Plateau Biology, Chinese Academy of Science, Xining 810008, China; Yanglucun@nwipb.cas.cn
3   Research Institute of nature protected Area, Chinese Academy of Forestry, Beijing 100091, China
4   Qinghai Key Laboratory of Qing-Tibet Biological Resources, Xining 810008, China
*   Correspondence: zhougy@nwipb.cas.cn; Tel.: +86-971-6159630; Fax: +86-971-6143282

Received: 17 September 2019; Accepted: 4 November 2019; Published: 5 November 2019

Abstract: Litter is an important component of terrestrial ecosystems and plays a significant role in carbon cycles. Quantifying regional-scale patterns of litter standing crop distribution will improve our understanding of the mechanisms of the terrestrial carbon cycle, and thus enable accurate predictions of the responses of the terrestrial carbon cycle to future climate change. In this study, we aimed to estimate the storage and climatic controlling factors of litter standing crop carbon in the Tibetan Plateau shrublands. We investigated litter standing crop carbon storage and its controlling factors, using a litter survey at 65 shrublands sites across the Tibetan Plateau from 2011–2013. Ordinary least squares regression analyses were conducted to estimate the relationships between litter standing crop carbon, longitude, and latitude. Multiple linear regressions were used to evaluate relationships among litter standing crop carbon, mean annual temperature (MAT), mean annual precipitation (MAP), and aboveground biomass. The litter standing crop carbon storage was 10.93 Tg C, 7.40 Tg C, and 3.53 Tg C in desert shrublands and alpine shrublands, respectively. Litter standing crop carbon decreased with longitude, and was stable with increasing latitude. Most (80%) of the litter standing crop was stored in branches, with only 20% stored in foliage in the shrublands on the Tibetan Plateau. The conversion coefficient was 0.44 for litter standing crop to litter standing crop carbon, and 0.39 and 0.45 for foliage and branch litter standing crop to foliage and branch litter standing crop carbon, respectively. Aboveground biomass can accelerate more inputs of litter and has a positive effect on litter standing crop carbon. MAT had a positive effect on litter standing crop carbon due to stimulating more input of aboveground biomass. However, MAP had a negative relationship with litter standing crop carbon by enhancing litter decomposition.

Keywords: litter standing crop carbon; conversion coefficient; climatic factors; Tibetan Plateau; shrublands

---

1. Introduction

Litter is an important component of terrestrial ecosystems. Its production and accumulation contribute to carbon sequestration and soil fertility [1]. The litter standing crop depends on the rates of litter production and litter decomposition. If the litter production rate is greater than the litter decomposition rate, litter becomes a sink of atmosphere CO_2; if the litter decomposition rate is greater than the litter production rate, it becomes a carbon source [2]. Although soil acts as the largest carbon pool [3], the litter standing crop pool also accounts for a non-ignorable amount of emissions; the global litter standing crop carbon pool has been estimated to be 43 Pg C for all forest ecosystem carbon stocks [4], and the litter standing crop carbon pool in both forest and grassland ecosystems in China has been estimated as being 0.52 Pg C [5], which means litter plays a significant role in carbon cycles [1,6,7]. Most researchers have focused on litter production and litter decomposition [8,9], while few studies have considered litter standing crop carbon in shrublands [2], especially on the Tibetan Plateau.

As a result of global climatic changes, the Tibetan Plateau has been experiencing warming, and the mean annual temperature (MAT) has increased by 0.05 °C every year [10]. Meanwhile the mean annual precipitation (MAP) has increased by 10.2 mm every ten years [10]. Furthermore, climatic conditions, such as changes in precipitation and temperature, are predicted to not only continue, but also perhaps to accelerate in the future [11,12]. It has been demonstrated that climatic factors such as MAP and MAT greatly affect litter standing crop carbon [5]. Climatic factors also indirectly influence litter decomposition through their effects on biotic factors [2]. It has been demonstrated that MAT and MAP have positive effects on litter production in forests in China [13]. At the regional scale, a similar trend was observed in Tiantong national forest park [14]. However, other studies have indicated that litter production in northeastern China's forests is influenced by temperature, but not significantly affected by precipitation [15]. Therefore, the effects of climatic factors on litter production remain uncertain. Compared with climatic effects on litter production, its effects on litter standing crop have been less researched in the shrublands on the Tibetan Plateau.

The conversion coefficient of litter standing crop to litter standing crop carbon is an important parameter in the carbon budget [16]. Generally, researchers have used 0.50 to convert biomass to biomass carbon [17]. However, the parameter of 0.50 overestimated carbon by 22% when estimating the litter carbon budget in shrublands across south China [16]. Furthermore, different components play important roles in shaping the magnitude of the conversion coefficient [18]. Although the conversion coefficient of litter standing crop to litter standing crop carbon has been estimated in shrublands of southern China [16], components such as foliage and branch standing crop to foliage and branch standing crop carbon have not been considered. Furthermore, although biomass and soil organic carbon have been explored in the Tibetan Plateau shrublands [19–21], it was difficult to assess the total carbon storage in shrublands ecosystem without litter standing crop carbon. It has been demonstrated that the biophysical processes that control soil carbon on the Tibetan Plateau may differ from those in other regions [20,22]. We hypothesized that the conversion coefficient in the Tibetan Plateau shrublands was different from that in the southern China shrublands. We also aimed to explore the litter standing crop carbon responses to changes of MAT, MAP and aboveground biomass in the shrublands on the Tibetan Plateau.

2. Materials and Methods

2.1. Study Area

This study investigated shrublands on the northeastern Tibetan Plateau (Figure 1). Shrublands are one of the most widely distributed ecosystems on the plateau [23]. Shrublands compose woody vegetation, with coverage of more than 30% and a mean height below 5 m [24] and are primarily classified as desert shrublands or alpine shrublands [23]. Desert shrublands are mostly distributed in drier areas, such as the Qaidam Basin, and are characterized by species that can endure severe drought, such as *Kalidium foliatum* (Pall.) Moq, *Salsola abrotanoides* Bunge, and *Sympegma regelii* Bunge. Compared with alpine shrublands, only a few types of super-xerophytic herbs grow in desert shrublands. Brown desert soil and grey brown desert soil are mainly soil types in desert shrublands ecosystems, while soil in desert shrublands is very thin and more infertile than that in alpine shrublands [23]. In contrast, alpine shrublands are located in mountains with a cold and semiarid environment. Chestnut soil, grey cinnamon soil, and alpine shrubby meadow soil are mainly soil types in alpine shrublands ecosystems [23]. Representative species include *Rhododendron thymifolium* Maxim, *R. capitatum* Maxim, and *Sibiraea laevigata* (Linn.) Maxim. Mesophyte herbs, including *Kobresia* spp., *Carex* spp., and

Oxytropis spp., are located in alpine desertblands. MAP and MAT in the study regions range from 17.6 to 764.4 mm and −5.6 to 8.9 °C, respectively [25].

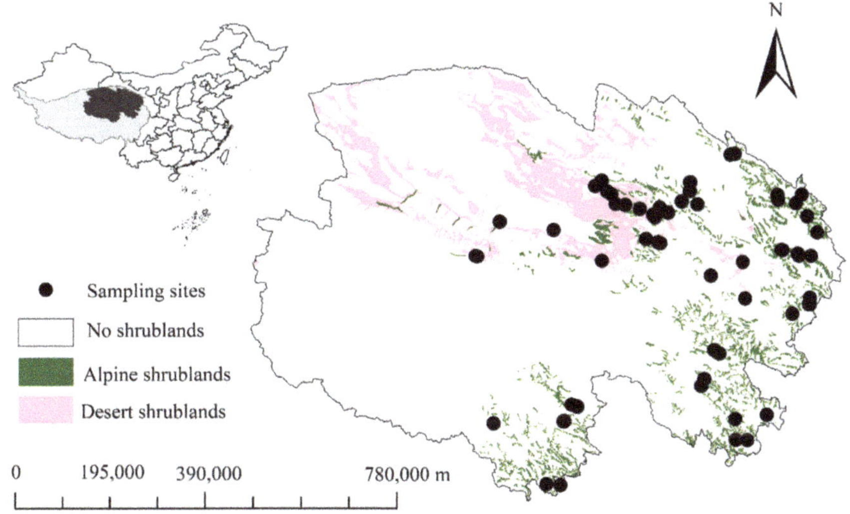

Figure 1. Distribution of 65 sampling sites in the shrublands of the Tibetan Plateau, which was based on the vegetation of China (1:100 0000) [26].

2.2. Field Investigation and Laboratory Measurements

In line with China's vegetation atlas (1:1000000) [26] and field investigations, we systematically selected 195 plots in 65 sites across the northeastern Tibetan Plateau shrublands in July and August from 2011 to 2013 (Figure 1). We selected three plots (1 m × 1 m) at each site and the distance of two plots was between 5 m and 50 m. At each site, all of the litter, including older litter and decomposed litter in the three plots, was collected to determine the litter standing crop [27]. Also, all shrubland plants in three plots were harvested to measure aboveground biomass. The samples of litter and aboveground biomass were oven dried at 65 °C to a constant weight, and weighed to the nearest 0.01 g. The litters were primarily foliage or branches. Carbon content of litter was determined using an elemental analyzer (2400 II CHNS; Perkin-Elmer, Boston, MA, USA).

2.3. Climate Data

To investigate the potential effects of climate factors on the litter standing crop carbon, MAT and MAP were extracted for each site from the WorldClim database (http://www.Worldclim.org/) with a spatial resolution of 1 × 1 km^2 [28,29].

2.4. Data Analysis

Ordinary least squares regression analyses were used to estimate the relationships between litter standing crop carbon, longitude, and latitude. Multiple linear regressions were conducted to evaluate relationships among litter standing crop carbon, MAT, MAP, and aboveground biomass. All analyses were conducted using SPSS 22.0 (SPSS Inc., Chicago, USA). Graphs were created in SigmaPlot 12.5 (Systat Software, Inc., Point Richmond, CA, USA).

3. Results

3.1. Size and Storage of Litter Standing Crop in Tibetan Plateau Shrublands

The litter standing crop ranged from 0.04 kg m^{-2} to 0.48 kg m^{-2}. The branch standing crop ranged from 0.02 kg m^{-2} to 0.48 kg m^{-2}, and foliage was no more than 0.25 kg m^{-2}. The mean litter standing crop was 0.23 kg m^{-2} in the shrublands overall (Table 1), and it was 0.25 kg m^{-2} and 0.19 kg m^{-2} in desert shrublands and alpine shrublands, respectively (Table 1). The mean values for foliage and branch standing crops in the shrublands were 0.04 kg m^{-2}, and 0.19 kg m^{-2}, respectively (Table 1). Most litter standing crop was stored in the branch standing crop, accounting for 80% of all litter standing crop overall (Figure 2a), and 76% and 87% in alpine shrublands and desert shrublands, respectively (Figure 2b,c).

Table 1. Size, storage and carbon storage of foliage, branch, and litter standing crops in the shrublands of the Tibetan Plateau.

Shrubland Type	Area (10^4 km^2)	Litter Standing Crop Density (kg m^{-2})			Litter Standing Crop Storage (Tg)			Litter Standing Crop Carbon Storage (Tg C)		
		Foliage	Branch	Litter	Foliage	Branch	Litter	Foliage	Branch	Litter
Desert shrublands	6.69	0.03	0.22	0.25	2.11	14.63	16.74	0.82	6.58	7.40
Alpine shrublands	4.29	0.05	0.14	0.19	1.94	6.15	8.09	0.76	2.77	3.53
Overall	10.98	0.04	0.19	0.23	4.05	20.78	24.83	1.58	9.35	10.93

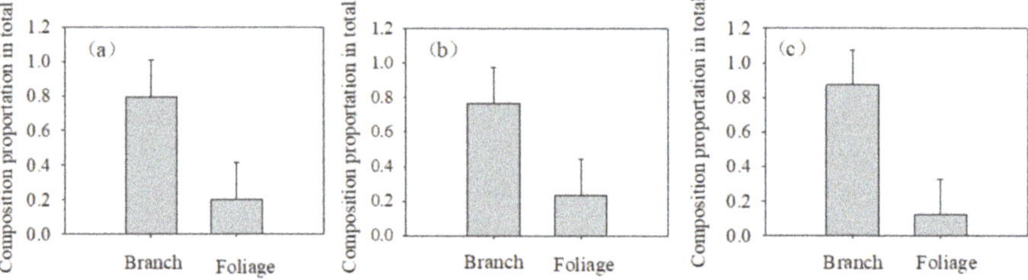

Figure 2. The proportions of branch and foliage standing crop in the litter standing crop on the Tibetan Plateau, (**a**) all shrublands, (**b**) alpine shrublands, (**c**) desert shrublands.

Overall, the litter standing crop storage in the shrublands was 24.83 Tg; 16.74 Tg and 8.09 Tg in desert shrublands and alpine shrublands, respectively. Storages in the foliage and branch standing crops were 4.05 Tg and 20.78 Tg, respectively (Table 1).

3.2. Conversion Coefficient, Storage, and Distribution of Litter Standing Crop Carbon

The conversion coefficient of litter standing crop to litter standing crop carbon was 0.44 in the shrublands of the Tibetan Plateau, which supported our hypothesis that the conversion coefficient was different from that in the shrublands of southern China. The conversion coefficients for foliage and branch standing crop to foliage and branch standing crop carbon were 0.39 and 0.45, respectively. Litter standing crop carbon was 10.93 Tg C in the northeastern Tibetan Plateau shrublands, and they were 7.40 Tg C and 3.53 Tg C in desert shrublands and alpine shrublands, respectively (Table 1). Furthermore, the conversion coefficients of branch standing crop to branch standing crop carbon were 0.42 and 0.46 in desert shrublands and alpine shrublands, respectively. Meanwhile, the conversion coefficients of foliage standing crop to foliage standing crop carbon were 0.27 and 0.40 in desert shrublands and alpine shrublands, respectively.

Spatially, the litter standing crop carbon was stable with increased latitude (Figure 3a), but it decreased with longitude in total shrublands (Figure 3b). Similarly, the significant trend also existed in the alpine shrublands (Figure 3c). In the desert shrublands, the range of longitude was between 94.51° and 98.73°. The litter standing crop carbon has a decreasing trend with an increasing longitude in desert shrublands (Figure 3d). Specifically, the highest litter standing crop carbon was in a low longitude (94.51°–95.91°), while the lowest litter standing crop carbon was in high MAT (97.32°–98.73°). In medium MAT (95.91°–97.32°), the litter standing crop carbon was moderate (Figure 3d).

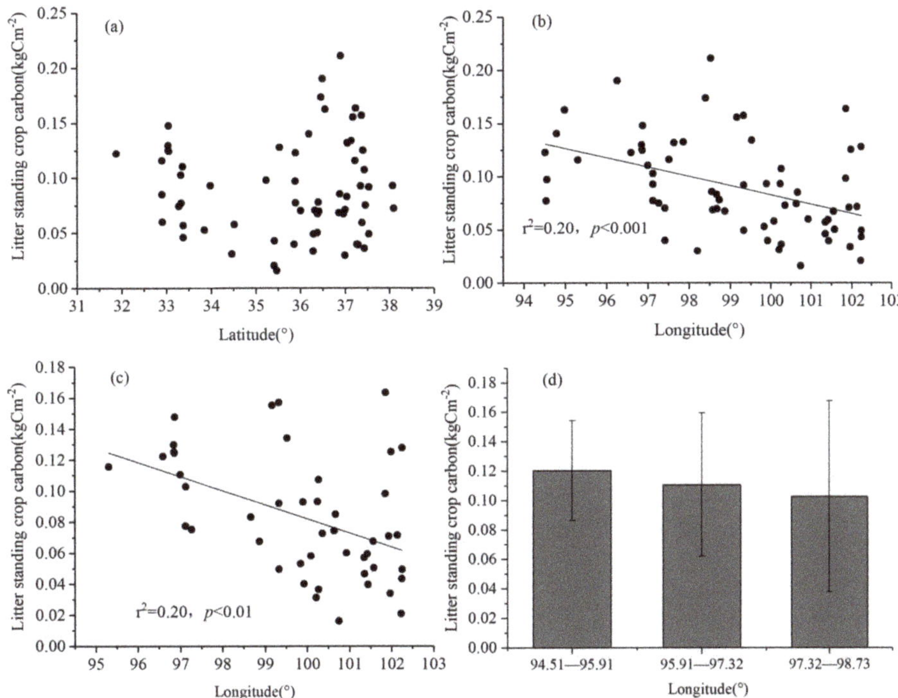

Figure 3. Relationships between litter standing crop carbon and latitude and longitude, (**a**) and (**b**) total shrublands, (**c**) alpine shrublands, (**d**) desert shrublands.

3.3. Effects of MAP and MAT on Litter Standing Crop Carbon

Litter standing crop carbon significantly decreased with MAP (Figure 4a) ($p < 0.01$). In the alpine shrublands, the range of MAP was between 160.83 mm and 572.50 mm. The litter standing crop carbon showed a negative trend with MAP in alpine shrublands. Specifically, the highest litter standing crop carbon was in low MAP (160.83 mm–298.06 mm), while the lowest litter standing crop carbon in high MAP (435.28–572.50 mm), and in medium MAP (298.06 mm–435.28 mm), the litter standing crop carbon was moderate (Figure 4b). Similarly, and the range of MAP was between 35.83 mm and 265.00 mm in desert shrublands. The highest litter standing crop carbon was in low MAP (35.83 mm–112.22 mm), while the lowest litter standing crop carbon in high MAP (188.61 mm–265.00 mm), and in medium MAP (112.22 mm–188.61 mm), the litter standing crop carbon was moderate (Figure 4c).

Litter standing crop carbon increased with MAT ($p < 0.01$) (Figure 4d). In the alpine shrublands, the range of MAT was between −2.83 °C and 3.20 °C. The litter standing crop carbon showed a positive trend with MAT in alpine shrublands. Specifically, the highest litter standing crop carbon was in high MAT (1.18 °C–3.20 °C), while the lowest litter standing crop carbon in low MAT(−2.85 °C−−0.83 °C), and in medium MAT (−0.83 °C–1.18 °C), the litter standing crop carbon was moderate (Figure 4e).

Similarly to desert shrublands, and the range of MAT was between −1.14 °C and 5.47 °C. The litter standing crop carbon showed a positive trend with MAT in desert shrublands. Specifically, the highest litter standing crop carbon was in high MAT (3.65 °C–5.47 °C), while the lowest litter standing crop carbon was in low MAT (−1.14 °C–0.42 °C), and in medium MAP (0.42 °C–3.65 °C), the litter standing crop carbon was moderate (Figure 4f). Generally, multiple linear regressions showed that in total shrublands, litter standing crop carbon = −0.001 MAP + 0.01 MAT + 0.25 ($r = 0.40$, $p < 0.01$).

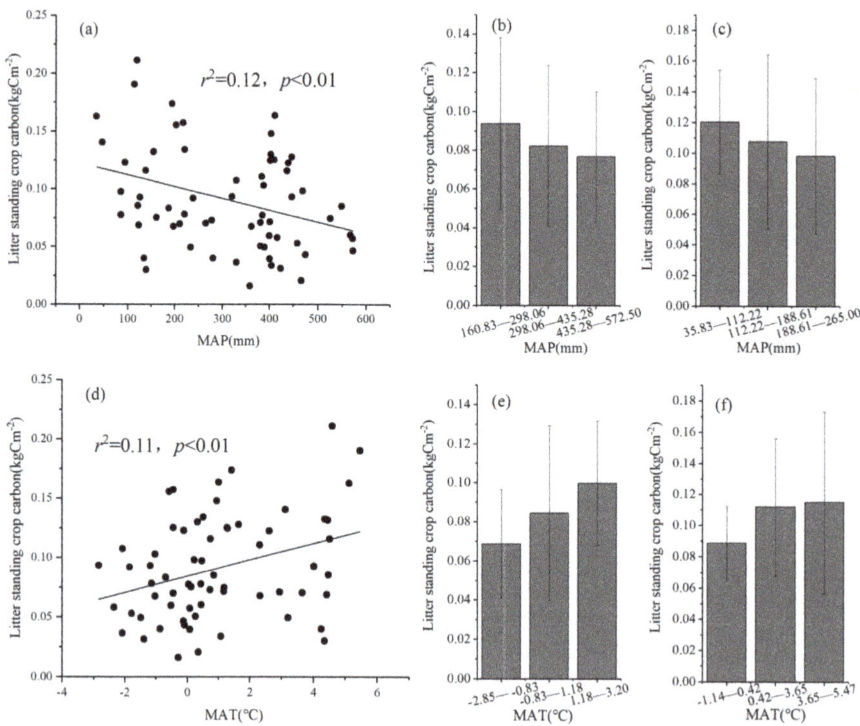

Figure 4. Relationships among mean annual precipitation (MAP) and mean annual temperature (MAT) and litter standing crop carbon, (**a**,**d**) total shrublands, (**b**,**e**) alpine shrublands, (**c**,**f**) desert shrublands.

3.4. Effects of Aboveground Biomass and Climatic Factors on Litter Standing Crop Carbon

Increasing aboveground biomass can stimulate litter standing crop carbon (Figure 5a) ($p < 0.01$), specifically, the positive trend also existed in both alpine shrublands and desert shrublands (Figure 5b). Generally, multiple linear regressions showed that litter standing crop carbon = −0.001 MAP + 0.01 MAT + 0.0001Aboveground biomass + 0.11 ($r = 0.42$, $p < 0.01$). Furthermore, the relationships among litter standing crop carbon, aboveground biomass, MAT, and MAP were shown in Figure S1.

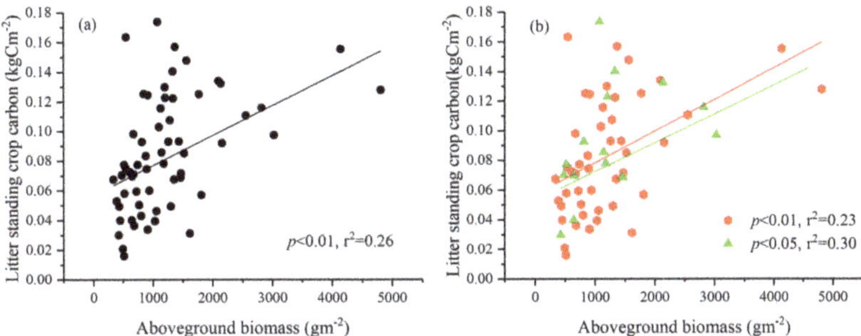

Figure 5. Relationships between aboveground biomass and litter standing crop carbon, (**a**) total shrublands, (**b**) red circle means alpine shrublands, green triangle means desert shrublands.

4. Discussion

4.1. Size of Litter Standing Crop

The density of the litter standing crop in the Tibetan Plateau shrublands was 0.23 kg m^{-2}, nearly half of the litter standing crop in the forests (0.45 kg m^{-2}) of the Tibetan Plateau [5], and was more than three times as much as the litter standing crop of the grasslands (0.05 kg m^{-2}) of the Tibetan Plateau [5]. The mean aboveground biomass in the shrublands was 1.10 kg m^{-2} [19], less than that of 4.22 kg m^{-2} in the forests [30], and greater than that of 0.08 kg m^{-2} in the grasslands of the plateau [31]. As aboveground biomass made the greatest contribution to litter standing crop in ecosystem [1], the litter standing crop in shrublands was in the range between forest and grasslands.

The litter standing crop in the shrublands of south China was 0.32 kg m^{-2} [16], larger than our result of 0.23 kg m^{-2} for the Tibetan Plateau. This difference may result from the greater amount of aboveground biomass in the shrublands across south China, compared to the Tibetan Plateau shrublands. Aboveground biomass can shape the input resource of litter standing crop [1] and significantly accumulate litter standing crop. It has been demonstrated that aboveground biomass in shrublands in the south region of China was 1.44 kg m^{-2} [32], greater than the 1.10 kg m^{-2} of the Tibetan Plateau shrublands [19].

4.2. Controlling Factors of Litter Standing Crop Carbon

Climatic factors can significantly shape litter standing crop [33,34]. Generally, precipitation not only stimulates vegetation growth in arid regions [35], and furthermore increases input of litter standing crop, but also affects the leaching of chemical substances from litter and consequently limits the litter decomposition rate [2]. So precipitation has been shown to have a positive effect on litter production [13] and stimulates accumulation of litter standing crop. However, precipitation has restrained effects on accumulation of litter standing crop carbon in the Tibetan Plateau shrublands. This result was different from those for other regions and biomes. For forests, precipitation did not have a significant effect on litter standing crop in forests of the eastern Inner Mongolia Plateau [18] or forests in China [5]. Litter standing crop carbon was stable with increasing precipitation in southern regions of China shrublands [16]. It should be noted that precipitation can also enhance litter decomposition by maximizing decomposer activity [11,36], which can reduce litter standing crop storage. Litter production and litter decomposition determine litter standing crop [37]. The restrain effects on the accumulation of litter standing crop carbon from precipitation may exceed the positive effects; consequently, increasing MAP decreases litter standing crop carbon in the shrublands, both alpine shrublands and desert shrublands, on the Tibetan Plateau.

Temperature is considered to be the most significant factor determining litter decomposition [38,39]. Rising temperature increases the litter decomposition rate [2]. In our study, MAT contributed to the accumulation of litter standing crop carbon in the shrublands on the Tibetan Plateau, which was different from other reported findings, specifically, that temperature had negative effects on accumulation of litter standing crop carbon in shrubland ecosystems in southern China [15] and in forests in China [5], whereas temperature was not a significant controlling factor for litter standing crop carbon in grasslands in China [5]. One of the most significant climate characteristics on the Tibetan Plateau is cold, and in the Tibetan Plateau shrublands, increasing temperature has been shown to play a significant role in shaping the amount of aboveground biomass [40]. Temperature contributes to the accumulation of aboveground biomass [40], which shapes amount of litter input, and stimulates more inputs into the litter standing crop; consequently, the litter standing crop in the shrublands of the Tibetan Plateau shows an increasing trend with MAT.

The climatic controlling factors on the litter standing crop carbon in the Tibetan Plateau shrublands, such as MAT and MAP, were different from those in other regions, which may indicate that the biophysical processes that control carbon on the Tibetan Plateau may differ from those in other regions [20,22]. In the global warming scenario, MAT is increasing by 0.05 °C every year [10,41], which, combined with the positive relationship between litter standing crop carbon and temperature, contributes to the carbon sink trend in the Tibetan Plateau shrublands.

4.3. Distribution and Conversion Coefficient of Litter Standing Crop to Litter Standing Crop Carbon

Spatially, the litter standing crop carbon decreased with longitude in shrublands, which may result from climatic changes with different regions. With an increasing longitude, the MAP has an increasing trend (Figure S2a), and consequently, increasing precipitation limited accumulation of litter standing crop carbon (Figure 4a–c). At the same time, MAT has a decreasing trend with longitude (Figure S2b), and decreasing temperature also limited the accumulation of litter standing crop carbon (Figure 4d–e). The climatic changes with longitude can explain the distribution of litter standing crop carbon in shrublands on the Tibetan Plateau.

It has been estimated that the conversion coefficient for litter standing crop to litter standing crop carbon is 0.41 in the shrublands of southern China [16], which is lower than our result of 0.44. Different vegetation types result in various rates of litter production and litter decomposition due to climatic conditions [33,34], and eventually, shape the conversion coefficient for litter standing crop to litter standing crop carbon. Cold weather is the most significant climatic characteristics of the Tibetan Plateau [42], due to its high elevation. On the one hand, lower temperature suppresses decomposer activity [11,43], which contributes to the weak decomposition of litter on the Tibetan Plateau. On the other hand, plants tend to produce organic matter with a higher carbon content [44–46]. Therefore, the conversion coefficient for litter standing crop to litter standing crop carbon in shrublands was smaller on the Tibetan Plateau compared with in shrublands of southern China. Researchers use a coefficient of 0.50 to convert biomass to biomass carbon in woody ecosystems such as forests [47,48] and shrublands [17], while the coefficient is smaller (0.45) in herbaceous ecosystems such as grasslands [49,50]. Our results reveal that the conversion coefficient for branch standing crop to branch standing crop carbon was 0.45, larger than the conversion coefficient of 0.36 for foliage standing crop to foliage standing crop carbon, which indicates that the conversion coefficient in woody ecosystems is larger than in herbaceous ecosystems. Studies of forest and grassland ecosystem support this trend, with a conversion coefficient of 0.46 in forests and 0.40 in grasslands in China [5]. Therefore, it is better to use specific conversion coefficients for different ecosystems to estimate litter standing crop carbon. Furthermore, even with in the same ecosystem, such as shrublands, it is better to use a specific parameter to assess carbon pools, specifically, 0.41 for the shrublands of southern China [16] and 0.44 for the Tibetan Plateau shrublands. To conduct a more accurate estimation of the carbon pool, it is necessary to use conversion coefficients for the different components of litter standing crop (in this case, 0.45 and 0.36, respectively, for branch and foliage standing crop in the shrublands of the Tibetan Plateau). Meanwhile, in the Tibetan Plateau, conversion coefficients of branch standing crop to branch standing crop carbon and foliage standing crop to foliage standing crop carbon were 0.46 and 0.42,

0.40, and 0.27 greater in alpine shrublands than in desert shrublands, respectively, which furthermore means that in the neighborhood regions (such as the Tibetan Plateau), the given conversion coefficient needs to be considered for more accurate estimation.

5. Conclusions

We used data from 65 sites to explore the storage, distribution, and environmental controls of litter standing crop carbon in the shrublands of the Tibetan Plateau. The litter standing crop carbon was 10.93 Tg C in the shrublands of the plateau. Most of the litter standing crop was stored in the branch standing crop, which accounted for 80% of all litter standing crop. The conversion coefficient was 0.44 for litter standing crop to litter standing crop carbon. Spatially, the litter standing crop showed a decreasing trend with increasing longitude, and was stable with latitude. Litter standing crop carbon decreased with increasing MAP, but MAT and aboveground biomass have had positive effects on litter standing crop carbon in the shrublands on the Tibetan Plateau.

Supplementary Materials: The following are available online at http://www.mdpi.com/1999-4907/10/11/987/s1, Figure S1: Relationships among aboveground biomass, litter standing crop carbon, and climatic factors, (a) mean annual temperature (MAT), (b) mean annual precipitation (MAP); Figure S2: Relationships among longitude, (a) mean annual precipitation (MAP), (b) mean annual temperature (MAT).

Author Contributions: Conceptualization, G.Z. and X.N.; Methodology, G.Z., X.N.; Investigation, L.Y.; Writing—Original Draft Preparation X.N. and D.W.; Project Administration, G.Z.

Acknowledgments: We thank Wenzhu Song, Zebing Zhong, Hechun Liu, and Yi Ning for facilitating our field surveys on the Tibetan Plateau and laboratory assistance. This study is funded by the National Program on Basic Work Project of China (2015FY11030001-5), Qinghai Province Natural Science Fund Project (2019-ZJ-910), Qinghai Province International Exchange and Cooperation Project (2019-HZ-807) and Strategic Priority Research Program of CAS (XDA0505030304).

Conflicts of Interest: The authors declare no conflict of interest. The founding sponsors had no role in the design of the study; in the collection, analyses, or interpretation of data; in the writing of the manuscript and in the decision to publish the results.

References

1. Descheemaeker, K.; Muys, B.; Nyssen, J.; Poesen, J.; Raes, D.; Haile, M.; Deckers, J. Litter production and organic matter accumulation in exclosures of the Tigray highlands, Ethiopia. *For. Ecol. Manag.* **2006**, *233*, 21–35. [CrossRef]

2. Peng, S.; Liu, Q. The dynamicis of forest litter and its responses to global warming. *Acta Ecol. Sin.* **2002**, *9*, 1534–1544.

3. Bojko, O.; Kabała, C. Loss-on-ignition as an estimate of total organic carbon in the mountain soils. *Pol. J. Soil Sci.* **2014**, *47*, 71–79.

4. Pan, Y.; Birdsey, R.A.; Fang, J.; Houghton, R.; Kauppi, P.E.; Kurz, W.A.; Phillips, O.L.; Shvidenko, A.; Lewis, S.L.; Canadell, J.G. A large and persistent carbon sink in the world's forests. *Science* **2011**, *333*, 988–993. [CrossRef] [PubMed]

5. Wen, D.; He, N. Spatial patterns of litter density and their controlling factors in forests and grasslands of China. *Acta Ecol. Sin.* **2016**, *36*, 2876–2884.

6. Marty, C.; Houle, D.; Gagnon, C. Variation in stocks and distribution of organic C in soils across 21 eastern Canadian temperate and boreal forests. *Forest Ecol. Manag.* **2015**, *345*, 29–38. [CrossRef]

7. Domke, G.M.; Perry, C.H.; Walters, B.F.; Woodall, C.W.; Russell, M.B.; Smith, J.E. Estimating litter carbon stocks on forest land in the United States. *Sci. Total Environ.* **2016**, *557*, 469–478. [CrossRef]

8. Liu, L.; Shen, G.; Chen, F.; Luo, L.; Xie, Z.; Yu, J. Dynamic characteristics of litterfall and nutrient return of four typical forests along the altitude gradients in Mt. Shennongjia, China. *Acta Ecol. Sin.* **2012**, *32*, 2142–2149. [CrossRef]

9. McLaren, J.R.; Buckeridge, K.M.; Weg, M.J.; Shaver, G.R.; Schimel, J.P.; Gough, L. Shrub encroachment in Arctic tundra: Betula nana effects on above-and belowground litter decomposition. *Ecology* **2017**, *98*, 1361–1376. [CrossRef]

10. Yang, Y. Ecological processes in alpine ecosystems under changing environment. *Chin. J. Plant Ecol.* **2018**, *42*, 1–5.

11. Taylor, P.G.; Cleveland, C.C.; Wieder, W.R.; Sullivan, B.W.; Doughty, C.E.; Dobrowski, S.Z.; Townsend, A.R. Temperature and rainfall interact to control carbon cycling in tropical forests. *Ecol. Lett.* **2017**, *20*, 779–788. [CrossRef] [PubMed]

12. Wohl, E.; Barros, A.; Brunsell, N.; Chappell, N.A.; Coe, M.; Giambelluca, T.; Goldsmith, S.; Harmon, R.; Hendrickx, J.M.; Juvik, J. The hydrology of the humid tropics. *Nat. Clim. Chang.* **2012**, *2*, 655. [CrossRef]

13. Jia, B.; Zhou, G.; Liu, Y.; Jiang, Y. Spatial pattern and environment controls of annual litterfall production in natural forest ecosystems in China. *Sci. Sin. Vitae* **2016**, *11*, 1304–1311.

14. Zhao, Y.; Wang, X.; Zhang, Y. Relationship between litterfall production and meteorological factors in evergreen broad-leaved forest in Tiantong national forest park. *J. Beijing Univ. Agric.* **2017**, *32*, 73–77.

15. Zhang, X.; Wang, X.; Zhu, B.; Zong, Z.; Peng, C.; Fang, J. Litter fall production in relation to environmental factions in northeast China's forests. *Chin. J. Plant Ecol.* **2008**, *32*, 1031–1040.

16. Ge, J.; Xiong, G.; Li, J.; Xu, W.; Zhao, C.; Lu, Z.; Li, Y.; Xie, Z. 2017 Litter standing crop of shrubland ecosytem in southern China. *Chin. J. Plant Ecol.* **2017**, *41*, 5–13.

17. Hu, H.; Wang, Z.; Liu, G.; Fu, B. Vegetation carbon storage of Major shrublands in China. *Chin. J. Plant Ecol.* **2006**, *30*, 539–544.

18. Chen, S.; Liu, H.; Guo, D. Litter stocks and chemical quality of natural birch forests along temperature and precipitatioan gradients in eastern Inner Mongolia China. *Chin. J. Plant Ecol.* **2010**, *34*, 1007–1015.

19. Nie, X.; Yang, Y.; Yang, L.; Zhou, G. Above- and Belowground Biomass Allocation in Shrub Biomes across the Northeast Tibetan Plateau. *PLoS ONE* **2016**, *11*, e0154251. [CrossRef]

20. Nie, X.; Peng, Y.; Li, F.; Yang, L.; Xiong, F.; Li, C.; Zhou, G. Distribution and controlling factors of soil organic carbon storage in the northeast Tibetan shrublands. *J. Soils Sediments* **2019**, *19*, 322–331. [CrossRef]

21. Nie, X.; Yang, L.; Li, F.; Xiong, F.; Li, C.; Zhou, G. Storage, patterns and controls of soil organic carbon in the alpine shrubland in the Three Rivers Source Region on the Qinghai-Tibetan Plateau. *Catena* **2019**, *178*, 154–162. [CrossRef]

22. Yang, Y.; Fang, J.; Tang, Y.; Ji, C.; Zheng, C.; He, J.; Zhu, B. Storage, patterns and controls of soil organic carbon in the Tibetan grasslands. *Glob. Chang. Biol.* **2008**, *14*, 1592–1599. [CrossRef]

23. Zhou, X.; Wang, Z.B.; Du, Q. *Vegetation of Qinghai Province*; Qinghai People's Publishing House: Xining, China, 1987; pp. 53–72.

24. Wu, Z. *Vegetation of China*; Science Press: Beijing, China, 1980; p. 430.

25. Zhang, Z. *Geography of Qinghai Plateau*; Science Press: Beijing, China, 2009; pp. 25–26.

26. Chinese Academy of Science. *Vegetation Altas of China*; Science Press: Beijing, China, 2001.

27. Technical Manual Writing Group of Ecosystem Carbon Sequestration Project. *Observation and Investigation for Carbon Sequestration in Terrestrial Ecosystems*; Science Press: Beijing, China, 2015; pp. 153–155.

28. Yang, X.; Tang, Z.; Ji, C.; Liu, H.; Ma, W.; Mohhamot, A.; Shi, Z.; Sun, W.; Wang, T.; Wang, X.; et al. Scaling of nitrogen and phosphorus across plant organs in shrubland biomes across Northern China. *Sci. Rep.* **2014**, *4*, 5448. [CrossRef] [PubMed]

29. Nie, X.; Xiong, F.; Yang, L.; Li, C.; Zhou, G. Soil nitrogen storage, distribution, and associated controlling factors in the northeast Tibetan Plateau shrublands. *Forests* **2017**, *8*, 416. [CrossRef]

30. He, H. *Modeling Forest NPP Patterns on the Tibetan Plateau and Its Responses to Cliamte Change*; Chinese Academy of Forestry: Beijing, China, 2008.

31. Yang, Y. Carbon and Nitrogen in Alpine Grasslands on the Tibetan Plateau. Ph.D. Thesis, Peking University, Beijing, China, 2008.

32. Li, J. Biodiversity and Its Relationship to Ecosystem Production in Shrubland across Subtropical Region in China. Ph.D. Thesis, Institute of Botany, Chinese Academy of Sciences, Beijing, China, 2015.

33. Zhou, X.; Talley, M.; Luo, Y. Biomass, litter, and soil respiration along a precipitation gradient in southern Great Plains, USA. *Ecosystems* **2009**, *12*, 1369–1380. [CrossRef]

34. Chave, J.; Navarrete, D.; Almeida, S.; Alvarez, E.; Aragão, L.E.; Bonal, D.; Châtelet, P.; Silva-Espejo, J.; Goret, J.Y.; Hildebrand, P. Regional and seasonal patterns of litterfall in tropical South America. *Biogeosciences* **2010**, *1*, 43–55. [CrossRef]

35. Wynn, J.G.; Bird, M.I.; Vellen, L.; Grand-Clement, E.; Carter, J.; Berry, S.L. Continental-scale measurement of the soil organic carbon pool with climatic, edaphic, and biotic controls. *Glob. Biogeochem. Cycles* **2006**, *20*, GB1007. [CrossRef]

36. Wieder, W.R.; Cleveland, C.C.; Townsend, A.R. Controls over leaf litter decomposition in wet tropical forests. *Ecology* **2009**, *90*, 3333–3341. [CrossRef]

37. Hilli, S.; Stark, S.; Derome, J. Litter decomposition rates in relation to litter stocks in boreal coniferous forests along climatic and soil fertility gradients. *Appl. Soil Ecol.* **2010**, *46*, 200–208. [CrossRef]

38. Hobbie, S.E. Temperature and plant species control over litter decomposition in Alaskan tundra. *Ecol. Monogr.* **1996**, *66*, 503–522. [CrossRef]

39. Fierer, N.; Craine, J.M.; McLauchlan, K.; Schimel, J.P. Litter quality and the temprtature sensitivity of decomposition. *Ecology* **2005**, *86*, 320–326. [CrossRef]

40. Nie, X.; Yang, L.; Xiong, F.; Li, C.; Fan, L.; Zhou, G. Aboveground biomass of the alpine shrub ecosystems in Three-River Source Region of the Tibetan Plateau. *J. Mt. Sci.* **2018**, *15*, 357–363. [CrossRef]

41. Duan, K.; Yao, T.; Wang, K.; Tian, L.; Xu, B. The Difference in precipitation variabil ity between the north and south Tibetan Plateaus. *J. Glaciol. Geocryol.* **2008**, *30*, 726–732.

42. Ding, J.; Li, F.; Yang, G.; Chen, L.; Zhang, B.; Liu, L.; Fang, K.; Qin, S.; Chen, Y.; Peng, Y.; et al. The permafrost carbon inventory on the Tibetan Plateau: A new evaluation using deep sediment cores. *Glob. Chang. Biol.* **2016**, *22*, 2688–2701. [CrossRef] [PubMed]

43. Cleveland, C.C.; Wieder, W.R.; Reed, S.C.; Townsend, A.R. Experimental drought in a tropical rain forest increases soil carbon dioxide losses to the atmosphere. *Ecology* **2010**, *91*, 2313–2323. [CrossRef] [PubMed]

44. Vergutz, L.; Manzoni, S.; Porporato, A.; Novais, R.F.; Jackson, R.B. Global resorption efficiencies and concentrations of carbon and nutrients in leaves of terrestrial plants. *Ecol. Monogr.* **2012**, *82*, 205–220. [CrossRef]

45. Schreeg, L.A.; Mack, M.C.; Turner, B.L. Nutrient-specific solubility patterns of leaf litter across 41 lowland tropical woody species. *Ecology* **2013**, *94*, 94–105. [CrossRef]

46. Sun, X.; Kang, H.; Chen, H.; Berg, B.; Bartels, S.F.; Liu, C. Biogeographic patterns of nutrient resorption from Quercus variabilis Blume leaves across China. *Plant Biol.* **2016**, *18*, 505–513. [CrossRef]

47. Fang, J.; Chen, A.; Peng, C.; Zhao, S.; Ci, L. Changes in forest biomass carbon storage in China between 1949 and 1998. *Science* **2001**, *292*, 2320–2322. [CrossRef] [PubMed]

48. Guo, Z.; Hu, H.; Li, P.; Li, N.; Fang, J. Spatio-temporal changes in biomass carbon sinks in China's forests from 1977 to 2008. *Sci. China Life Sci.* **2013**, *56*, 661–671. [CrossRef]

49. Piao, S.; Fang, J.; Zhou, L.; Tan, K.; Tao, S. Changes in biomass carbon stocks in China's grasslands between 1982 and 1999. *Glob. Biogeochem. Cycles* **2007**, *21*, GB2002. [CrossRef]

50. Ma, W.; Fang, J.; Yang, Y.; Mohammat, A. Biomass carbon stocks and their changes in northern China's grasslands during 1982–2006. *Sci. China Life Sci.* **2010**, *53*, 841–850. [CrossRef] [PubMed]

 *forests*

Article

Divergent Responses of Foliar N:P Stoichiometry During Different Seasons to Nitrogen Deposition in an Old-Growth Temperate Forest, Northeast China

Dongxing Yang, Hongrui Mao and Guangze Jin *

Center for Ecological Research, Northeast Forestry University, Harbin 150040, China;
ydx19930719@sina.com (D.Y.); damao731@126.com (H.M.)
* Correspondence: kwpine@nefu.edu.cn; Tel.: +86-451-82191823

Received: 4 February 2019; Accepted: 10 March 2019; Published: 13 March 2019

Abstract: Atmospheric nitrogen (N) deposition has rapidly increased during the last few decades; however, the seasonal responses of leaf N:P stoichiometry to N deposition remain unclear. In 2008, a simulated N deposition experiment (0, 30, 60, and 120 kg·N·ha^{-1}·yr^{-1}) was conducted in an old-growth temperate forest in Northeast China. In 2014, the leaves of 17 woody species and soil were sampled in spring, summer, and autumn in each treatment, and N:P stoichiometry was assessed. Community N and P in summer were significantly lower than that in spring and autumn. Unlike broadleaved species, conifers showed no significant variation among the three seasons. N addition significantly enhanced community N and soil available P but decreased soil total P in summer and autumn, and decreased community P, as well as the P concentration of three life forms (conifer, tree, and shrub), in autumn. Our results emphasize the importance of multiple sampling across seasons in temperate forests. Arguing against the traditional consensus, the productivity of the old-growth temperate forests is limited by both N and P.

Keywords: nitrogen deposition; N and P colimitation; leaf N:P stoichiometry; soil N:P stoichiometry; seasonal variations

---

1. Introduction

Increased atmospheric nitrogen (N) deposition due to the combustion of fossil fuels and the increase in agricultural production during the last few decades, has captured the attention of many ecologists [1–3]. N deposition also has rapidly increased in China since the 1980s because of the country's rapid industrial development [4,5]. In addition, N and P are commonly considered the two most limiting elements of net primary productivity (NPP) in terrestrial ecosystems [6,7], increased N deposition may result in an imbalance in the input of N vs. P in ecosystems [8,9] and further alter foliar N:P, the community nutrient limitation, and species composition [10–12].

Although positive effects of increased foliar N by N deposition are often reported [8,13], neutral and even inhibitory effects have also been observed [14,15]. Studies of foliar P under N deposition are relatively less common than those of foliar N [16], and foliar N:P ratio variations under N deposition remain uncertain. In fact, the determination of a limiting status is important for predicting how a change in foliar N:P stoichiometry under N deposition occurs. The effects of N deposition on community production and on N:P stoichiometry occur in three stages [16–18]. During the first stage, plant growth under N limitation will show a rapid increase [19]. During the second stage, the response of plant growth will decrease compared to that during the first stage as the status of N saturation is approached [17]. When N limitation changes to P limitation during the third stage, increased N input may result in the loss of soil available nutrients, and plant growth will be inhibited in response to N deposition [14]. In addition, P is a major limiting nutrient in tropical regions with old-aged soils, whereas temperate regions in the Northern Hemisphere with younger soils are often limited by N [7]. Arguing against the traditional consensus, Elser et al. [6] suggested that P limitation has an equivalent status to that of N limitation across terrestrial systems. Through a fertilization experiment in 13 boreal forests, Goswami et al. [20] found evidence that most middle-aged and mature forests exhibit P limitation rather than N limitation. Other studies have described a shift from N limitation to P limitation with succession, as N-cycling properties recover and the dominance of a conservative P cycle re-emerges, in mature and old-growth forests [21,22]. Overall, as the critical forest type of the northern region of China, further research on the mixed broadleaved-Korean pine (*Pinus koraiensis*) forest, an old-growth temperate forest, regarding its nutrient limitation status is needed.

Initially, fertilization experiments are widely employed to assess community nutrient status and N or P limitation [23–26]. Based on the strong relationship between N and P, Koerselman and Meuleman [27] also proposed a simple method that utilizes the foliar N:P ratio to evaluate the limiting status. However, a one-time sampling may not fully reflect the stoichiometric characteristics of a plant community [28]. Because of the change in foliar development stages and the existence of a retranslocation mechanism in plant leaves, foliar N and P may undergo marked variations through two seasons in just a few months, even in mature green leaves [19,29,30]. Thus, considering the effect of seasonal change is important when studying patterns of change in foliar N:P stoichiometry and the response of foliar N:P stoichiometry to N deposition. To our knowledge, there is no research focusing on the effects of N deposition on foliar N:P stoichiometry in different seasons.

Because plant traits affect internal and external nutrient cycling, different responses to N deposition and seasonal change may appear among species and plant functional groups [7,31]. Additionally, mycorrhizae, as a type of key classification for plants, are also attracting the attention of many biologists and ecologists [20,32,33]. Indeed, mycorrhizal symbiosis has a strong effect on plant growth and plays a key role in nutrient cycling [34,35], and various mycorrhizal types possess different mechanisms and capacities for influencing the process in which plants acquire nutrients from the soil. For example, the hyphae of mycorrhizal fungi can search for limited nutrients, including nitrates and phosphates, to support plant growth, but only ectomycorrhizal (EM) and ericoid mycorrhizal fungi can obtain organically bound nutrients [36]. As plants of old-growth temperate forests are associated with both arbuscular mycorrhizal (AM) and ectomycorrhizal (EM) species, the plants associated with differential mycorrhizal types may have diverse responses to altered surroundings.

This study investigated the effects of seasonal changes on the response pattern of leaf N:P stoichiometry to N deposition through a simulated N addition experiment. We sought to determine (1) how leaf stoichiometric characteristics change with seasonal change, (2) whether the responses of the leaf N:P stoichiometry to N deposition show differences in three seasons, and (3) the nature of the nutrient limitation status of an old-growth temperate forest in Northeast China.

2. Materials and Methods

2.1. Site Description

The study site is in the Heilongjiang Liangshui National Nature Reserve (47°10′50″ N, 128°53′20″ E) in Dailing, Heilongjiang Province, Northeast China. The annual mean temperature is −0.3 °C, and the annual mean maximum temperature is 7.5 °C. The annual mean minimum temperature is −6.6 °C. The ≥ 0 °C accumulated temperature is 2200 °C–2600 °C. The annual mean precipitation is 676 mm, and >90% of the precipitation falls during the growing season from May to October. There are 100–120 frost-free days and 130–150 snow-cover days. The average elevation is 400 m, and the difference in the relative elevation is 100–200 m. The zonal soil is a dark brown forest soil. The mixed broadleaved-Korean pine (*Pinus koraiensis*) forest is the zonal vegetation of the reserve. The stand age is between 200 and 300 years, with some individual ages older than 400 years.

2.2. Experimental Design and Measurements

The N addition experiment began during May 2008 in the mixed broadleaved-Korean pine (*Pinus koraiensis*) forest, an old-growth temperate forest. Twelve 20-m × 20-m sequential plots were established in a random manner. To prevent mutual interference, 10-m barriers between adjacent plots were retained. In every plot, species with a diameter at breast height (DBH, 1.3 m) >2 cm were surveyed; species, DBH, and coordinates were recorded and labeled in May 2008 and surveyed again in July 2014. The background conditions of each plot surveyed in 2008 are shown in Table 1. Using urea $(CO(NH_2)_2)$ as the N source, four treatment groups were established, with three replicates—control (N0, 0 $kg \cdot ha^{-1} \cdot yr^{-1}$), low N (N1, 30 $kg \cdot ha^{-1} \cdot yr^{-1}$), medium N (N2, 60 $kg \cdot ha^{-1} \cdot yr^{-1}$), and high N (N3, 120 $kg \cdot ha^{-1} \cdot yr^{-1}$). The background N deposition (inorganic and organic N) was 12.9 $kg \cdot N \cdot ha^{-1} \cdot yr^{-1}$, as recorded by a dry–wet deposition collector (New Star Environmental LLC, Roswell, GA, USA). N additions were applied at the beginning of June, July, and August every year and began in June 2008. The $CO(NH_2)_2$ for each treatment was dissolved in 20 L of water, and the solutions were sprayed onto the soil surface using a backpack sprayer. The control group was sprayed with the same amount of water.

Table 1. Natural conditions in each plot before the N addition experiment in 2008.

Plots	Treatments	Average N Addition $(kg \cdot ha^{-1} \cdot yr^{-1})$	Repeat	Density $(trees \cdot ha^{-1})$	Mean Breast Diameter (cm)	Slope Aspect	Slope Degree
1	N0	0	N0-1	1325	10.70	W	<6°
2	N0	0	N0-2	1625	8.60	W	<6°
3	N1	30	N1-1	1575	8.46	W	<6°
4	N3	120	N3-1	2925	7.68	W	<6°
5	N3	120	N3-2	1825	9.48	W	<6°
6	N3	120	N3-3	1100	14.02	W	<6°
7	N0	0	N0-3	1650	9.43	W	<6°
8	N1	30	N1-2	1550	11.47	W	<6°
9	N1	30	N1-3	1575	10.41	W	<6°
10	N2	60	N2-1	2300	7.62	W	<6°
11	N2	60	N2-2	1800	11.44	W	<6°
12	N2	60	N2-3	1525	11.45	W	<6°

When collecting samples from tall trees, a branch with many leaves and sufficient sunlight was cut by a person specialized in climbing trees. The healthy and green leaves were selected and placed into a plastic vacuum packaging bag. Seventeen species were collected randomly in 3 plots for each treatment in the middle of May, July, and September during 2014. However, insufficient material was collected for a proportion of the species because of natural and anthropogenic reasons; for example, most of the *Fraxinus mandshurica* individuals had not renewed their foliage in May, and their leaves had already fallen by September, or a few trees were so high that samples could not be obtained. In this study, May, July, and September represented spring, summer, and autumn, respectively. The 17 species were of three types— deciduous broad-leaved trees, deciduous broad-leaved shrubs and evergreen conifers (hereafter trees, shrubs and conifers, respectively). Conifers were assigned their own group because they have unique traits compared to those of other trees [7]. The mycorrhizal types of species in this study were according to Shi et al. [37], and species belonging to the same family were classified into the same mycorrhizal type. The detailed sampling information is listed in Table 2.

Soil samples were also collected at the same time as leaf samples. In each plot, soils were randomly sampled by taking three 5-cm-diameter soil cores from depths of 0–10, 10–20, 20–30, and 30–40 cm. The litter on the soil surface was removed before sampling. The samples were placed in plastic bags and transported to the lab after being air-dried.

Table 2. Sampling condition of each species in 2014.

Scientific Name of Species	Life Form	Ectomycorrhizal (EM) or Arbuscular Mycorrhizal (AM)	The important value (%) under Control Group in 2014	Uncollected Samples in Different Treatments		
				Spring	Summer	Autumn
Pinus koraiensis	Conifer	EM	27.67	N3		
Acer mono	Tree	EM	11.19		N1	
Corylus mandshurica	Shrub	EM	9.60			
Tilia amurensis	Tree	EM	8.04	N1		N1
Betula costata	Tree	EM	7.05	N1 N2 N3		N0 N1 N2
Ulmus laciniata	Tree	EM	5.66			
Ulmus japonica	Tree	EM	4.13	N2		
Acer tegmentosum	Tree	AM	3.75		N1	N0 N1 N2 N3
Fraxinus mandshurica	Tree	AM	3.03	N1 N2 N3		N1 N2
Prunus padus	Tree	AM	1.98			
Syringa reticulate var. _mandshurica_	Tree	AM	1.54			
Acer ukurunduense	Tree	EM	1.19	N1	N1	
Euonymus pauciflorus	Shrub	AM	1.13			
Picea koraiensis	Conifer	EM	0	N2		
Deutzia gladata	Shrub	AM	0			
Lonicera chrysantha	Shrub	AM	0			
Ribes mandschuricum	Shrub	AM	0			

Each leaf sample was ground to a fine powder, passed through a 0.149-mm mesh, and enveloped. The sample was oven-dried at 65 °C to a constant weight for further analysis. The soil sample was passed through a 2-mm mesh; a portion was conserved in an envelope, and the remaining material was conserved after passing through a 0.149-mm mesh. In October 2014, leaf N concentration was analyzed using a Kjeltec KTM 2300 analyzer (Foss Teactor AB, Hoganas, Sweden) after H_2SO_4 + K_2SO_4 + $CuSO_4$ digestion. Soil total N (TN) in every layer was analyzed using the same method but with H_2SO_4 + H_2O_2 digestion. Leaf and soil total P (TP) were measured using the molybdenum antimony colorimetric spectrophotometer method after H_2SO_4 + H_2O_2 digestion. These samples were oven-dried again at 65 °C for 2 h before digestion. For soil available P (AP), 2-mm of soil from every layer was pooled and fully mixed, and the air-dried soil was extracted using 0.05 $mol·L^{-1}$ HCl + 0.025 $mol·L^{-1}$ H_2SO_4. The P concentration in the extract was measured using the molybdenum antimony colorimetric spectrophotometer method. N and P concentrations are expressed as $mg·g^{-1}$, and a mass basis was used to calculate stoichiometric ratios.

2.3. Statistical Analyses

The important value (*IV*) of *i*-th species in each treatment group in 2008 and 2014 were calculated using the following equation:

$$IV_{(i)} = (\text{relative density}_{(i)} + \text{relative frequency}_{(i)} + \text{relative prominence}_{(i)}) / 3$$

The change in *IVs* for each species was directly calculated using the following equation:

$$\text{The changed value}_{(i)} = IV_{(i)(2014)} - IV_{(i)(2008)}$$

Two-way analysis of variance (ANOVA) with season and treatment as fixed factors was applied to examine the effects on leaf and soil N, P and the N:P ratio based on averages. Community N:P stoichiometry (N:P stoichiometry means N, P, and the N:P ratio, community N:P stoichiometry means N:P stoichiometry of the plant community) weighted by $IV_{(i)(2014)}$ was also examined in the same manner (hereafter, community N:P stoichiometry is the leaf N:P stoichiometry weighted by the $IV_{(i)(2014)}$). Meanwhile, the marginal means of the main factors were calculated and recorded. Three-way ANOVA with season, treatment, and different groups (broadleaves vs. conifers, trees vs. shrubs, ectomycorrhizal species (EM) vs. arbuscular mycorrhizal species (AM), broadleaves including broadleaved trees and shrubs; the data for conifers were not included in tree and EM groups) as fixed factors were used to analyze effects on leaf N, P, and the N:P ratio. Nonsignificant interactions were removed from the analysis. Leaf P, the leaf N:P ratio and soil N:P stoichiometry in our data were apparently of right-skewed distribution, and they were naturally logarithmically transformed before performing ANOVA; untransformed data are shown in the figures.

The data were separated according to (1) different seasons and (2) different seasons and different classified groups. Least squares discrimination (LSD) one-way ANOVA was applied to assess differences between treatment groups, and $IV_{(i)(2014)}$ was calculated as weightiness. Differences between N:P stoichiometry averages in our study and previous studies were evaluated using the one-sample *t*-test. The Kolmogorov–Smirnov and Levene's tests were employed to assess the normality of the residuals and the homogeneity of variance, respectively. All figures were created using Origin 9.0, and all data were analyzed using SPSS 20 (SPSS, Inc., Chicago, IL, USA). Significance was determined at the 0.05 level.

3. Results

3.1. N, P, and the N:P Ratio in the Three Seasons Under Ambient (without N Addition) Conditions

Community N, average N, and soil N showed a significant increase in September after an initial decrease from spring to summer (Figure 1a,d and Figure 2a). A significant increase from spring to autumn was found for soil N:P, which exhibited the same trend as soil N and average N:P (Figures 1f and 2a,d). Community N:P in autumn was significantly lower than that in summer but was significantly higher than that in spring. Community P and average P in spring were significantly higher than those in the other two seasons (Figure 1b,c,e,f). Although soil TP showed no significant variation during the three seasons, soil AP had a similar trend to that of community P, tending to increase again in autumn after a significant decrease in summer (Figures 1b and 2b,c). The leaf N:P stoichiometry of conifers was significantly lower than that of trees and shrubs ($p < 0.05$), and the conifer P and N:P ratios displayed no significant seasonal variation, different from the patterns of the trees and shrubs. Moreover, a nearly identical seasonal pattern between the trees and shrub N:P stoichiometry was observed (Figure 3a–i).

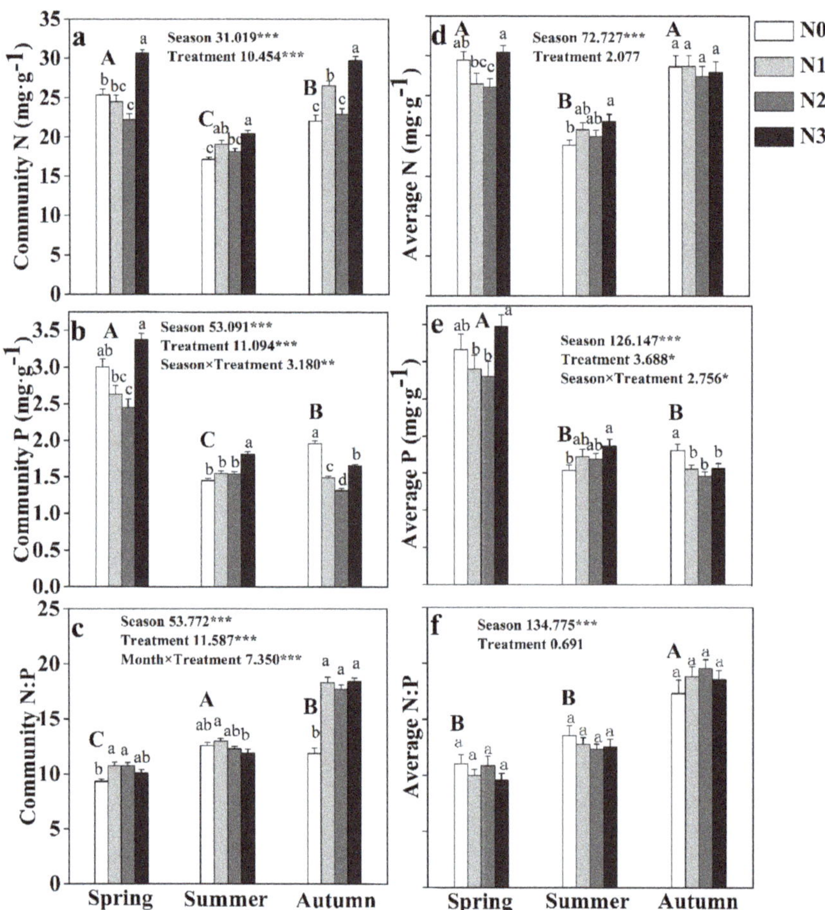

Figure 1. Effects of N addition on community and average N (**a,d**), P (**b,e**) and the N:P ratio (**c,f**) in three seasons. The results of two-way ANOVAs are shown in each panel and interactions without significance were removed from the analysis. The results of least squares discrimination (LSD) one-way ANOVA were shown above the bars. Different uppercase letters indicate a significant difference among the three seasons under ambient conditions. Different lowercase letters indicate a significant difference among treatments in each month. Data are shown as the mean + *SE*. * $p < 0.05$, ** $p < 0.01$, *** $p < 0.001$.

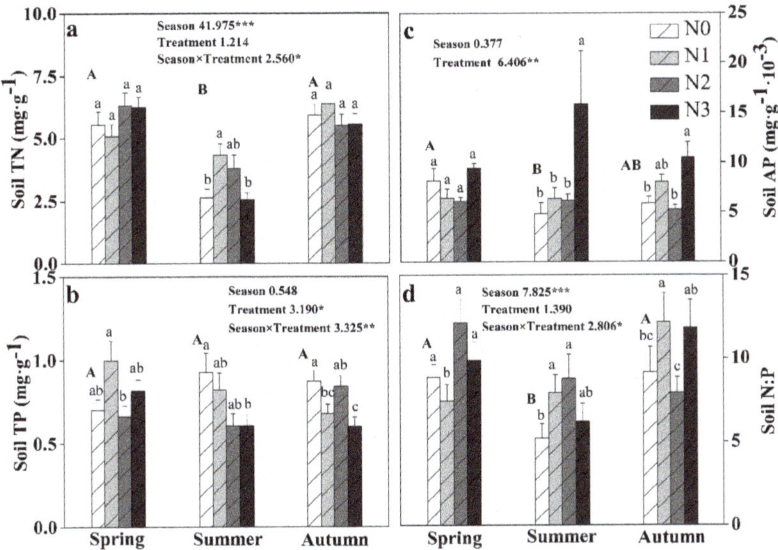

Figure 2. Effects of N addition on soil total N (TN) (**a**), soil total P (TP) (**b**), soil available P (AP) (**c**) and soil total N:P ratio (**d**) in the three seasons. The results of two-way ANOVAs are shown in each panel and interactions without significance were removed from the analysis. The results of least squares discrimination (LSD) one-way ANOVA were shown above the bars. Different uppercase letters indicate a significant difference among the three seasons under ambient conditions. Different lowercase letters indicate a significant difference among the treatments in each month. Data are shown as the mean + *SE*. * $p < 0.05$, ** $p < 0.01$, *** $p < 0.001$.

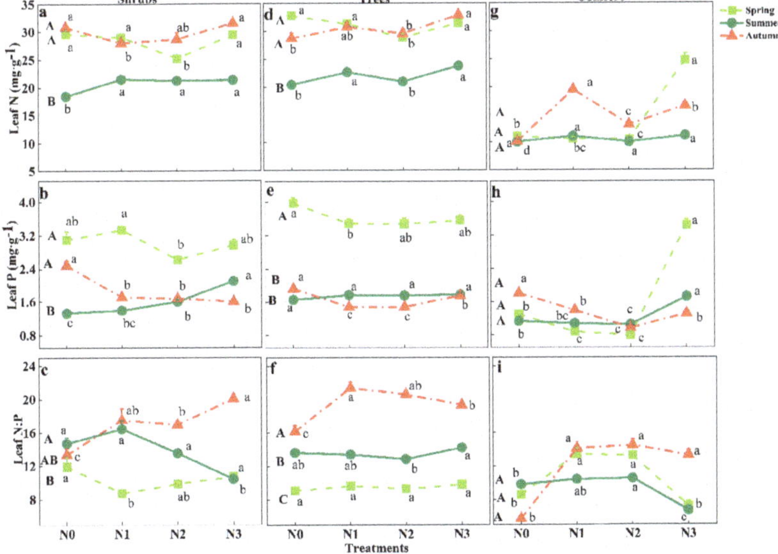

Figure 3. Effects of N addition on leaf N (**a,d,g**), leaf P (**b,e,h**) and the leaf N:P ratio (**c,f,i**) of three life forms in three seasons. Different uppercase letters indicate a significant difference among the three seasons under ambient. Different lowercase letters indicate a significant difference among the treatments in each month. Data are shown as the mean + *SE*.

Summer was the peak season in the survey region. During this season, leaf average N and P in our study was similar to the average for plants in China, though N:P was significantly lower. Leaf P was significantly lower than reported in two global studies, and leaf N was significantly lower than that in the study of Elser et al. [38]. We also found some inconsistent results for leaf N:P stoichiometry in spring and autumn compared to that in summer (Table 3).

Table 3. Comparison of the average leaf N, P, and N:P in our study with China and the world.

Stoichiometry and Seasons	N			P			N:P		
	Spring	Summer	Autumn	Spring	Summer	Autumn	Spring	Summer	Autumn
Han et al. [39]		20.2			1.46			16.3	
(China)	+ *	−	+ *	+ *	+	+ *	− *	− *	+
Reich and Oleksyn [7]		20.1			1.77			13.8	
(World)	+ *	−	+ *	+ *	− *	+	− *	−	+ *
Elser et al. [38]		20.6			1.99			12.7	
(World)	+ *	− *	+ *	+ *	− *	−	−	+	+ *

"+" means the average in this study is higher than average of China and the world. "−" means the value in this study is lower than the tested value. * $p < 0.05$.

3.2. Effect of N Addition on Community and Soil N:P Stoichiometry during Different Seasons

Community N and P showed a similar variation to that of average N and P under N addition in spring and summer, with a similar change pattern for N to that of P (Figure 1a,b,d,e). During spring, community N and average N were significantly increased under N3 after a decreasing trend (Figure 1a,d). Significant increases in community N:P were found for N1 and N2 treatments, but there was no significant variation between the different treatment groups with regard to average N:P (Figure 1c,f). In summer, an upward trend in community and average N and P was observed (Figure 1a,b,d,e); community N:P under N3 was significantly lower than that under N1, and there was no significant effect of N deposition on average N:P (Figure 1c,f). In autumn, N addition significantly increased community N, P, and N:P and significantly decreased average P; however, there was no significant variation in average N and N:P (Figure 1a–f). A two-way ANOVA showed that seasonal change had a significant impact on the community and average N:P stoichiometry (Figure 1a–f). Treatment also had a significant impact on community N:P stoichiometry but no significant impact on average N and N:P (Figure 1c,d,f). In addition, their interaction had a significant influence on community P, N:P and average P (Figure 1b,c,e).

During summer, soil N and N:P showed an increasing trend, with a decrease under N3. In addition, the N3 treatment significantly decreased soil P and increased soil AP. The patterns of variation in soil P and soil AP in autumn were similar to those in summer; soil N:P increased under N1 and N3, and soil N showed no significant change (Figure 2a–d). A two-way ANOVA showed that seasonal change had a significant impact on soil N and soil N:P, and treatment had a significant impact on soil P and soil AP; however, their interaction had no significant impact on soil AP (Figure 2a–d). In addition, the marginal means of main factors (season and treatment) are listed in the Table S1.

3.3. Response of Leaf N:P Stoichiometry in Different Groups to Seasonal Change and N Addition

A three-way ANOVA showed that although only a small variation in the response of N:P stoichiometry between shrubs and trees to N addition was found in the three seasons, the response was significantly different than that of conifers (Table 4; Figure 3a–i). In spring, the leaf N of shrubs, trees and conifers all exhibited a significant decrease under N2 that was increased under N3 (Figure 3a,d,g). A similar decrease under N2 in leaf P was found (Figure 3b,e,h). In summer, N addition significantly increased N in trees and shrubs, with no remarkable effect on conifer N (Figure 3a,d,g). In autumn, increased leaf N was found in the leaves of the trees and conifers, but their leaf P showed a trend opposite to leaf N under N addition (Figure 3d,e,g,h). N addition significantly increased the N:P of shrub, trees, and conifers in autumn (Figure 3c,f,i).

A three-way ANOVA showed that different mycorrhizal types displayed notable differences in leaf N:P stoichiometry, and there were significant interactions between mycorrhizal type and other factors (Table 4; Figure 4a–f). In particular, EM species in autumn showed a nearly opposite trend to that of AM species in terms of N:P stoichiometry (Figure 4a–f). From 2008 to 2014, the changed *IV* of AM species showed an increasing trend with higher available N. In contrast, the changed *IV* of EM showed a decreased trend under increasing N addition (Figure 5).

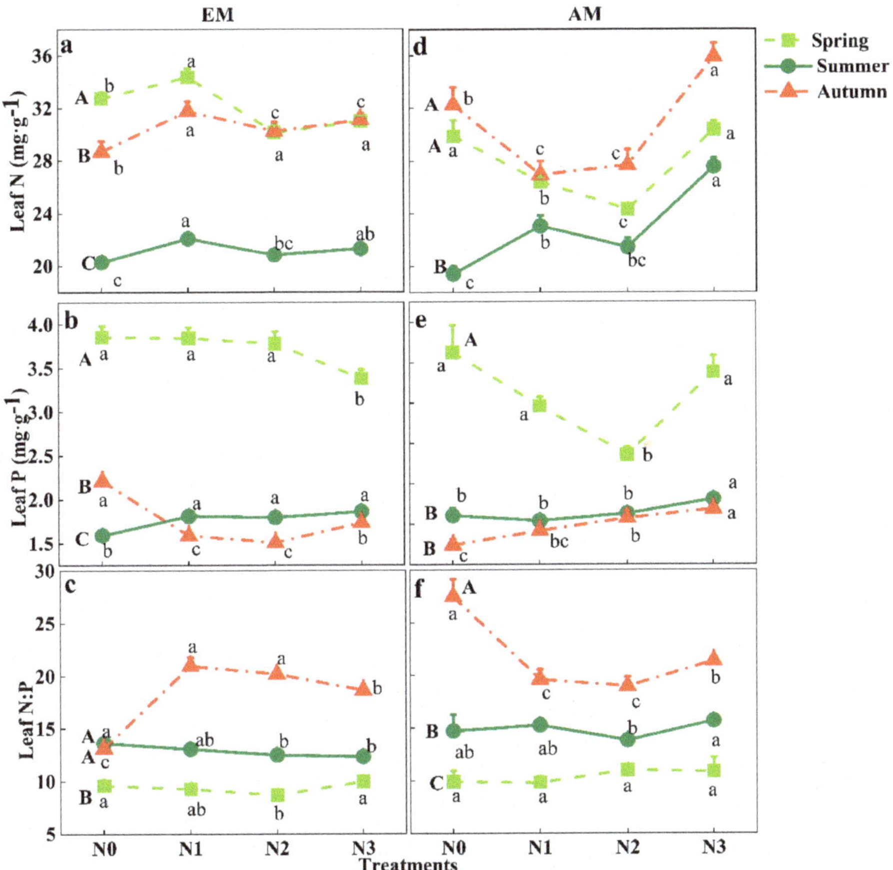

Figure 4. Effect of N addition on leaf N (**a,d**), leaf P (**b,e**) and the leaf N:P ratio (**c,f**) of ectomycorrhizal species (EM) vs. arbuscular mycorrhizal species (AM) in the three seasons. Different uppercase letters indicate a significant difference among the three seasons under ambient conditions. Different lowercase letters indicate a significant difference among the treatments in each month. Data are shown as the mean + *SE*.

Table 4. Results (*F* values) of three-way ANOVA on the effects of leaf N:P stoichiometry in different groups.

Different Groups	Factors	Season	Treatment	Group	Season × Treatment	Season × Group	Treatment × Group	Season × Treatment × Group
Broadleaves vs. conifers	N	53.480 ***	6.269 ***	408.073 ***	2.857 *	8.434 ***		2.220 *
	P	25.034 ***	20.112 ***	134.599 ***	6.762 ***	18.939 ***	9.209 ***	3.286 **
	N:P	33.11 ***	21.895 **	39.339 ***	11.868 **	14.017 **	9.716 **	3.817 **
Trees vs. Shrubs	N	106.743 ***	4.307 **	5.477 *				
	P	104.666 ***	2.415	0.906	2.458 *	3.691 *		
	N:P	131.018 ***	3.085 *	0.005	2.581 *			
EM vs. AM	N	80.135 ***	7.201 ***	1.092		8.889 ***	7.089 ***	
	P	142.893 ***	1.639	17.898 ***				2.250 **
	N:P	127.291 ***	0.433	14.060 ***				2.798 ***

Interactions without significance were removed from the analysis, and the date are not shown in the table. * $p < 0.05$, ** $p < 0.01$, *** $p < 0.001$.

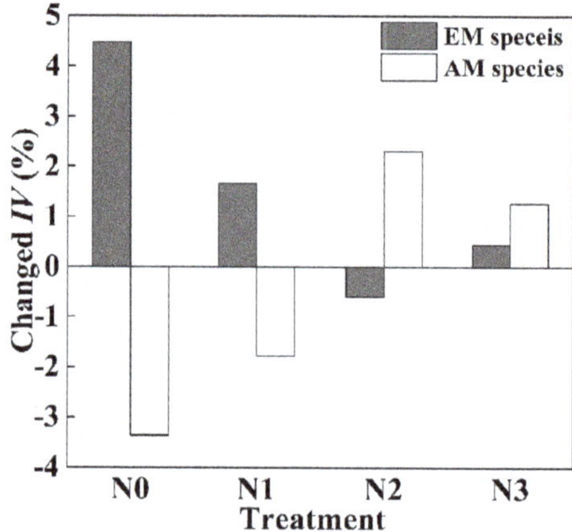

Figure 5. Changed *IV* (%) of AM species and EM species from 2008 to 2014 under four treatments.

4. Discussion

4.1. Seasonal Change in Leaf N:P Stoichiometry

Consistent with previous study [19], among the three seasons, community N and P were highest in spring, which is due to the rapid growth rate of leaves [28,29]. N is the main element of proteins, and P is the main element of P-rich ribosomal RNA (rRNA). In spring, leaves are in the early stage of their growth period, the foliar size is relatively small, and foliar cells show a high degree of splitting and selection to absorb considerable N and P to support their rapid growth. Therefore, foliar nutrient concentrations are relatively high [18,40]. Summer is the peak season: The rate of foliar nutrient absorption is lower than that of growth, leaves are larger than they are in spring, and most of the broad leaves have fully expanded. Thus, the concentrations of nutrients are diluted, which results in rapidly decreasing nutrient concentrations in summer [41]. However, when leaves are no longer growing and are fully expanded, the dilution process stops, and foliar nutrient concentrations increase again because of continuous nutrient absorption; accordingly, community N and P in autumn are higher than in summer.

Because of the rapid decrease in foliar P compared to foliar N in summer, community N:P is of a significantly higher level than that in spring. The sharp decrease in foliar P from the early growing season to the late growing season has also been found in previous studies, but without explanation [19,42]. A possible reason is the decreased demand for P. Because plants are older in summer than in spring, active meristems (e.g., inflorescences, shoot tips, young leaves) will restrict plant growth; in contrast, mature leaves no longer grow but are still photosynthetically active, which results in a greater decrease in the P requirement for rRNA synthesis compared to the decrease in N for protein synthesis [43]. Similarly, soil AP significantly decreases, by 41.2%, from spring to summer, and the percentage of soil AP of soil TP during Spring, Summer, and Autumn is 1.17%, 0.52%, and 0.67%, respectively. Wang and Moore [30] also found a strong increase in available N and K but no change in soil AP in summer and autumn compared to available N, P, and K in spring.

When we divided the foliar data into shrubs, trees, and conifers, remarkably different seasonal patterns and nutrient concentrations between coniferous and broadleaved species were found. Indeed, leaves with high nutrient concentrations (both N and P) tend to be shorter lived and have a high photosynthetic capacity, whereas leaves with low nutrient concentrations tend to have longer lifespans and lower metabolic capacities [44,45] (Evergreen-Deciduous hypothesis). Conifers (mainly consisting of *Pinus koraiensis*, the *IV* of which under ambient conditions is 27.67%) maintain a low leaf nutrient status that may constitute a strategy related to canopy dominance and space occupancy [46]. In addition, unlike broadleaved species that translocate N and P to leaves during the early growing season, conifers gradually translocate these nutrients to new leaves throughout the growing season to maintain a relatively stable nutrient concentration and stoichiometry, thereby decreasing the effect of the external environment [29]. This finding confirms our results that conifers exhibit less seasonal variation than the other two plant groups examined in this study. It was also noted that the weighted N:P stoichiometry of shrubs and trees demonstrated a pattern similar to that of the average N:P stoichiometry, although this was not the case for conifers, which explains the reduced variation in community N:P stoichiometry from summer to autumn compared to the variation in the average N:P stoichiometry. These findings suggest that compared to average values, weighted values may better reflect the N:P stoichiometry at a community level.

Similar seasonal pattern of foliar nutrient concentrations has also been exhibited in the previous study [30]: foliar N concentration of deciduous species showed an increased trend from July to September after a sharp decrease from May to June, though the increase was not statistically significant and there was a sharp decrease again in October due to nutrient resorption. As mentioned earlier, the seasonal pattern of N concentration exhibits the strong effects of foliar development stage on nutrient concentrations of leaves, and exposes the limitation of calculating foliar nutrients based on the mass ratio. In fact, it is a well-known phenomenon in agriculture that nutrient concentrations decrease with ontogenetic development of individual plants, and this decrease is mainly because of the imbalanced rate between increased nutrients and accumulated dry matter [19]. However, the application of mass ratio is still widely accepted in the field of plant stoichiometry [16,19,43], because we are accustomed to thinking about organism body mass or estimating total biomass of species or components in ecosystems [40]. Nonetheless, our study suggests that compared to the foliar concentration (mg·g^{-1}), the calculation based on the foliar mass (mass·leaf^{-1}) may be a better expression when considered the seasonal effects, because it can directly express the nutrient contents in leaves and ignore the effects of foliar development stage [29].

4.2. Seasonal Variation in the Leaf N:P Stoichiometry Response to N Addition

Seasonal change had a highly significant effect on community N:P stoichiometry. Furthermore, leaf nutrients responded differently to N addition in the three seasons. In fact, although N addition increases the availability of N and promotes its absorption, it can at the same time stimulate the growth of plants, which is accompanied by dilution of nutrient concentrations [19,28]. Thus, the differences among treatments may reflect the relative magnitude of the foliar growth rate and the nutrient absorption rate. During the early growth stage, N addition increases the growth rate of leaves more rapidly than it increased N absorption rate. Accumulated N is diluted by rapidly growing leaves, which results in a decreasing trend under low and medium N addition. When the increase in absorption rate is faster than the increase in growth rate, the N concentration shows an increasing trend, for example: the community N under high N addition during spring. Uniformly, the foliar growth rate gradually slowed or even stopped, but the effect of N addition on the nutrient absorption rate was still effective, which resulted in the increasing trend under N addition during summer.

Studies have suggested that imbalanced N-P loading affects P cycling and may enhance P limitation [12,45,46]. This effect is exhibited in the decreased soil TP in summer and autumn and increased community N:P in autumn under N addition. In the short term, N-induced P limitation may be alleviated by enhanced P absorption due to enhanced activity of soil and root phosphatases or by increased P conservation due to increased P resorption [47–51]. In this study, a similar change pattern between community P and community N during spring and summer was found, and soil AP rose significantly as a result of high N addition, confirming the alleviation of N-induced P limitation. Mitigation of N-induced P limitation under N deposition has also been reported in temperate plantations by Deng et al. [16] and in a semiarid grassland by Long et al. [52], reflecting the capacity of plants to regulate P acquired to maintain homeostasis (*H*) [43,53].

Nonetheless, community P as well as the P concentration of the three life forms (shrub, tree, and conifer) all showed a decreasing trend under N addition in autumn. There are two possible reasons for this variation. One, is that these regulatory mechanisms may not be able to quantitatively supply sufficient P to balance the increase in N in plants under continuous N input [26]; however, the limited variations across seasons in soil TP, increased community P under ambient conditions and decreased P demand of leaves during this stage suggest that this explanation may not be viable. Another, more plausible reason, is internal nutrient retranslocation, i.e., the process by which plants transform nutrients in leaves into other tissues [29,49]. The recycling mechanism occurs not only in senescing leaves but also during the entire life cycle of leaves [54] as a response to N-induced P limitation. Plants thus transfer more P from mature green leaves, which have a relatively low P demand in autumn, to other tissues to support their growth. Unfortunately, we did not collect more data for nutrients in other tissues to support this explanation, and more research needs to be completed in the future. In the long term, the mitigating effect of N addition on P limitation may not be persistent, and the degree of P limitation will still be enhanced by N addition because of the reduced P concentration in leaves, which will slow the P recycling rate and further decrease the amount of available P [12,47,55].

Mycorrhizal associations play a critical role in regulating the acquisition of limited nutrients [34,35]. Mycorrhizal associations exist in nearly all ecosystems, with mycorrhizal plants dominating in many habitats [56,57]. AM-associated plants are often dominant in low-latitude areas, whereas EM-associated plants are often dominant in high-latitude areas. Overall, compared to EM, AM associations result in more effective P acquisition [31,58]. Consistent with Deng et al. [16], who reported an increased AM:EM ratio in a young stand under N addition, we also found that the changed *IV* of AM species increased but that of EM species decreased with increasing N addition, which suggests once again an enhanced P limitation under N addition as well as the regulation of community N-induced P limitation. Furthermore, we did not observe significant differences in responses to N addition between shrubs and trees, but AM species and EM species did display such differences. Studies have shown that EM fungi tend to specialize in N acquisition, whereas AM species might experience secondary N limitation under the enhanced P limitation induced by N addition [20,59]. This may explain the diverse variation in N concentration under N addition between the two mycorrhizal types. Similar to the variation in community P, the P concentration among EM species also decreased under N addition. Zhang et al. [33] found a significantly greater P resorption rate for EM-associated trees than for AM trees in a boreal forest, which may from another perspective bolster our explanation for the decrease in P during autumn. Moreover, the increased P level in AM species demonstrates a stronger capacity to acquire P. In addition, other properties are also important in driving the growth of a plant community, for example: The evolution of soil organic to humic matter, which may interact in the response of foliar N:P stoichiometry to N deposition [60,61].

4.3. Status of Nutrient Limitation

How to assess nutrient limitation in primary productivity is always a concern of ecologists, and a fertilization experiment is a frequently used method that defines nutrient limitation as a requirement for additional nutrients [26,47,62]. Accordingly, the improved foliar N level via N addition across the three seasons showed that N may be a constraint in this forest. Nevertheless, we cannot conclude that N is the sole limiting nutrient. In contrast, we infer that P is also a critical limiting nutrient for this forest because of the initiation of some adaptive strategies that aim to alleviate P limitation under N addition.

In addition to fertilization experiments, community foliar nutrient concentrations and ratios based on averages are widely applied for assessing nutrient limitation. Koerselman and Meuleman [27] found that a foliar N:P ratio <14 indicated that the plant was limited by N alone, that N:P >16 indicated limitation solely by P, and that 14 < N:P < 16 indicated colimitation. Later, Güsewell et al. [63] proposed a more conservative threshold value; i.e., N:P < 10 and N:P > 20 solely represented N limitation and P limitation, respectively. Recently, community N:P weighted by the *IV* or biomass of species is also being applied in some research [13,22]. In this study, the values of N:P across three seasons were all between 10 and 20, regardless of whether they were calculated using the weighted or average value. In addition, through a comparison with previous research during the peak season, our results show that the foliar P in the study region is lower than the worldwide average and that the foliar N:P is lower than the average for China. These results support the opinion of Han et al. [39], who found that foliar P across most areas in China was lower than the global average and that northern regions in China were more limited by N. More evidence regarding the existence of P limitation is reflected in the low-supplied soil AP. Through investigation of pools and distributions of soil P in China, Zhang et al. [64] found that dark brown forest soil, which is the zonal soil in this region, has a relatively high soil TP but a quite low soil AP compared to other soil types. The highest soil AP:TP ratio in this study was approximately 0.012 under ambient conditions. These results further prove our inference that N and P limitation coexist in this forest.

Studies have shown that the relative status of N vs. P limitation changes with succession [21,22]. Inputs of N and P have different sources; P is derived from rock weathering, whereas N can accumulate from the atmosphere through biological N fixation and atmospheric N deposition [65]. Therefore, young ecosystems lacking the input of atmospheric elements are usually limited by the supply of N, and with the development of succession, P will decrease to a limiting level because of the accumulation of soil organic carbon and soil TN [26]. The old-growth temperate forest in the present study has a long history of succession, which may explain the reason for the observed N and P colimitation. Wang and Moore [30] also recently reported colimitation of N and P in an ombrotrophic peatland in eastern Ontario, Canada, and Goswami et al. [20] found evidence of P limitation in most middle-aged and mature stands in central New Hampshire, USA. These two regions are all at latitudes similar to that of the region in this study.

5. Conclusions

This is the first study that explores the divergent responses of foliar N:P stoichiometry during different seasons to N deposition. In this study, plant leaf stoichiometric characteristics showed different variations with seasonal changes, and responses to N deposition in three seasons also differed. Thus, significant variations in leaf stoichiometric characteristics across these three seasons over just four months may be worth considering. Our research emphasizes the importance of multiple sampling across seasons in a temperate forest. Broadleaved species vs. conifers and AM species vs. EM species showed significant differences in seasonal variation and in response to N addition. Arguing against the traditional consensus, the net primary productivity of this mixed broadleaved-Korean pine (*Pinus koraiensis*) forest, an old-growth temperate forest and a critical forest type in Northeast China, is under N and P colimitation. The level of P limitation is enhanced by N deposition, and this N-induced

P limitation may be alleviated by the adjustment of strategies for plant adaption such as nutrient retranslocation and mycorrhizal association.

Supplementary Materials: The following are available online at http://www.mdpi.com/1999-4907/10/3/257/s1, Table S1: Marginal means of main factors.

Author Contributions: G.J. conceived and designed the experiments; D.Y. and H.M. performed the experiments; D.Y. analyzed the data; D.Y. contributed reagents/materials/analysis tools; D.Y. and G.J. wrote the paper.

Funding: This work was funded by Fundamental Research Funds for Central Universities (2572017EA02).

Acknowledgments: We thank three anonymous reviewers and editor for constructive suggestions to improve the quality of the manuscript.

Conflicts of Interest: The authors declare no conflict of interest.

References

1. Galloway, J.N.; Dentener, F.J.; Capone, D.G.; Boyer, E.W.; Howarth, R.W.; Seitzinger, S.P.; Asner, G.P.; Cleveland, C.C.; Green, P.A.; Holland, E.A.; et al. Nitrogen cycles: Past, present, and future. *Biogeochemistry* **2004**, *70*, 153–226. [CrossRef]
2. Neff, J.C.; Townsend, A.R.; Gleixner, G.; Lehman, S.J.; Turnbull, J.; Bowman, W.D. Variable effects of nitrogen additions on the stability and turnover of soil carbon. *Nature* **2002**, *419*, 915–917. [CrossRef] [PubMed]
3. Phoenix, G.K.; Emmett, B.A.; Britton, A.J.; Caporn, S.J.M.; Dise, N.B.; Helliwell, R.; Jones, L.; Leake, J.R.; Leith, I.D.; Sheppard, L.J.; et al. Impacts of atmospheric nitrogen deposition: Responses of multiple plant and soil parameters across contrasting ecosystems in long-term field experiments. *Glob. Chang. Biol.* **2012**, *18*, 1197–1215. [CrossRef]
4. Liu, X.; Duan, L.; Mo, J.; Du, E.; Shen, J.; Lu, X.; Zhang, Y.; Zhou, X.; He, C.; Zhang, F. Nitrogen deposition and its ecological impact in China: An overview. *Environ. Pollut.* **2011**, *159*, 2251–2264. [CrossRef] [PubMed]
5. Lu, X.; Mao, Q.; Gilliam, F.S.; Luo, Y.; Mo, J. Nitrogen deposition contributes to soil acidification in tropical ecosystems. *Glob. Chang. Biol.* **2014**, *20*, 3790–3801. [CrossRef] [PubMed]
6. Elser, J.J.; Bracken, M.E.; Cleland, E.E.; Gruner, D.S.; Harpole, W.S.; Hillebrand, H.; Ngai, J.T.; Seabloom, E.W.; Shurin, J.B.; Smith, J.E. Global analysis of nitrogen and phosphorus limitation of primary producers in freshwater, marine and terrestrial ecosystems. *Ecol. Lett.* **2007**, *10*, 1135–1142. [CrossRef] [PubMed]
7. Reich, P.B.; Oleksyn, J. Global patterns of plant leaf N and P in relation to temperature and latitude. *Proc. Natl. Acad. Sci. USA* **2004**, *101*, 11001–11006. [CrossRef] [PubMed]
8. Crowley, K.F.; McNeil, B.E.; Lovett, G.M.; Canham, C.D.; Driscoll, C.T.; Rustad, L.E.; Denny, E.; Hallett, R.A.; Arthur, M.A.; Boggs, J.L.; et al. Do nutrient limitation patterns shift from nitrogen toward phosphorus with increasing nitrogen deposition across the northeastern United States? *Ecosystems* **2012**, *15*, 940–957. [CrossRef]
9. Parfitt, R.L.; Ross, D.J.; Coomes, D.A.; Richardson, S.J.; Smale, M.C.; Dahlgren, R.A. N and P in new zealand soil chronosequences and relationships with foliar N and P. *Biogeochemistry* **2005**, *75*, 305–328. [CrossRef]
10. Clark, C.M.; Tilman, D. Loss of plant species after chronic low-level nitrogen deposition to prairie grasslands. *Nature* **2008**, *451*, 712–715. [CrossRef] [PubMed]
11. Peñuelas, J.; Sardans, J.; Rivas-ubach, A.; Janssens, I.A. The human-induced imbalance between C, N and P in earth's life system. *Glob. Chang. Biol.* **2012**, *18*, 3–6. [CrossRef]
12. Penuelas, J.; Poulter, B.; Sardans, J.; Ciais, P.; van der Velde, M.; Bopp, L.; Boucher, O.; Godderis, Y.; Hinsinger, P.; Llusia, J.; et al. Human-induced nitrogen-phosphorus imbalances alter natural and managed ecosystems across the globe. *Nat. Commun.* **2013**, *4*, 2934. [CrossRef] [PubMed]
13. Han, X.; Sistla, S.A.; Zhang, Y.H.; Lü, X.T.; Han, X.G. Hierarchical responses of plant stoichiometry to nitrogen deposition and mowing in a temperate steppe. *Plant Soil* **2014**, *382*, 175–187. [CrossRef]
14. Magill, A.H.; Aber, J.D.; Currie, W.S.; Nadelhoffer, K.J.; Martin, M.E.; McDowell, W.H.; Melillo, J.M.; Steudler, P. Ecosystem response to 15 years of chronic nitrogen additions at the Harvard forest LTER, Massachusetts, USA. *For. Ecol. Manag.* **2004**, *196*, 7–28. [CrossRef]
15. Mayor, J.R.; Wright, S.J.; Turner, B.L.; Austin, A. Species-specific responses of foliar nutrients to long-term nitrogen and phosphorus additions in a lowland tropical forest. *J. Ecol.* **2014**, *102*, 36–44. [CrossRef]

16. Deng, M.; Liu, L.; Sun, Z.; Piao, S.; Ma, Y.; Chen, Y.; Wang, J.; Qiao, C.; Wang, X.; Li, P. Increased phosphate uptake but not resorption alleviates phosphorus deficiency induced by nitrogen deposition in temperate *Larix principis-rupprechtii* plantations. *New Phytol.* **2016**, *212*, 1019–1029. [CrossRef] [PubMed]

17. Aber, J.D.; Nadelhoffer, K.J.; Steudler, P.; Melillo, J.M. Nitrogen saturation in northern forest ecosystem. *Bioscience* **1989**, *39*, 378–386. [CrossRef]

18. Elser, J.J.; Fagan, W.F.; Kerkhoff, A.J.; Swenson, N.G.; Enquist, B.J. Biological stoichiometry of plant production: Metabolism, scaling and ecological response to global change. *New Phytol.* **2010**, *186*, 593–608. [CrossRef] [PubMed]

19. Ågren, G.I. Stoichiometry and nutrition of plant growth in natural communities. *Annu. Rev. Ecol. Evol. Syst.* **2008**, *39*, 153–170. [CrossRef]

20. Goswami, S.; Fisk, M.C.; Vadeboncoeur, M.A.; Garrison-Johnston, M.; Yanai, R.D.; Fahey, T.J. Phosphorus limitation of aboveground production in northern hardwood forests. *Ecology* **2018**, *99*, 438–449. [CrossRef]

21. Davidson, E.A.; de Carvalho, C.J.; Figueira, A.M.; Ishida, F.Y.; Ometto, J.P.; Nardoto, G.B.; Saba, R.T.; Hayashi, S.N.; Leal, E.C.; Vieira, I.C.; et al. Recuperation of nitrogen cycling in amazonian forests following agricultural abandonment. *Nature* **2007**, *447*, 995–998. [CrossRef]

22. Zhang, W.; Zhao, J.; Pan, F.; Li, D.; Chen, H.; Wang, K. Changes in nitrogen and phosphorus limitation during secondary succession in a karst region in southwest China. *Plant Soil* **2015**, *391*, 77–91. [CrossRef]

23. Chapin, F.S. The mineral nutrition of wild plants. *Ann. Rev. Ecol. Syst.* **1980**, *11*, 233–260. [CrossRef]

24. Chapin, F.S.; Vitousek, P.M.; Van Cleve, K. The nature of nutrient limitation in plant communities. *Am. Nat.* **1986**, *127*, 48–58. [CrossRef]

25. Vitousek, P.M.; Howarth, R.W. Nitrogen limitation on land and in the sea: How can it occur? *Biogeochemistry* **1991**, *13*, 87–115. [CrossRef]

26. Vitousek, P.M.; Porder, S.; Houlton, B.Z.; Chadwick, O.A. Terrestrial phosphorus limitation: Mechanisms, implications, and nitrogen–phosphorus interactions. *Ecol. Appl.* **2010**, *20*, 5–15. [CrossRef]

27. Koerselman, W.; Meuleman, A.F.M. The vegetation N:P ratio: A new tool to detect the nature of nutrient limitation. *J. Appl. Ecol.* **1996**, *33*, 1441–1450. [CrossRef]

28. Townsend, A.R.; Cleveland, C.C.; Asner, G.P.; Bustamamte, M.M.C. Controls over foliar N:P ratios in tropical forests. *Ecology* **2007**, *88*, 107–118. [CrossRef]

29. Fife, D.N.; Nambiar, E.K.S.; Saur, E. Retranslocation of foliar nutrients in evergreen tree species planted in a mediterranean environment. *Tree Physiol.* **2008**, *28*, 187–196. [CrossRef]

30. Wang, M.; Moore, T.R. Carbon, nitrogen, phosphorus, and potassium stoichiometry in an ombrotrophic peatland reflects plant functional type. *Ecosystems* **2014**, *17*, 673–684. [CrossRef]

31. Lambers, H.; Raven, J.A.; Shaver, G.R.; Smith, S.E. Plant nutrient-acquisition strategies change with soil age. *Trends Ecol. Evol.* **2008**, *23*, 95–103. [CrossRef]

32. Jeffries, P.; Gianinazzi, S.; Perotto, S.; Turnau, K.; Barea, J.M. The contribution of arbuscular mycorrhizal fungi in sustainable maintenance of plant health and soil fertility. *Biol. Fertil. Soils* **2003**, *37*, 1–16.

33. Zhang, H.Y.; Lu, X.T.; Hartmann, H.; Keller, A.; Han, X.G.; Trumbore, S.; Phillips, R.P. Foliar nutrient resorption differs between arbuscular mycorrhizal and ectomycorrhizal trees at local and global scales. *Glob. Ecol. Biogeogr.* **2018**, *27*, 875–885. [CrossRef]

34. Smith, S.E.; Smith, F.A. Roles of arbuscular mycorrhizas in plant nutrition and growth: New paradigms from cellular to ecosystem scales. *Annu. Rev. Plant Biol.* **2011**, *62*, 227–250. [CrossRef]

35. Van der Heijden, M.G.; Martin, F.M.; Selosse, M.A.; Sanders, I.R. Mycorrhizal ecology and evolution: The past, the present, and the future. *New Phytol.* **2015**, *205*, 1406–1423. [CrossRef]

36. Read, D.J.; Perez-Moreno, J. Mycorrhizas and nutrient cycling in ecosystems—A journey towards relevance? *New Phytol.* **2003**, *157*, 475–492. [CrossRef]

37. Shi, W.; Wang, Z.Q.; Liu, J.L.; Gu, J.C.; Guo, D.L. Fine root morphology of twenty hardwood species in maoershan natural secondary forest northeastern China. *J. Plant Ecol.* **2008**, *32*, 1217–1226.

38. Elser, J.J.; Sterner, R.W.; Gorokhova, E.; Fagan, W.F.; Markow, T.A.; Cotner, J.B.; Harrison, J.F.; Hobbie, S.E.; Odell, G.M.; Weider, L.J. Biological stoichiometry from genes to ecosystems. *Ecol. Lett.* **2000**, *3*, 540–550. [CrossRef]

39. Han, W.; Fang, J.; Guo, D.; Zhang, Y. Leaf nitrogen and phosphorus stoichiometry across 753 terrestrial plant species in China. *New Phytol.* **2005**, *168*, 377–385. [CrossRef]

40. Sterner, R.W.; Elser, J.J. *Ecological Stoichiometry: The Biology of Elements from Molecules to the Biosphere*; Princeton University Press: Princeton, NJ, USA, 2002.

41. Aronsson, A.; Elowson, S.; Persson, T. Effects of irrigation and fertilization on mineral nutrients in scots pine needles. *Ecol. Bull.* **1980**, *32*, 219–228.

42. Son, Y.; Lee, I.K.; Ryu, S.R. Nitrogen and phosphorus dynamics in foliage and twig of pitch pine and Japanese larch plantations in relation to fertilization. *J. Plant Nutr.* **2000**, *23*, 697–710. [CrossRef]

43. Güsewell, S. N:P ratios in terrestrial plants: Variation and functional significance. *New Phytol.* **2004**, *164*, 243–266. [CrossRef]

44. Reich, P.B.; Uhl, C.; Walters, M.B.; Ellsworth, D.S. Leaf lifespan as a determinant of leaf structure and function among 23 amazonian tree species. *Oecologia* **1991**, *86*, 16–24. [CrossRef]

45. Reich, P.B.; Walters, M.B.; Ellsworth, D.S. From tropics to tundra: Global convergence in plant functioning. *Proc. Natl. Acad. Sci. USA* **1997**, *94*, 13730–13734. [CrossRef]

46. Westoby, M.; Falster, D.S.; Moles, A.T.; Vesk, P.A.; Wright, I.J. Plant ecological strategies: Some leading dimensions of variation between species. *Annu. Rev. Ecol. Syst.* **2002**, *33*, 125–159. [CrossRef]

47. Li, Y.; Niu, S.; Yu, G. Aggravated phosphorus limitation on biomass production under increasing nitrogen loading: A meta-analysis. *Glob. Chang. Biol.* **2016**, *22*, 934–943. [CrossRef]

48. Perring, M.P.; Hedin, L.O.; Levin, S.A.; McGroddy, M.; de Mazancourt, C. Increased plant growth from nitrogen addition should conserve phosphorus in terrestrial ecosystems. *Proc. Natl. Acad. Sci. USA* **2008**, *105*, 1971–1976. [CrossRef]

49. Cleveland, C.C.; Houlton, B.Z.; Smith, W.K.; Marklein, A.R.; Reed, S.C.; Parton, W.; Del Grosso, S.J.; Running, S.W. Patterns of new versus recycled primary production in the terrestrial biosphere. *Proc. Natl. Acad. Sci. USA* **2013**, *110*, 12733–12737. [CrossRef]

50. Phoenix, G.K.; Booth, R.E.; Leake, J.R.; Read, D.J.; Grime, J.P.; Lee, J.A. Simulated pollutant nitrogen deposition increases P demand and enhances root-surface phosphatase activities of three plant functional types in a calcareous grassland. *New Phytol.* **2003**, *161*, 279–290. [CrossRef]

51. Treseder, K.K. A meta-analysis of mycorrhizal responses to nitrogen, phosphorus, and atmospheric CO_2 in field studies. *New Phytol.* **2004**, *164*, 347–355. [CrossRef]

52. Long, M.; Wu, H.H.; Smith, M.D.; La Pierre, K.J.; Lü, X.T.; Zhang, H.Y.; Han, X.G.; Yu, Q. Nitrogen deposition promotes phosphorus uptake of plants in a semi-arid temperate grassland. *Plant Soil* **2016**, *408*, 475–484. [CrossRef]

53. Yu, Q.; Chen, Q.; Elser, J.J.; He, N.; Wu, H.; Zhang, G.; Wu, J.; Bai, Y.; Han, X. Linking stoichiometric homoeostasis with ecosystem structure, functioning and stability. *Ecol. Lett.* **2010**, *13*, 1390–1399. [CrossRef] [PubMed]

54. Nambiar, E.K.S.; Fife, D.N. Nutrient retranslocation in temperate conifers. *Tree Physiol.* **1991**, *9*, 185–207. [CrossRef]

55. Knorr, M.; Frey, S.D.; Curtis, P.S. Nitrogen additions and litter decomposition: A meta-analysis. *Ecology* **2005**, *86*, 3252–3257. [CrossRef]

56. Brundrett, M.C. Mycorrhizal associations and other means of nutrition of vascular plants: Understanding the global diversity of host plants by resolving conflicting information and developing reliable means of diagnosis. *Plant Soil* **2009**, *320*, 37–77. [CrossRef]

57. Read, D.J.; Lewis, D.H.; Fitter, A.H.; Alexander, I.J.; Battley, E.H. Mycorrhizas in ecosystems. *Experientia* **1991**, *47*, 376–391. [CrossRef]

58. Averill, C.; Turner, B.L.; Finzi, A.C. Mycorrhiza-mediated competition between plants and decomposers drives soil carbon storage. *Nature* **2014**, *505*, 543–545. [CrossRef]

59. Smith, S.; Read, D. Mycorrhizal symbiosis. *Q. Rev. Biol.* **2008**, *3*, 273–281.

60. Pizzeghello, D.; Nicolini, G.; Nardi, S. Hormone-like activity of humic substances in *Fagus sylvaticae* forests. *New Phytol.* **2001**, *151*, 647–657. [CrossRef]

61. Pizzeghello, D.; Nicolini, G.; Nardi, S. Hormone-like activities of humic substances in different forest ecosystems. *New Phytol.* **2002**, *155*, 393–402. [CrossRef]

62. Gilbert, J.B.; Lawesj, H. Agricultural, botanical, and chemical results of experiments on the mixed herbage of permanent meadow, conducted for more than twenty years in succession on the same land. Part I. *Philos. Trans. R. Soc. Lond.* **1880**, *171*, 289–416.

63. Güsewell, S.; Koerselman, W.; Verhoeven, J.T.A. Biomass N:P ratios as indicators of nutrient limitation for plant populations in wetland. *Ecol. Appl.* **2003**, *13*, 372–384. [CrossRef]
64. Zhang, C.; Tian, H.Q.; Liu, J.Y.; Wang, S.Q.; Liu, M.L.; Pan, S.F.; Shi, X.Z. Pools and distributions of soil phosphorus in China. *Glob. Biogeochem. Cycles* **2005**, *19*. [CrossRef]
65. Walker, T.W.; Syers, J.K. The fate of phosphorus during pedogenesis. *Geoderma* **1976**, *15*, 1–19. [CrossRef]

Article

Leaf Nitrogen and Phosphorus Stoichiometry of *Cyclocarya paliurus* across China

Yang Liu [1,2], Qingliang Liu [1], Tongli Wang [2] and Shengzuo Fang [1,3,*]

[1] College of Forestry, Nanjing Forestry University, Nanjing 210037, China; lyang_188@sina.com (Y.L.); zhuanshag@163.com (Q.L.)
[2] Department of Forest and Conservation Sciences, University of British Columbia, 3041-2424 Main Mall, Vancouver, BC V6T 1Z4, Canada; tongli.wang@ubc.ca
[3] Co-Innovation Center for Sustainable Forestry in Southern China, Nanjing Forestry University, Nanjing 210037, China
* Correspondence: fangsz@njfu.edu.cn; Tel.: +86-25-8542-7797

Received: 16 November 2018; Accepted: 13 December 2018; Published: 13 December 2018

Abstract: Leaf stoichiometry (nitrogen (N), phosphorus (P) and N:P ratio) is not only important for studying nutrient composition in forests, but also reflects plant biochemical adaptation to geographic and climate conditions. However, patterns of leaf stoichiometry and controlling factors are still unclear for most species. In this study, we determined leaf N and P stoichiometry and their relationship with soil properties, geographic and climate variables for *Cyclocarya paliurus* based on a nation-wide dataset from 30 natural populations in China. The mean values of N and P concentrations and N:P ratios were 9.57 mg g^{-1}, 0.91 mg g^{-1} and 10.51, respectively, indicating that both leaf N and P concentrations in *C. paliurus* forests were lower than those of China and the global flora, and almost all populations were limited in N concentration. We found significant differences in leaf N and P concentrations and N:P ratios among the sampled *C. paliurus* populations. However, there were no significant correlations between soil properties (including organic C, total N and P concentrations) and leaf stoichiometry. The pattern of variation in leaf N concentration across the populations was positively correlated with latitude (24.46° N–32.42° N), but negatively correlated with mean annual temperature (MAT); meanwhile, leaf N concentration and N:P ratios were negatively correlated with mean temperature in January (MT$_{min}$) and mean annual frost-free period (MAF). Together, these results suggested that temperature-physiological stoichiometry with a latitudinal trend hold true at both global and regional levels. In addition, the relationships between leaf stoichiometry and climate variables provided information on how leaf stoichiometry of this species may respond to climate change.

Keywords: leaf stoichiometry; *Cyclocarya paliurus*; geographic variations; natural populations; climate variables; nitrogen; phosphorus; N:P ratio

---

1. Introduction

Patterns of leaf stoichiometry play a vital role in studying biological nutrient dynamics, biological symbiosis relationship, microbial nutrition, judgment of restrictive elements, consumer-driven nutrient cycle, and global C, N, P biogeochemical cycles [1–5]. The mechanisms of leaf stoichiometry in forests and their relationship to the environment conditions have attracted the attention of many scholars in recent years [6–10]. It is demonstrated that leaf stoichiometry is correlated with both geographic and climate variables such as latitude, temperature and precipitation, of which several hypotheses have been developed [11,12]. One famous hypothesis is the plant physiology hypothesis, which proposes that the developmental processes of plants are temperature sensitive, and plants will increase their nutrient concentrations (including leaf N and P) to compensate for the decreases in the growth rate that happen in lower-temperature or higher-latitude regions [6,11]. Another hypothesis is the biogeochemical hypothesis. This assumes that soil nutrient conditions, which are influenced by precipitation through leaching effects, drives the variation of plant nutrient (e.g., N, P concentrations) [13,14].

Previous studies of leaf stoichiometry at the global or regional levels have revealed a non-linear relation between leaf N concentration and climate factors [6,7,9,15,16]. Based on data across North America, Yin reported that leaf N in forests increased from boreal to temperate regions, and then decreased towards subtropical area [17]. Reich and Oleksyn's study at the global level also showed a similar pattern, where leaf N concentration increased from cold regions (mean annual temperature (MAT): -10 °C), peaked at temperate regions (MAT: 15 °C), and then tended to decrease in areas of high temperature (MAT: 30 °C) [6]. Recently, studies have been focused on the patterns of leaf stoichiometry in individual families [6], genus [9,18], and also species [14]. However, information about leaf stoichiometry at species level is still limited (e.g., among natural forests of a given species), and whether the mechanisms are consistent across different scales is unknown [19–21]. The relationships between leaf stoichiometry and climate factors have been found to differ among plants due to their dissimilarities caused by the ranges of different habitats. For example, Reich and Oleksyn reported that leaf N and P concentrations decreased with MAT in *Calamagrostis*, increased with MAT in birch (*Betula*), but showed a convex curve with MAT in maple (*Acer*) [6]. Wu et al. also demonstrated that leaf N and P concentrations of *Quercus* species across China decreased with mean temperature in January (MT_{min}) [9]. However, in Scots pine (*Pinus sylvestris* Linn.), leaf N concentration appears to decrease with latitude across Europe regions [22]. Thus, the response of leaf stoichiometry to climate change in plants may need to be differentiated by individual species.

Cyclocarya paliurus Batal. is a multiple function woody plant native to China, with a wide distribution from the warm temperate to the sub-tropical areas [23,24] and from the plain to highlands (e.g., up to about 2000 m in altitude in Guangxi province) [24]. Such a wide-range distribution (a broad gradient of both altitude and temperature) provides an opportunity to validate the hypothesis of temperature–plant physiological stoichiometry [25,26]. The nutritional status of *C. paliurus* in both plantations and natural forests has been investigated due to its importance in providing food and drug ingredient for the treatment of diabetes mellitus and hypertension [27–30]. In recent years, cultivation techniques including optimizing soil (NPK fertilizer used) and light environment have also been carried out to improve plant growth and yield of targeted health-promoting substances [31–34]. However, no information on the variation in leaf stoichiometry is available for *C. paliurus* forests or plantations. Therefore, our major objectives in this study were to determine the variation of leaf stoichiometry in *C. paliurus* sampled from different populations across China and the relationships between leaf stoichiometry of *C. paliurus* and geographic origin and climate factors. Soil organic C, total N and total P concentrations were analyzed at the same time to determine whether the differences of leaf stoichiometry were linked to the soil properties (concentrations). The findings from this study not only provide information for the characterization of the pattern of variation in leaf stoichiometry of this species, but also help to understand how leaf stoichiometry of *C. paliurus* populations may respond to climate change.

2. Materials and Methods

2.1. Study Areas and Materials

We investigated a total of 30 *C. paliurus* populations across the major distribution areas in 10 provinces of China (Figure 1). These populations range from 290 to 1798 m a.s.l. in altitude, 24.46° N–32.42° N in latitude and 103.78° E–121.79° E in longitude. Leaf sampling was carried out in September 2014, because leaf nutrients are relatively stable at this stage [9]. The longitude, latitude and altitude were measured by GPS on the spot. The climate data of the populations was obtained from ClimateAP for the historical period 1991–2014, including mean annual precipitation (MAP), mean annual temperature (MAT), mean temperature in July (MT_{max}), mean temperature in January (MT_{min}) and mean annual frost-free period (MAF) (Table 1, http://climateap.net/) [35,36]. The detailed method of sample collection and pre-treatment was described as Liu et al. (2018) [24]. Briefly, each sample consisted of about 400 g mature leaves collected from six to thirty average-sized trees (>20 years) in each population. All samples were dried to a constant weight at 70 °C and ground into fine powder in the lab. Soil samples (0–20 cm, n = 3) were collected from each location at the same time, sealed in polythene bags and brought back to laboratory.

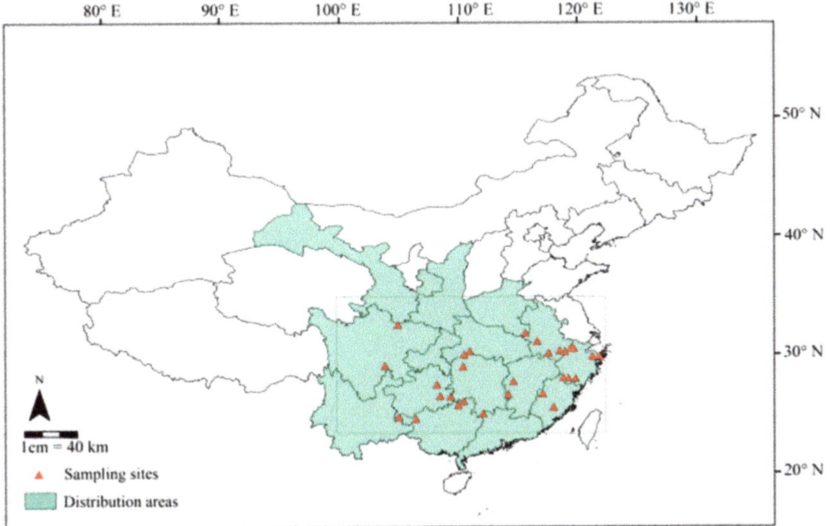

Figure 1. Natural distribution of *C. paliurus* (colored area in line box) and locations of 30 populations sampled (red triangle).

Table 1. List of the 30 sampled populations and their geographic and climate variables, as well as N and P concentrations and N:P ratios in the leaf of *C. paliurus*.

Sample ID	Locations	Longitude (°)	Latitude (°)	Altitude (m)	MAP (mm)	MAT (°C)	MT_{max} (°C)	MT_{min} (°C)	MAF (d)	Leaf N (mg g^{-1})	Leaf P (mg g^{-1})	N:P
HF	Hefeng, Hubei	110.42	29.88	1150	1669	12.7	23.3	1.3	292	8.43 ± 1.38 h~n	0.96 ± 0.03 fg	8.78 ± 1.21 j~o
WF	Wufeng, Hubei	110.90	30.19	959	1434	13.4	24.2	2	298	9.57 ± 0.46 e~i	0.93 ± 0.13 f~h	10.40 ± 1.40 f~k
LS	Longsheng, Guangxi	109.89	25.62	606	1704	14.7	25	3.6	320	8.58 ± 0.49 g~m	0.79 ± 0.01 ij	10.88 ± 0.65 e~h
ZY	Ziyuan, Guangxi	110.38	25.92	820	1915	13.4	23.8	2.3	300	10.47 ± 1.11 c~f	1.07 ± 0.03 cde	9.77 ± 1.04 g~n
BS	Baise, Guangxi	106.34	24.46	1798	1850	15.9	22.6	7.4	352	7.76 ± 0.22 i~o	0.86 ± 0.01 hi	9.04 ± 0.35 h~n
JZS	Jinzhongshan, Guangxi	104.95	24.61	1798	1568	16.2	22.2	7.7	351	9.41 ± 1.18 f~j	1.00 ± 0.01 ef	9.40 ± 1.08 g~n
JGS	Jinggangshan, Jiangxi	114.10	26.51	970	2001	13.8	23.1	3.3	312	7.75 ± 0.82 i~o	1.10 ± 0.04 a~d	7.04 ± 0.65 o
FY	Fenyi, Jiangxi	114.53	27.63	450	1834	17.4	27.6	6.4	343	8.97 ± 0.67 f~l	1.13 ± 0.01 abc	7.92 ± 0.56 no
GNJ	Guniujiang, Anhui	117.53	30.02	1100	2371	14.4	25.7	1.7	306	11.98 ± 1.04 c	0.76 ± 0.01 j	15.81 ± 1.44 b
JD	Jingde, Anhui	118.45	30.23	610	1777	15.2	27	2.4	312	14.48 ± 0.75 ab	1.18 ± 0.01 a	12.32 ± 0.56 cde
SC	Sucheng, Anhui	116.54	31.02	769	1679	13.6	25.4	1.1	275	13.58 ± 1.91 b	1.01 ± 0.06 ef	13.53 ± 1.16 c
QLF	Qingliangfeng, Anhui	118.89	30.15	680	1691	13.2	24.4	0.9	289	14.40 ± 1.68 ab	0.86 ± 0.03 hi	16.77 ± 1.54 b
QC	Qingchuan, Sichuan	104.86	32.42	1383	1093	11.7	21.6	0.7	270	15.18 ± 1.16 a	1.16 ± 0.03 ab	13.09 ± 0.95 cd
MC	Muchuan, Sichuan	103.78	28.97	1100	1448	15.7	24.5	5.7	348	11.15 ± 0.59 cde	0.80 ± 0.09 ij	13.86 ± 1.32 c
LP	Liping, Guizhou	109.24	26.34	727	1322	15.9	25.8	5.1	339	9.21 ± 0.75 f~k	1.10 ± 0.02 bcd	8.38 ± 0.58 l~o
LGS	Leigongshan, Guizhou	108.38	26.37	1178	1574	15.3	24.1	4.8	341	9.89 ± 0.17 e~h	1.16 ± 0.03 ab	8.53 ± 0.07 k~o
BWS	Bawanshan, Guizhou	108.38	26.37	1156	1564	15.3	24.1	4.8	341	10.22 ± 0.39 d~g	0.95 ± 0.02 fg	10.77 ± 0.32 e~i
SQ	Shiqian, Guizhou	108.11	27.35	1239	1268	14.6	23.9	4	332	7.64 ± 0.16 k~o	0.94 ± 0.04 fg	8.11 ± 0.28 mno
YS	Yongshun, Hunan	110.33	28.88	680	1556	15.3	25.9	4	328	9.60 ± 0.28 e~i	0.90 ± 0.02 gh	10.62 ± 0.44 e~j
JH	Jianhe, Hunan	112.03	24.92	845	1740	16.2	26.8	4.8	333	7.31 ± 0.88 l~p	0.89 ± 0.02 gh	8.18 ± 0.95 mno
PC	Pucheng, Fujian	118.76	27.93	650	1921	16.8	26.3	6.8	342	6.67 ± 0.37 op	0.76 ± 0.02 j	8.75 ± 0.48 j~o
NML	Niumulin, Fujian	117.93	25.43	500	1705	18.5	26.4	9.5	354	7.11 ± 1.04 m~p	0.66 ± 0.03 k	10.75 ± 1.19 e~i
MX	Mingxi, Fujian	117.01	26.59	574	1812	17.6	26.5	7.4	340	7.91 ± 1.00 i~o	0.92 ± 0.02 gh	8.63 ± 0.98 k~o
SCH	Shangcheng, Henan	115.55	31.72	910	1906	13.4	25.7	0.7	259	8.35 ± 0.86 h~o	0.76 ± 0.03 j	10.92 ± 0.77 e~h
WC	Wencheng, Zhejiang	119.79	27.88	959	1911	15.3	24.5	6	337	5.89 ± 0.17 p	0.50 ± 0.07 l	11.74 ± 0.83 def
FH	Fenghua, Zhejiang	121.22	29.76	513	1641	15.7	27.3	4.1	329	6.82 ± 0.88 n~p	0.67 ± 0.08 k	10.21 ± 0.71 f~l
LQ	Longquan, Zhejiang	119.19	27.91	1216	2224	14.1	23.3	4.3	323	8.79 ± 0.34 f~m	0.79 ± 0.03 ij	11.13 ± 0.25 efg
NB	Ningbo, Zhejiang	121.79	29.80	290	1452	16.4	27.2	6	344	10.33 ± 0.77 def	1.04 ± 0.01 de	9.92 ± 0.80 f~m
MW	Meiwu, Zhejiang	119.64	30.41	774	1643	13.8	25	2.3	289	8.02 ± 0.11 i~o	0.90 ± 0.08 gh	8.96 ± 0.81 i~n
LWS	Laowangshan, Zhejiang	119.41	30.50	540	1555	14.6	26.4	2.6	297	11.55 ± 0.26 b~f	0.63 ± 0.02 k	18.39 ± 0.39 a

MAP: mean annual precipitation (mm), MAT: mean annual temperature (°C), MT_{max}: mean temperature in July (°C), MT_{min}: mean temperature in January (°C), MAF: mean annual frost-free period (d). Sample ID was defined from the abbreviation of the location name. Different letters indicate significant differences (p <0.05 by Tukey's test) between populations.

2.2. Measurements

All samples were sieved through a mesh screen (1 mm) before measurements of leaf N and P concentrations. Leaf N concentration was detected following the method of Wu et al., using an auto analyzer (Kjeltec 2300 Analyzer Unit, Foss Tecator, Hoganas, Sweden) (n = 3) [9]. Leaf P concentration was measured according to the ammonium molybdate method described by the General Administration of Quality Supervisionin of China (reference code: GBW08513) (n = 3). Soil samples were air dried at 70 °C, grounded, and then sieved through a 2-mm mesh before analysis. Soil organic C, total N and total P concentrations in each sample were calculated following the method of Jiao et al. [37].

2.3. Data Analysis

One-way analysis of variance (ANOVA) was conducted to detect the quantitative differences in leaf stoichiometry and soil properties among different populations followed by Tukey's multiple range tests. All data were expressed as means ± standard deviation (SD). Scatter plots were used to show the relationships between leaf N and P concentrations and N:P ratios and factors studied (including longitude, latitude, altitude, MAT, MT_{max}, MT_{min}, MAP, and MAF), with appropriate regression equations developed. Relationships between leaf stoichiometry and soil properties were evaluated using the Pearson's correlation analysis. All statistical analyses were performed by using SPSS 19.0 software (SPSS, Chicago, IL, USA).

3. Results

3.1. Variation in Leaf Stoichiometry among C. paliurus Populations

Significant differences were found in leaf N and P concentrations and N:P ratios among different populations of *C. paliurus* (Table 1). For the 30 populations studied, the mean values of N and P concentrations and N:P ratios were 9.57 mg g^{-1}, 0.91 mg g^{-1} and 10.51, respectively, with a range of 5.89 (Wencheng, Zhejiang)–15.18 (Qingchuan, Sichuan) mg g^{-1}, 0.50 (Wencheng, Zhejiang)–1.18 (Jingde, Anhui) mg g^{-1}, and 7.04 (Jinggangshan, Jiangxi)–18.39 (Laowangshan, Zhejiang) (Table 1), respectively. Moreover, leaf N concentrations were significantly correlated with leaf P concentrations (R^2 = 0.1815, p = 0.0189) (Figure 2).

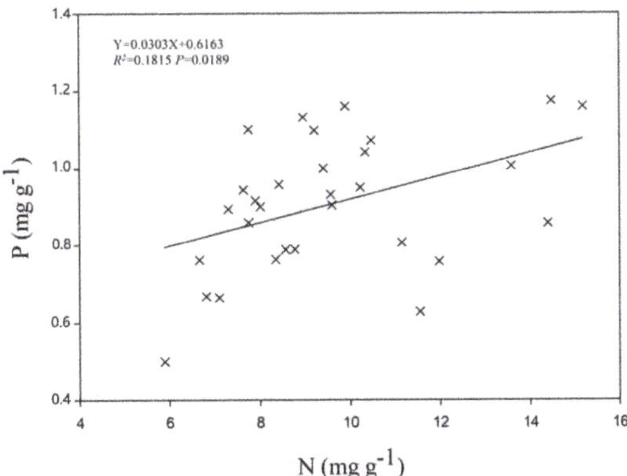

Figure 2. The relationship between leaf N and P concentrations for the 30 *C. paliurus* populations.

3.2. Soil Properties and Their Relationships with Leaf Stoichiometry

Soil organic C, total N and total P concentrations varied significantly among the natural populations, ranging from 24.04 (Jianhe, Hunan) to 75.82 (Leigongshan, Guizhou) mg g^{-1}, 1.94 (Yongshun, Hunan) to 5.84 (Leigongshan, Guizhou) mg g^{-1}, and 0.76 (Baise, Guangxi) to 3.52

(Shangcheng, Henan) mg g^{-1} (Table 2), respectively. Table 3 shows that there were no significant correlations between leaf stoichiometry and the soil properties studied (including soil C, N and P concentrations).

Table 2. Soil organic C, total N and total P concentrations for the 30 *C. paliurus* populations.

Sample ID	Locations	Organic C (mg g^{-1})	Total N (mg g^{-1})	Total P (mg g^{-1})
HF	Hefeng, Hubei	28.41 ± 1.01 ijk	2.07 ± 0.06 gh	1.45 ± 0.23 c~f
WF	Wufeng, Hubei	47.46 ± 6.79 d~h	2.72 ± 0.10 e~h	1.30 ± 0.24 d~g
LS	Longsheng, Guangxi	50.53 ± 1.74 c~f	4.21 ± 0.62 bcd	1.29 ± 0.13 d~g
ZY	Ziyuan, Guangxi	31.05 ± 4.62 ijk	3.80 ± 0.84 b~e	0.88 ± 0.08 efg
BS	Baise, Guangxi	51.33 ± 6.22 c~f	3.48 ± 0.21 c~g	0.76 ± 0.13 g
JZS	Jinzhongshan, Guangxi	52.54 ± 1.97 b~f	4.44 ± 0.82 abc	3.51 ± 0.97 a
JGS	Jinggangshan, Jiangxi	45.64 ± 3.57 e~h	4.94 ± 0.78 ab	1.67 ± 0.16 cde
FY	Fenyi, Jiangxi	45.57 ± 5.57 e~h	3.93 ± 0.75 b~e	1.13 ± 0.02 d~g
GNJ	Guniujiang, Anhui	36.03 ± 2.10 h~k	2.14 ± 0.16 gh	0.80 ± 0.16 fg
JD	Jingde, Anhui	37.49 ± 4.33 g~j	2.17 ± 0.26 fgh	1.83 ± 0.22 bcd
SC	Sucheng, Anhui	49.76 ± 1.43 c~g	4.13 ± 0.27 b~e	2.18 ± 0.16 bc
QLF	Qingliangfeng, Anhui	sm	sm	sm
QC	Qingchuan, Sichuan	31.55 ± 4.91 ijk	2.79 ± 0.24 d~h	0.95 ± 0.05 efg
MC	Muchuan, Sichuan	35.73 ± 2.11 h~k	2.67 ± 0.17 e~h	0.80 ± 0.19 fg
LP	Liping, Guizhou	58.22 ± 3.91 bcd	3.90 ± 0.58 b~e	1.57 ± 0.10 c~f
LGS	Leigongshan, Guizhou	75.82 ± 5.73 a	5.84 ± 0.66a	1.37 ± 0.24 d~g
BWS	Bawanshan, Guizhou	56.21 ± 2.39 b~e	3.98 ± 0.42 b~e	1.23 ± 0.06 d~g
SQ	Shiqian, Guizhou	54.58 ± 2.27 b~e	3.80 ± 0.38 b~e	1.86 ± 0.25 bcd
YS	Yongshun, Hunan	25.99 ± 2.52 jk	1.94 ± 0.19 h	0.96 ± 0.10 efg
JH	Jianhe, Hunan	24.04 ± 0.52 k	1.97 ± 0.23 h	1.49 ± 0.29 c~f
PC	Pucheng, Fujian	47.28 ± 1.40 d~h	3.92 ± 0.25 b~e	1.17 ± 0.15 d~g
NML	Niumulin, Fujian	40.55 ± 5.23 f~i	3.34 ± 0.13 c~h	0.92 ± 0.22 efg
MX	Mingxi, Fujian	55.25 ± 3.32 b~e	4.12 ± 0.24 b~e	1.12 ± 0.05 d~g
SCH	Shangcheng, Henan	61.30 ± 3.38 bc	4.64 ± 0.16 abc	3.52 ± 0.20 a
WC	Wencheng, Zhejiang	64.82 ± 4.20 ab	3.45 ± 0.23 c~g	0.85 ± 0.03 fg
FH	Fenghua, Zhejiang	47.97 ± 2.28 d~h	4.05 ± 0.75 b~e	2.57 ± 0.25 b
LQ	Longquan, Zhejiang	36.21 ± 1.20 h~k	2.71 ± 0.07 e~h	0.79 ± 0.06 fg
NB	Ningbo, Zhejiang	58.83 ± 5.47 bcd	4.23 ± 0.50 bcd	1.18 ± 0.19 d~g
MW	Meiwu, Zhejiang	64.10 ± 0.86 ab	4.79 ± 0.57 abc	2.17 ± 0.17 bc
LWS	Laowangshan, Zhejiang	61.36 ± 2.41 bc	3.60 ± 0.46 b~f	1.39 ± 0.02 def

Note: QLF sample missing (sm). Sample ID was defined from the abbreviation of the location name. Different letters indicate significant differences (p <0.05 by Tukey's test) between populations.

Table 3. Pearson correlation coefficients between leaf stoichiometry and soil properties (n = 90).

Leaf Stoichiometry	Soil Properties		
	Organic C (mg g^{-1})	Total N (mg g^{-1})	Total P (mg g^{-1})
Leaf N (mg g^{-1})	−0.222	−0.245	−0.063
Leaf P (mg g^{-1})	−0.102	0.152	0.065
Leaf N:P	−0.073	−0.337	−0.140

3.3. Leaf Stoichiometry in Relation to Geographic and Climate Variables

Leaf stoichiometry of *C. paliurus populations* was significantly related to both geographic and climate variables of their origins. Leaf N concentration was positively correlated with latitude (R^2 = 0.2747 and p = 0.0030 for the linear fit) (Figure 3A), but negatively correlated with mean annual temperature (MAT) (R^2 = 0.2130 and p = 0.0103 for the linear fit) (Figure 4A), mean temperature in January (MT$_{min}$) (R^2 = 0.3077 and p = 0.0015 for the linear fit) (Figure 4C), and mean annual frost-free period (MAF) (R^2 = 0.2319 and p = 0.0071 for the linear fit) (Figure 5B). Leaf P concentration was negatively related to longitude (R^2 = 0.1660 and p = 0.0255 for the linear fit) (Figure 3E). Leaf N:P ratios were positively correlated with latitude (R^2 = 0.2921 and p = 0.0020 for linear fit) (Figure 3G), mean temperature in January (MT$_{min}$) (R^2 = 0.1741 and p = 0.0218 for the linear fit) (Figure 4I), and mean annual frost-free period (MAF) (R^2 = 0.1655 and p = 0.0257 for the linear fit) (Figure 5F).

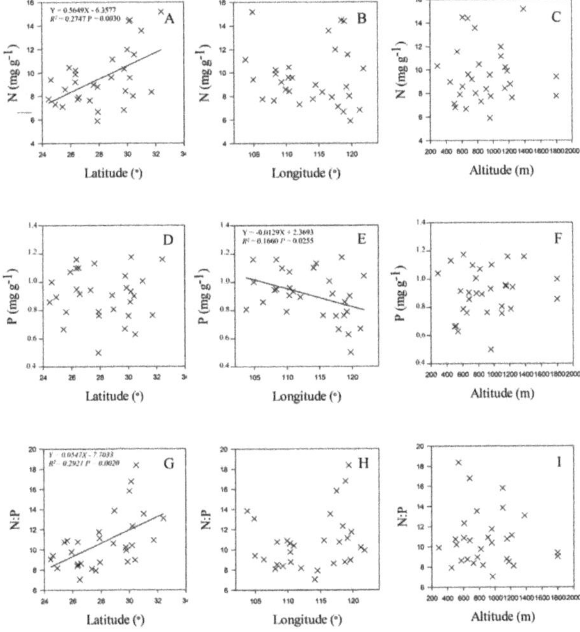

Figure 3. Relationships between leaf N (**A–C**), P (**D–F**), N:P ratio (**G–I**) and geographic origin (latitude, longitude and altitude) of *C. paliurus* populations.

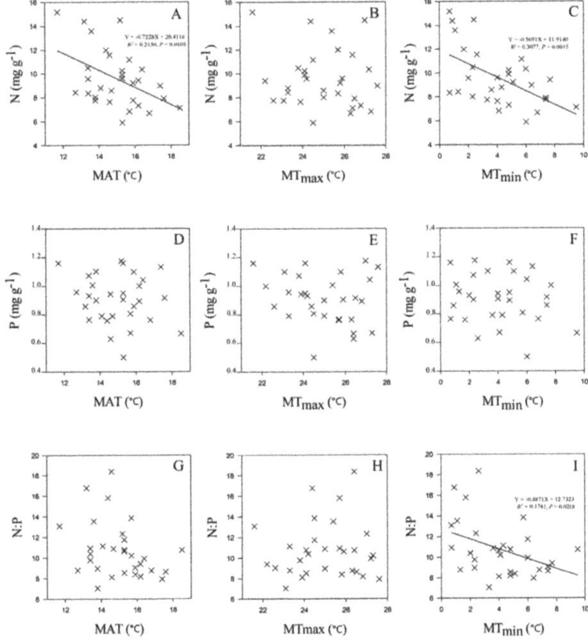

Figure 4. Relationships between leaf N (**A–C**), P (**D–F**), N:P ratio (**G–I**) and temperature (MAT: mean annual temperature, MT$_{max}$: mean temperature in July, and MT$_{min}$: mean temperature in January) for *C. paliurus* populations.

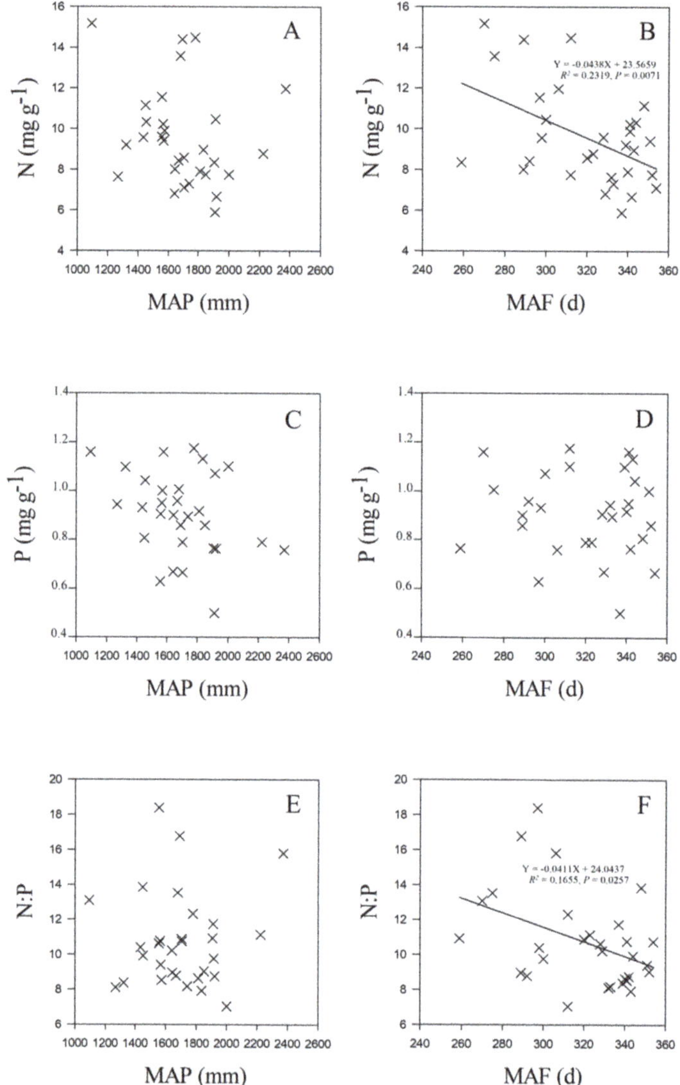

Figure 5. Relationships between leaf N (**A**,**B**), P (**C**,**D**), N:P ratio (**E**,**F**) and mean annual precipitation (MAP), and mean annual frost-free period (MAF) of *C. paliurus* populations.

4. Discussion

4.1. Patterns of Leaf Stoichiometry of C. paliurus across China

The ranges of both leaf N and P concentrations and N:P ratio of *C. paliurus* populations were smaller than those species of Juglandaceae family that have been studied previously (*Carya cathayensis* Sarg. in China and *Juglans nigra* Linn. in the USA) [38,39], and are also smaller than those of China and the global flora (Table 4) [6,7]. Leaf mean N concentration was also smaller than that of China and the global flora, which might be due to the relatively lower nitrogen resorption capacity in woody plants than that of herbs [7,40]. Previous studies have reported that leaf P concentration in China was much lower than that at the global level, which was confirmed in this study [6,7]. Lower leaf P concentrations in China have been demonstrated to be related to the lower soil available P concentration (3.83 mg kg^{-1} for China vs 7.65 mg kg^{-1} for the globe) [7,41,42]. Thus, soil conditions in *C. paliurus* populations, especially soil available P, needs to be further studied to confirm this internal relationship. Leaf N concentrations were significantly and positively correlated with leaf P concentrations in *C. paliurus* populations, which was inconsistent with previous studies [43]. This is mainly because N and P elements have similar biochemical pathways in plants, of which both are necessary in the basic biological processes such as protein synthesis, cell division and photosynthesis [44].

Table 4. Leaf N and P concentrations and N:P ratio for *C. paliurus* in the present study and other species of Juglandaceae family studied for China's and the global flora.

Data Source	Leaf N (mg g^{-1})		Leaf P (mg g^{-1})		N:P		References
	Range	Mean	Range	Mean	Range	Mean	
Present study	5.89–15.18	9.57	0.50–1.18	0.91	7.04–18.39	10.51	
Carya cathayensis Sarg. in China	13.10–24.65	19.92	1.24–2.70	1.75	8.83–15.31	11.92	ref. [38]
Juglans nigra Linn. in the USA	-	15.90	-	1.70	-	-	ref. [39]
China's flora	6.25–52.61	20.24	0.05–10.27	1.45	3.28–78.89	16.35	ref. [7]
The global flora	4.10–59.9	20.10	0.10–6.99	1.77	2.60–111.80	13.80	ref. [6]

Leaf N:P is an important index that reflects the limitation of nutrition in plants and ecosystem. For different plants and vegetation in different regions, the threshold of element restriction shown by N:P value is always distinct [9,45]. Based on the effects of N:P ratios on chlorophyll yield, Koerselman and Meuleman have proposed that a leaf N:P <14 indicates N limitation, and leaf N:P >16 indicates P limitation in plants [46]. The mean value of leaf N:P (10.51) over all *C. paliurus* populations suggested that in general *C. paliurus* forests were limited by N supply. However the LWS (Laowangshan) site had an leaf N:P ration of 18.39, which may indicate that it was limited by the supply of P although the concentration of total P in soil was not low compared to other sites (Table 2). The unusually high N:P ration at the LWS site might be due to the differences of plant growth rate caused by microclimate in *C. paliurus* forests.

4.2. Leaf N and P Stoichiometry in Relation to Soil Properties, Geographic Origins and Climate Factors

The results from our study showed that leaf stoichiometry of *C. paliurus* populations displayed significant relationships with geographic and climate factors. Meanwhile, results indicated that there were no significant correlations between leaf stoichiometry and the soil properties studied (including soil C, N and P concentrations). Our findings were quite different from those of Aerts and Chapin; they demonstrated that there was a significant correlation between leaf P and soil P concentrations [47]. This might be due to the differences in the species studied and their geographical distributions. Leaf N concentration increased with increasing latitude and decreasing MAT (Figures 3 and 4), which followed similar trends found in China's and the global flora [6,7,43]. Our study further demonstrated that leaf N concentration and N:P ratios increased with decreasing mean temperature in January (MT$_{min}$) (Figure 4C) and mean annual frost-free period (MAF) (Figure 5B). Together, these results confirmed that temperature-physiological stoichiometry with a latitudinal trend hold true at both global and regional levels [6]. Plants from colder habitats usually have higher values of leaf nutrient (N and P concentrations included) [11], which has been considered an adaptation mechanism to enhance their growth rates under lower temperatures [48,49].

A longitude-P relationship was also observed in *C. paliurus* populations across China (Figure 3E). Our study indicated that leaf P concentration increased with decreasing longitude. This result confirmed the biogeochemical hypothesis, which assumes that soil nutrient conditions, which are affected by leaching effects, drive the variation of plant nutrients [14]. Soil P concentration in China has been reported to decrease from northwest to southeast due to the variation of water distribution caused by precipitation [50,51]. Leaf N and P concentrations have been found strongly correlated with MAP at the global level [12]. However, a weak relationship between leaf P concentration and MAP was found in our study, which might be due to the small range of MAP (1093–2371 mm) in *C. paliurus* populations. Wu et al. also demonstrated that leaf N concentration and N:P ratios of *Quercus acutissima* Caruth. increased with increasing longitude in China [52]; however, this was not observed in our study. This difference further reveals that the responses of leaf stoichiometry to climate factors in plants are species specific. Overall, these results indicated that variation in leaf stoichiometry for a given species was more complicated, and these differences were mainly driven by mean annual temperature and water availability associated with geographic variation.

5. Conclusions

Our study demonstrated significant differences in leaf stoichiometry among *C. paliurus* populations, with the pattern of variation being related to geographic and climate variables. The leaf N concentration showed an increasing trend with increasing latitude, decreasing MAT, MT_{min} and MAF, which supported the hypothesis of temperature-plant physiological stoichiometry. These findings help to understand how the leaf stoichiometry of this species responds to climate variables, which might be useful for predicting the impact of climate change on the leaf stoichiometry of this species in the future. Meanwhile, we found that leaf P concentration increased with decreasing longitude, which might be due to the soil P concentration affected by leaching effects. The relatively low value of mean leaf N:P for *C. paliurus* populations, compared to other species, also suggested *C. paliurus* forests were generally limited by N supply. However, more tests about the relationship between leaf stoichiometry, soil conditions and leaf morphology need to be carried out in the future.

Author Contributions: S.F. conceived and designed the experiments; Q.L. collected the leaf samples, and performed the experiments; Y.L. analyzed the data, and wrote the manuscript; S.F. and T.W. participated in writing the manuscript.

Funding: We acknowledge the financial support of the Forestry Science and Technology Promotion Project from the State Forestry Administration of China (2017(08)), and the National Natural Science Foundation of China (No.31470637), which were funded by the Priority Academic Program Development of Jiangsu Higher Education Institutions (PAPD) and the Doctorate Fellowship Foundation of Nanjing Forestry University.

Conflicts of Interest: The authors declare that there are no conflicts of interest.

References

1. Allen, A.P.; Gillooly, J.F. Towards an integration of ecological stoichiometry and the metabolic theory of ecology to better understand nutrient cycling. *Ecol. Lett.* **2009**, *12*, 369–384. [CrossRef] [PubMed]
2. Elser, J.J.; Sterner, R.W.; Gorokhova, E.A.; Fagan, W.F.; Markow, T.A.; Cotner, J.B.; Weider, L.W. Biological stoichiometry from genes to ecosystems. *Ecol. Lett.* **2000**, *3*, 540–550. [CrossRef]
3. Venterink, H.O.; Güsewell, S. Competitive interactions between two meadow grasses under nitrogen and phosphorus limitation. *Funct. Ecol.* **2010**, *24*, 877–886. [CrossRef]
4. Pang, D.; Wang, G.; Li, G.; Sun, Y.; Liu, Y.; Zhou, J. Ecological stoichiometric characteristics of two typical plantations in the karst ecosystem of southwestern China. *Forests* **2018**, *9*, 56. [CrossRef]
5. Zhang, P.; Wang, H.; Wu, Q.; Yu, M.; Wu, T. Effect of wind on the relation of leaf N, P stoichiometry with leaf morphology in *Quercus* species. *Forests* **2018**, *9*, 110. [CrossRef]
6. Reich, P.B.; Oleksyn, J. Global patterns of plant leaf N and P in relation to temperature and latitude. *Proc. Natl. Acad. Sci.* **2004**, *101*, 11001–11006. [CrossRef] [PubMed]
7. Han, W.; Fang, J.; Guo, D.; Zhang, Y. Leaf nitrogen and phosphorus stoichiometry across 753 terrestrial plant species in China. *New Phytol.* **2005**, *168*, 377–385. [CrossRef] [PubMed]
8. Rentería, L.Y.; Jaramillo, V.J. Rainfall drives leaf traits and leaf nutrient resorption in a tropical dry forest in Mexico. *Oecologia* **2011**, *165*, 201–211. [CrossRef] [PubMed]
9. Wu, T.; Dong, Y.; Yu, M.; Wang, G.G.; Zeng, D.H. Leaf nitrogen and phosphorus stoichiometry of *Quercus* species across China. *For. Ecol. Manag.* **2012**, *284*, 116–123. [CrossRef]
10. Jiang, P.; Chen, Y.; Cao, Y. C: N: P stoichiometry and carbon storage in a naturally-regenerated secondary *Quercus variabilis* forest age sequence in the Qinling Mountains, China. *Forests* **2017**, *8*, 281. [CrossRef]

11. Chen, Y.; Han, W.; Tang, L.; Tang, Z.; Fang, J. Leaf nitrogen and phosphorus concentrations of woody plants differ in responses to climate, soil and plant growth form. *Ecography* **2013**, *36*, 178–184. [CrossRef]

12. Ordoñez, J.C.; Van Bodegom, P.M.; Witte, J.P.M.; Wright, I.J.; Reich, P.B.; Aerts, R. A global study of relationships between leaf traits, climate and soil measures of nutrient fertility. *Glob. Ecol. Biogeogr.* **2009**, *18*, 137–149. [CrossRef]

13. McGroddy, M.E.; Daufresne, T.; Hedin, L.O. Scaling of C:N:P stoichiometry in forests worldwide: Implications of terrestrial redfield-type ratios. *Ecology* **2004**, *85*, 2390–2401. [CrossRef]

14. Kang, H.; Zhuang, H.; Wu, L.; Liu, Q.; Shen, G.; Berg, B.; Liu, C. Variation in leaf nitrogen and phosphorus stoichiometry in *Picea abies* across Europe: An analysis based on local observations. *For. Ecol. Manag.* **2011**, *261*, 195–202. [CrossRef]

15. Sardans, J.; Rivas-Ubach, A.; Peñuelas, J. The C:N:P stoichiometry of organisms and ecosystems in a changing world: A review and perspectives. *Perspect. Plant Ecol.* **2012**, *14*, 33–47. [CrossRef]

16. Zhang, S.B.; Zhang, J.L.; Slik, J.F.; Cao, K.F. Leaf element concentrations of terrestrial plants across China are influenced by taxonomy and the environment. *Glob. Ecol. Biogeogr.* **2012**, *21*, 809–818. [CrossRef]

17. Yin, X. Variation in foliar nitrogen concentration by forest type and climatic gradients in North America. *Can. J. For. Res.* **1993**, *23*, 1587–1602. [CrossRef]

18. Gotelli, N.J.; Mouser, P.J.; Hudman, S.P.; Morales, S.E.; Ross, D.S.; Ellison, A.M. Geographic variation in nutrient availability, stoichiometry, and metal concentrations of plants and pore-water in ombrotrophic bogs in New England, USA. *Wetlands* **2008**, *28*, 827–840. [CrossRef]

19. Townsend, A.R.; Cleveland, C.C.; Asner, G.P.; Bustamante, M.M. Controls over foliar N:P ratios in tropical rain forests. *Ecology* **2007**, *88*, 107–118. [CrossRef]

20. Elser, J.J.; Fagan, W.F.; Kerkhoff, A.J.; Swenson, N.G.; Enquist, B.J. Biological stoichiometry of plant production: Metabolism, scaling and ecological response to global change. *New Phytol.* **2010**, *186*, 593–608. [CrossRef] [PubMed]

21. Hall, E.K.; Maixner, F.; Franklin, O.; Daims, H.; Richter, A.; Battin, T. Linking microbial and ecosystem ecology using ecological stoichiometry: a synthesis of conceptual and empirical approaches. *Ecosystems* **2011**, *14*, 261–273. [CrossRef]

22. Oleksyn, J.; Reich, P.B.; Zytkowiak, R.; Karolewski, P.; Tjoelker, M.G. Nutrient conservation increases with latitude of origin in European *Pinus sylvestris* populations. *Oecologia* **2003**, *136*, 220–235. [CrossRef] [PubMed]

23. Fang, S.; Wang, J.; Wei, Z.; Zhu, Z. Methods to break seed dormancy in *Cyclocarya paliurus* (Batal) Iljinskaja. *Sci. Hortic.* **2006**, *110*, 305–309. [CrossRef]

24. Liu, Y.; Fang, S.; Zhou, M.; Shang, X.; Yang, W.; Fu, X. Geographic variation in water-soluble polysaccharide content and antioxidant activities of *Cyclocarya paliurus* leaves. *Ind. Crop. Prod.* **2018**, *121*, 180–186. [CrossRef]

25. Ladanai, S.; Ågren, G.I. Temperature sensitivity of nitrogen productivity for Scots pine and Norway spruce. *Trees* **2004**, *18*, 312–319. [CrossRef]

26. Briceno-Elizondo, E.; Garcia-Gonzalo, J.; Peltola, H.; Matala, J.; Kellomäki, S. Sensitivity of growth of Scots pine, Norway spruce and silver birch to climate change and forest management in boreal conditions. *For. Ecol. Manag.* **2006**, *232*, 152–167. [CrossRef]

27. Liu, Y.; Cao, Y.; Fang, S.; Wang, T.; Yin, Z.; Shang, X.; Fu, X. Antidiabetic effect of *Cyclocarya paliurus* leaves depends on the contents of antihyperglycemic flavonoids and antihyperlipidemic triterpenoids. *Molecules* **2018**, *23*, 1042. [CrossRef]

28. Liu, Y.; Chen, P.; Zhou, M.; Wang, T.; Fang, S.; Shang, X.; Fu, X. Geographic variation in the chemical composition and antioxidant properties of phenolic compounds from *Cyclocarya paliurus* (Batal) Iljinskaja leaves. *Molecules* **2018**, *23*, 2440. [CrossRef]

29. Xie, J.H.; Dong, C.J.; Nie, S.P.; Li, F.; Wang, Z.J.; Shen, M.Y.; Xie, M.Y. Extraction, chemical composition and antioxidant activity of flavonoids from *Cyclocarya paliurus* (Batal.) Iljinskaja leaves. *Food Chem.* **2015**, *186*, 97–105. [CrossRef]

30. Fang, S.; Yang, W.; Chu, X.; Shang, X.; She, C.; Fu, X. Provenance and temporal variations in selected flavonoids in leaves of *Cyclocarya paliurus*. *Food Chem.* **2011**, *124*, 1382–1386. [CrossRef]

31. Deng, B.O.; Shang, X.; Fang, S.; Li, Q.; Fu, X.; Su, J. Integrated effects of light intensity and fertilization on growth and flavonoid accumulation in *Cyclocarya paliurus*. *J. Agric. Food Chem.* **2012**, *60*, 6286–6292. [CrossRef] [PubMed]

32. Liu, Y.; Fang, S.; Yang, W.; Shang, X.; Fu, X. Light quality affects flavonoid production and related gene expression in *Cyclocarya paliurus*. *J. Photochem. Photobiol. B Biol.* **2018**, *179*, 66–73. [CrossRef] [PubMed]

33. Liu, Y.; Qian, C.; Ding, S.; Shang, X.; Yang, W.; Fang, S. Effect of light regime and provenance on leaf characteristics, growth and flavonoid accumulation in *Cyclocarya paliurus* (Batal) Iljinskaja coppices. *Bot. Stud.* **2016**, *57*, 28. [CrossRef] [PubMed]

34. Liu, Y.; Wang, T.; Fang, S.; Zhou, M.; Qin, J. Responses of morphology, gas exchange, photochemical activity of photosystem II, and antioxidant balance in *Cyclocarya paliurus* to light spectra. *Front. Plant Sci.* **2018**, *9*, 1704. [CrossRef] [PubMed]

35. Wang, T.; Wang, G.; Innes, J.L.; Seely, B.; Chen, B. ClimateAP: An application for dynamic local downscaling of historical and future climate data in Asia Pacific. *Front. Agric. Sci. Eng.* **2017**, *4*, 448–458. [CrossRef]

36. Wang, T.; Campbell, E.M.; O'Neill, G.A.; Aitken, S.N. Projecting future distributions of ecosystem climate niches: Uncertainties and management applications. *For. Ecol. Manag.* **2012**, *279*, 128–140. [CrossRef]

37. Jiao, F.; Shi, X.R.; Han, F.P.; Yuan, Z.Y. Increasing aridity, temperature and soil pH induce soil CNP imbalance in grasslands. *Sci. Rep.* **2016**, *6*, 19601. [CrossRef] [PubMed]

38. Yan, D.L.; Huang, Y.J.; Jin, S.H.; Huang, J.Q. Temporal variation of C, N, P stoichiometric in functional organs rootlets, leaves of *Carya cathayensis* and forest soil (in Chinese). *J. Soil Water Conserv.* **2013**, *5*, 256–259.

39. Niklas, K.J.; Cobb, E.D. Biomass partitioning and leaf N, P-stoichiometry: comparisons between tree and herbaceous current-year shoots. *Plant Cell Environ.* **2006**, *29*, 2030–2042. [CrossRef] [PubMed]

40. Yuan, Z.Y.; Li, L.H.; Han, X.G.; Huang, J.H.; Jiang, G.M.; Wan, S.Q.; Chen, Q.S. Nitrogen resorption from senescing leaves in 28 plant species in a semi-arid region of northern China. *J. Arid Environ.* **2005**, *63*, 191–202. [CrossRef]

41. Batjes, N.H. *A Homogenized Soil Data File for Global Environmental Research: A Subset of FAO, ISRIC and NRCS Profiles (Version 1.0) (No. 95/10b)*; ISRIC: Wageningen, The Netherlands, 1995.

42. Han, W.X.; Fang, J.Y.; Reich, P.B.; Ian Woodward, F.; Wang, Z.H. Biogeography and variability of eleven mineral elements in plant leaves across gradients of climate, soil and plant functional type in China. *Ecol. Lett.* **2011**, *14*, 788–796. [CrossRef] [PubMed]

43. He, J.S.; Wang, L.; Flynn, D.F.; Wang, X.; Ma, W.; Fang, J. Leaf nitrogen: Phosphorus stoichiometry across Chinese grassland biomes. *Oecologia* **2008**, *155*, 301–310. [CrossRef] [PubMed]

44. Zhang, L.X.; Bai, Y.F.; Han, X.G. Application of N:P stoichiometry to ecology studies. *Acta Bot. Sin.* **2003**, *45*, 1009–1018.

45. Li, H.; Crabbe, M.J.C.; Xu, F.; Wang, W.; Niu, R.; Gao, X.; Chen, H. Seasonal variations in carbon, nitrogen and phosphorus concentrations and C:N:P stoichiometry in the leaves of differently aged *Larix principis-rupprechtii* Mayr. plantations. *Forests* **2017**, *8*, 373. [CrossRef]

46. Koerselman, W.; Meuleman, A.F. The vegetation N:P ratio: A new tool to detect the nature of nutrient limitation. *J. Appl. Ecol.* **1996**, *33*, 1441–1450. [CrossRef]

47. Aerts, R.; Chapin, F.S. The mineral nutrition of wild plants revisited: A reevaluation of processes and patterns. *Adv. Ecol. Res.* **2000**, *30*, 1–67.

48. Woods, H.A.; Makino, W.; Cotner, J.B.; Hobbie, S.E.; Harrison, J.F.; Acharya, K.; Elser, J.J. Temperature and the chemical composition of poikilothermic organisms. *Funct. Ecol.* **2003**, *17*, 237–245. [CrossRef]

49. Van de Waal, D.B.; Verschoor, A.M.; Verspagen, J.M.; van Donk, E.; Huisman, J. Climate-driven changes in the ecological stoichiometry of aquatic ecosystems. *Front. Ecol. Environ.* **2010**, *8*, 145–152. [CrossRef]

50. Zhang, C.; Tian, H.; Liu, J.; Wang, S.; Liu, M.; Pan, S.; Shi, X. Pools and distributions of soil phosphorus in China. *Glob. Biogeochem. Cycles* **2005**. [CrossRef]

51. Wendroth, O.; Vasquez, V.; Matocha, C.J. Field experimental approach to bromide leaching as affected by scale-specific rainfall characteristics. *Water Resour. Res.* **2011**. [CrossRef]

52. Wu, T.; Wang, G.G.; Wu, Q.; Cheng, X.; Yu, M.; Wang, W.; Yu, X. Patterns of leaf nitrogen and phosphorus stoichiometry among *Quercus acutissima* provenances across China. *Ecol. Complex.* **2014**, *17*, 32–39. [CrossRef]

Article

Stocks and Stoichiometry of Soil Organic Carbon, Total Nitrogen, and Total Phosphorus after Vegetation Restoration in the Loess Hilly Region, China

Hongwei Xu [1,3], Qing Qu [2], Peng Li [2], Ziqi Guo [2], Entemake Wulan [2] and Sha Xue [1,2,3],*

[1] State Key Laboratory of Soil Erosion and Dryland Farming on Loess Plateau, Institute of Soil and Water Conservation, Chinese Academy of Sciences and Ministry of Water Resources, Yangling 712100, China; xuhongwei16@mails.ucas.ac.cn
[2] Institute of Soil and Water Conservation, Northwest A&F University, Yangling 712100, China; ylxnqq@nwafu.edu.cn (Q.Q.); lipenglipeng1900@gmail.com (P.L.); Guoziqi100@outlook.com (Z.G.); 5141509166@mail.sjtu.edu.cn (E.W.)
[3] Institute of College of Natural Resources and Environment, University of Chinese Academy of Sciences, Beijing 100049, China
*   Correspondence: xuesha100@163.com; Tel.: +1-367-921-5717

Received: 17 November 2018; Accepted: 27 December 2018; Published: 3 January 2019

Abstract: The Loess Plateau is an important region for vegetation restoration in China; however, changes in soil organic carbon (SOC), soil nutrients, and stoichiometry after restoration in this vulnerable ecoregion are not well understood. Typical restoration types, including orchardland, grassland, shrubland, and forestland, were chosen to examine changes in the stocks and stoichiometry of SOC, soil total nitrogen (TN), and soil total phosphorus (TP) at different soil depths and recovery times. Results showed that SOC stocks first increased and then stabilized in orchardland, grassland, and shrubland at 0–30 cm depths, while in forestland, SOC stocks gradually increased. Soil TN stocks first increased and then decreased in orchardland, shrubland, and forestland with restoration age at 0–30 cm depths, while soil TP stocks showed little variation between restoration types; at the same time, the overall C:N, C:P, and N:P ratios increased with restoration age. In the later stages of restoration, the stocks of SOC and soil TN at 0–30 cm soil depths were still lower than those in natural grassland and natural forest. Additionally, the SOC, soil TN, and soil TP stocks and the C:N, C:P, and N:P ratios decreased with soil depth. The forestland had the highest rate of change in SOC and soil TN stocks, at 0–10 cm soil depth. These results indicate a complex response of SOC, soil TN, and soil TP stocks and stoichiometry to vegetation restoration, which could have important implications for understanding C, N, and P changes and nutrient limitations after vegetation restoration.

Keywords: soil stoichiometry; soil nutrient; nutrient limitations; natural grassland; natural forest

---

1. Introduction

Soil is an important component of terrestrial ecosystems and the main source of the nutrients required for plant growth and development. Soil organic carbon (SOC), soil total nitrogen (TN), and soil total phosphorus (TP) are the main structural and nutritional components of soil and are also the main limiting factors in terrestrial ecosystems [1]. Soil organic C, soil TN, and soil TP stocks reflect the potential of the soil to provide nutrients to vegetation. These elements continuously circulate between the layers of the soil (the biogeochemical cycles of C, N, and P), which ensures a smooth flow of energy and maintains the stability of ecosystems [2]. The availabilities of soil TN and soil TP are major factors regulating the carbon balance of the ecosystem. Elemental stoichiometry is an important indicator reflecting the C, N, and P cycles in soil and the accumulation and balance of nutrients in ecosystems, and it can help to determine the responses of ecological processes to global changes [3].

Vegetation restoration has received intensive interest because of its potential influence on global C and N cycling, soil quality improvement, land management, and regional economic development [4]. Land use change in the form of vegetation restoration plays an important role in improving the ecological environment and function of ecosystems and can also improve soil quality and soil nutrient cycling. Improved soil quality will, in turn, affect plant production and ecosystem function.

A large number of related studies have shown that stocks and stoichiometry of SOC, soil TN, and soil TP are closely related to land use type [5,6], and nutrient inputs and outputs are considered to be the main factors affecting soil nutrient content [7–9]. Some studies have found that vegetation restoration can promote photosynthesis, soil nutrient accumulation, and microbial activity [10–12], and increase the stoichiometry of SOC, soil TN, and soil TP [9]. However, other studies have indicated that land use change can lead to decreases in soil nutrient contents [13]. Studies estimating the impact of land use on the stocks and stoichiometry of SOC, TN, and TP have mainly focused on the topsoil (0–20 cm), as this is considered to be the most active soil layer in terms of natural and anthropogenic disturbances [14]. Recent studies have shown that the nutrient content of deep soils may also vary greatly with land use [14,15]. Therefore, understanding how C, N, and P stocks and stoichiometry change in soil with land use changes can clarify soil nutrient availability and nutrient cycling and balance mechanisms, and is of great significance for regional ecosystem health assessments.

The Loess Plateau, China, is located in a semi-arid/semi-humid climate zone which has undergone serious soil erosion. It is an ecologically fragile area and a key area for soil and water conservation efforts in China. Before the 1950s, extreme weather such as droughts, heavy rain, hailstorms, strong winds, and dust storms occurred frequently in this area, resulting in serious soil erosion. In addition, as a result of long-term and unsustainable land use, vegetation has been destroyed over large areas due to grazing and farming. The large-scale cultivation of sloping cropland further aggravated the soil erosion. The amount of nitrogen, phosphorus, and potassium lost from slope farmland has been estimated at 12.7 million tons per year [16]. Vegetation restoration was implemented in the 1970s in this region. In order to control soil erosion and improve ecosystem function, ecological restoration and environmental reconstruction work has been carried out in which slope cropland (slope > 25°) has been converted into orchardland, grassland, shrubland, and forestland. After decades of continuous efforts, vegetation coverage has increased, and the ecological environment has been greatly improved [17]. The sequestration and stoichiometry of SOC, soil TN, and soil TP varies among different vegetation types and restoration ages; therefore, its effect on soil physicochemical properties varies as well. It is important, therefore, to clarify annual and vertical variations in the SOC, soil TN, and soil TP stocks and stoichiometry in soils with different vegetation types in the Loess Hilly Region, China.

In order to better understand the SOC processes, carbon budget of the soil, and soil fertility issues after afforestation, we addressed the following questions: (1) How have stocks of SOC, soil TN, and soil TP, and their ratios, changed across the Loess Hilly Region, China, after decades of vegetation restoration? (2) Are these changes associated with soil depth? (3) How do these changes vary with restoration type? We further hypothesized that (1) as litter inputs to the soil increase with restoration age, stocks of SOC and soil TN increase, whereas soil TP stocks do not significantly change since P stocks are primarily affected by parent minerals. The change rate will be greatest for SOC, followed by soil TN, and then soil TP, causing an increase in the C:N, C:P, and N:P ratios with restoration age; (2) vegetation restoration affects stocks of SOC, soil TN, and soil TP at the soil surface more than at greater soil depths; and (3) due to differences in the litter produced by different vegetation, root secretions and soil microorganisms will also vary between restoration types, resulting in differences in stocks and ratios of SOC, soil TN, and soil TP.

2. Materials and Methods

2.1. Study Area

The study area was located in Ansai County, Shanxi Province, China in the center of the Loess Plateau (Figure 1). This region has a warm temperate and semi-arid climate with an annual average temperature of 8.8 °C. Annual precipitation is approximately 500 mm, 60% of which falls between July and September, and the frost-free period is 157 days. The soil is mainly composed of Huangmian soil developed on wind-deposited loessial parent material. This type of soil is characterized by weak cohesion, which has made it prone to severe soil erosion. The sand (2.00–0.02 mm grain size), silt (0.02–0.002 mm), and clay (<0.002 mm) contents are 65%, 24%, and 11%, respectively. The soil bulk density (BD) and soil pH of the tillage layer range from 1.15 to 1.35 g cm^{-3} and 8.4 to 8.6, respectively.

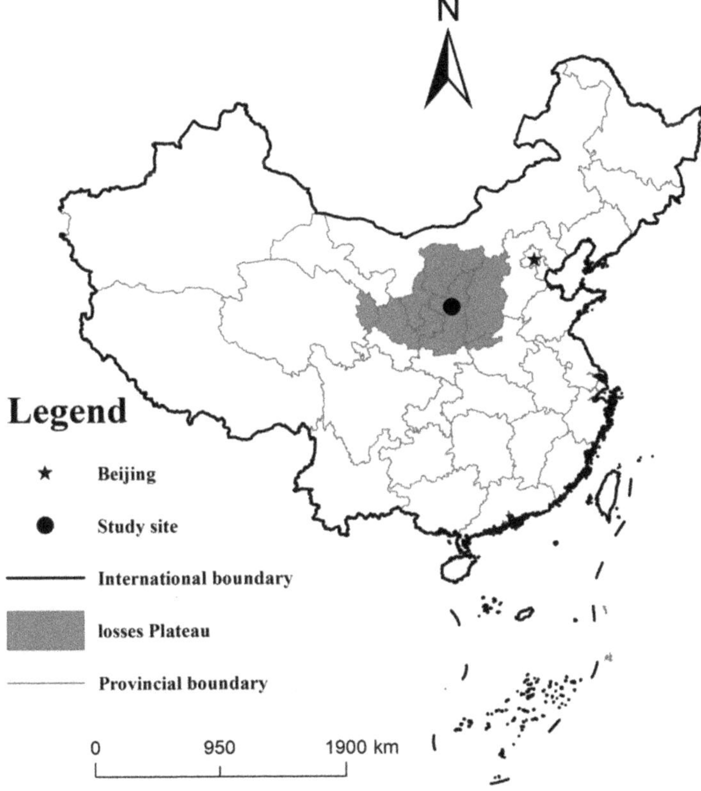

Figure 1. The locations of study sites in Ansai County, Shanxi Province, China.

2.2. Soil Sampling and Laboratory Analyses

With the aim to examine changes in the stocks and stoichiometry of soil organic carbon (SOC), soil total nitrogen (TN), and soil total phosphorus (TP) after vegetation restoration in different restoration types, this study adopted a "space for time" approach. A total of 82 sites representing four restoration types were selected based on vegetation type, topographic features, and restoration age, comprising 9 sites of orchardland (5, 10, and 20 years), 34 sites of grassland (2, 5, 8, 11, 15, 18, 26, and 30 years), 24 shrubland (5, 10, 20, 30, 36, and 47 years), and 15 forestland (5, 10, 20, 37, and 56 years). In addition, we selected three slope cropland sites, which were studied at 0 years. Because the four restoration areas were transformed from croplands, four natural grassland sites (age > 50 years) and nine natural forest sites (age > 100 years) were selected as controls (Table S1). Three 10 × 10 m plots were chosen in each orchardland, three 2 × 2 m plots in each grassland (grassland and natural grassland), three 10 × 10 m plots in each shrubland, and three 20 × 20 m plots in each forestland type (forestland and natural forest). Each plot was at least 50 m from the other plots. A total of 15 soil samples were collected from five soil depths (0–10, 10–20, 20–30, 30–50, and 50–100 cm) in a random sampling design using a soil drilling sampler (4 cm inner diameter). Soil samples from each plot from the same soil depth were mixed to form one sample. These soil samples were brought back to the laboratory and then divided into two parts. One part of the sample was naturally air-dried, plant roots and other impurities were removed, and then the SOC, soil TN, and soil TP were measured. The other part was stored in a refrigerator at 4 °C until further analysis of other indicators, which are not presented in this paper.

The soil bulk density (BD) of each depth was measured using the cutting ring method. The SOC was determined using the H_2SO_4–$K_2Cr_2O_7$ method [18]. The soil TN was measured using the Kjeldahl method [19], and the soil TP was determined colorimetrically using the ammonium molybdate method [20].

2.3. Calculation of SOC, Soil TN, and Soil TP Stocks

The stocks of SOC, soil TN, and soil TP from five soil depths of 0–10, 10–20, 20–30, 30–50, and 50–100 cm in different restoration types were selected in our study to research the carbon budget of the soil and soil fertility issues after afforestation. The SOC, soil TN, and soil TP stocks (Mg ha^{-1}) were calculated as follows:

$$SOC_i \text{ stock} = SOC_i \times BD_i \times D_i/10, \tag{1}$$

$$\text{Soil } TN_i \text{ stock} = \text{soil } TN_i \times BD_i \times D_i/10, \tag{2}$$

$$\text{Soil } TP_i \text{ stock} = \text{soil } TP_i \times BD_i \times D_i/10, \tag{3}$$

where SOC_i is the soil organic carbon content of the *i*th layer of soil (g kg^{-1}), soil TN_i is the soil total N content of the *i*th layer of soil (g kg^{-1}), soil TP_i is the soil total P content of the *i*th layer of soil (g kg^{-1}), BD_i is the soil bulk density of the *i*th layer of soil (g cm^{-3}), and D_i is the soil depth of the *i*th layer of soil (cm).

2.4. Statistical Analyses

Two-way ANOVAs were used to determine the effects of restoration age, soil depth, and their interaction on SOC, soil TN, and soil TP stocks and C:N, C:P, and N:P ratios. An independent samples *t*-test was used to compare the SOC, soil TN, and soil TP stocks and the C:N, C:P, and N:P ratios from sites grassland30, natural grassland, shrubland47, forestland56, and natural forest ($p < 0.05$). Before ANOVA analyses, we performed tests for normality and homogeneity of variance. In order to compare the effects of vegetation type, we selected SOC, soil TN, and soil TP contents and stoichiometry for the same or similar years from the four restoration types: orchardland (5 years, 10 years, 20 years), grassland (5 years, 11 years as 10 years, 18 as 20 years), shrubland (5 years, 10 years), and forestland (5 years, 10 years). In addition, we selected the SOC, soil TN, and soil TP contents from different soil depths as the rate of change with recovery years and compared the SOC, soil TN, and soil TP sequestration rates between different restoration types. All statistical analyses were conducted using SPSS 21.0 (SPSS Inc., Chicago, IL, USA). Figures were drawn using Origin 9.0.

3. Results

3.1. SOC, Soil TN, and Soil TP Stocks in Different Restoration Types

Restoration age had a significant effect on SOC and soil TN stocks ($p < 0.01$) (Table S2). In orchardland, the SOC stocks of the soil at 0–30 cm depth first increased and then stabilized after

5 years; the soil TN stock at the same depth first increased and then decreased with restoration age. The soil TP stocks at 0–100 cm depth showed no significant change with restoration age (Figure 2a–c). In grassland, the SOC and soil TN stocks at 0–30 cm depth first increased and then stabilized after 18 years, and the soil TP stocks at 0–100 cm depth showed no significant change with different restoration ages (Figure 2d–f). In shrubland, the SOC stocks at 0–50 cm depth and soil TN at 0–10 cm increased at first, peaked after 5 years, and then stabilized after 30 years; the soil TP stocks at 0–100 cm depth showed no significant change (Figure 2h–j). In forestland, the SOC at 0–50 cm gradually increased, and the soil TN stocks increased before peaking after 37 years, then decreased with restoration age; the soil TP stock at 0–100 cm depth showed no significant change (Figure 2k–m). In addition, SOC and soil TN stocks from the 0–30 cm soil layer of grassland30 were significantly lower than those in natural grassland (Figure 3a,b). SOC and soil TN stocks at 0–30 cm soil depth in shrubland47, as well as SOC stocks at 0–100 cm depth and soil TN stocks at 0–20 cm depth in forestland56, were significantly lower than those in natural forest (Figure 3d,e). The soil TP stocks at all soil depths showed no significant differences between grassland30 and natural grassland, shrubland47 and natural forest, or forestland56 and natural forest (Figure 3c,f).

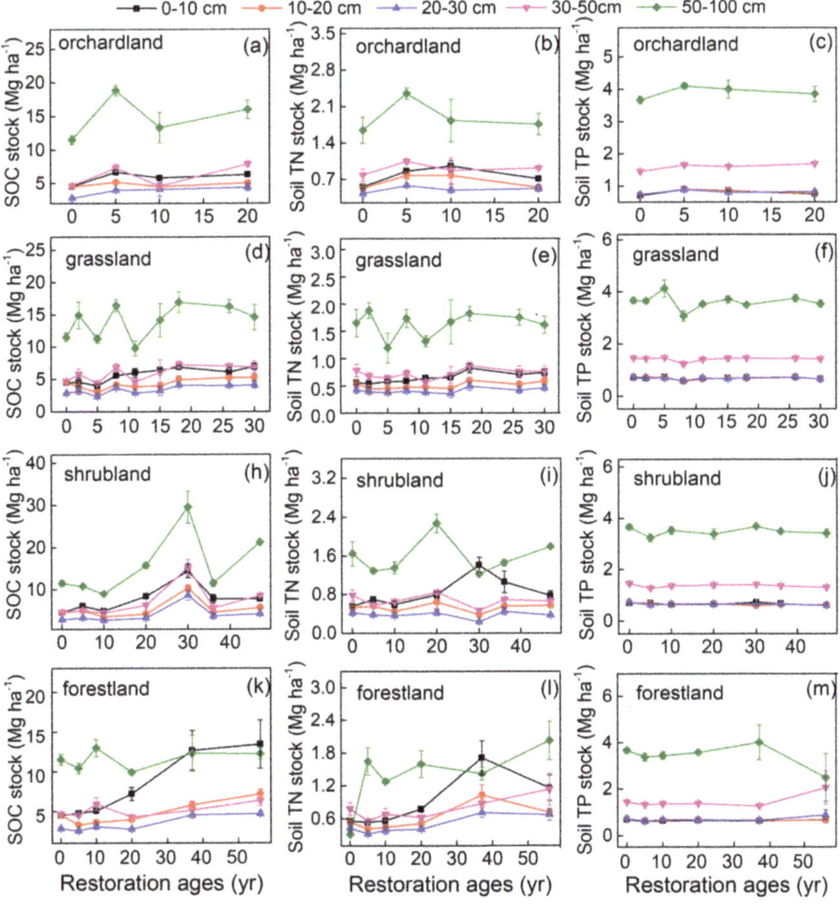

Figure 2. Changes in stocks of soil organic carbon (SOC), soil total nitrogen (TN), and soil total phosphorus (TP) with restoration age. Note: values are mean ± standard error.

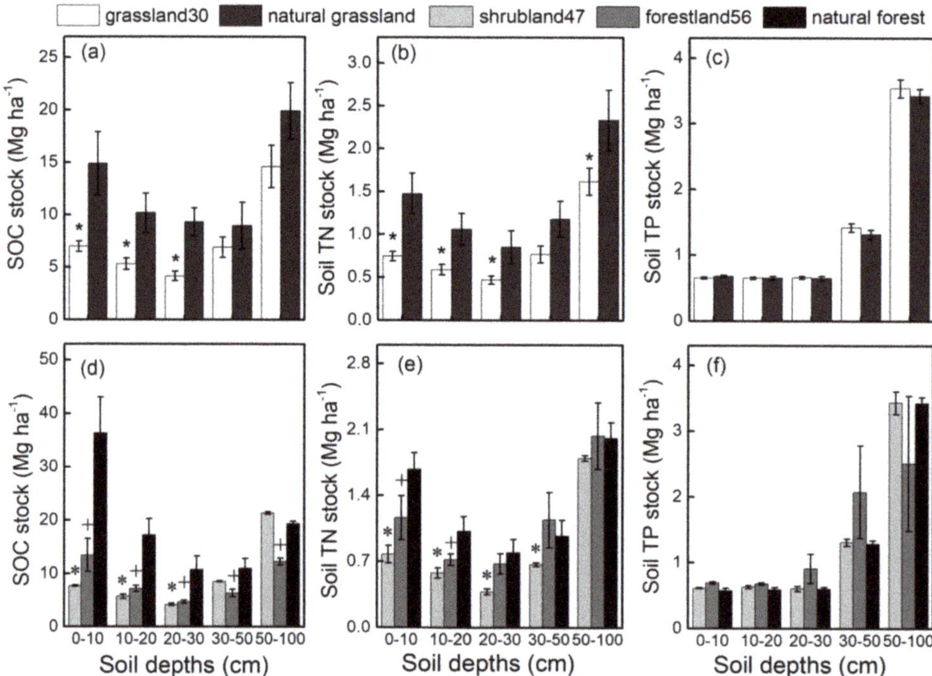

Figure 3. Stocks of soil organic carbon (SOC), soil total nitrogen (TN), and soil total phosphorus (TP) at different soil depths in grassland at 30 years compared to those of natural grassland, and those of shrubland at 47 years and forestland at 56 years compared to those of natural forest. Note: values are mean ± standard error. (**a–c**) means the SOC stock, Soil TN stock and Soil TP stock at different soil depths in grassland30 compared to those of natural grassland. (**d–f**) means the SOC stock, Soil TN stock and Soil TP stock at different soil depths in shrubland47 and forestland56 compared to those of natural forest * denotes significant differences between the natural grassland and grassland at 30 years after vegetation restoration ($p < 0.05$). * denotes significant differences between the natural forest and shrubland at 47 years after vegetation restoration ($p < 0.05$). + denotes significant differences between the natural forest and forestland at 56 years after vegetation restoration ($p < 0.05$).

3.2. Changes in SOC, Soil TN, and Soil TP Stoichiometry

In orchardland, the C:N ratio at 0–100 cm depth showed no obvious changes with restoration age, but there was an overall increasing trend (Figure 4a, Table S3). The C:P ratio at 0–30 cm increased, while the N:P ratio at 0–20 cm first increased and then decreased after 10 years of restoration (Figure 4b,c). In grassland, the C:N ratio at 0–100 cm showed an overall increasing trend with restoration age (Figure 4d, Table S3); the C:P and N:P ratios in soils of 0–30 cm depth gradually increased with restoration age (Figure 4e,f). In shrubland, the C:N ratio at 0–100 cm showed little variation with restoration age, but the overall trend was an increase (Figure 4h, Table S3); the C:P and N:P ratios at 0–10 cm depth first increased, then peaked at 30 years, before decreasing again with restoration age (Figure 4i,j). In forestland, the C:N ratio at 0–100 cm depth showed little change with restoration age, but the overall trend was an increase (Figure 4k, Table S3). The C:P ratio in soils of 0–100 cm depth gradually increased with restoration age (Figure 4l), and the N:P ratio at 0–50 cm first increased, then peaked at 37 years before decreasing again with restoration age (Figure 4m).

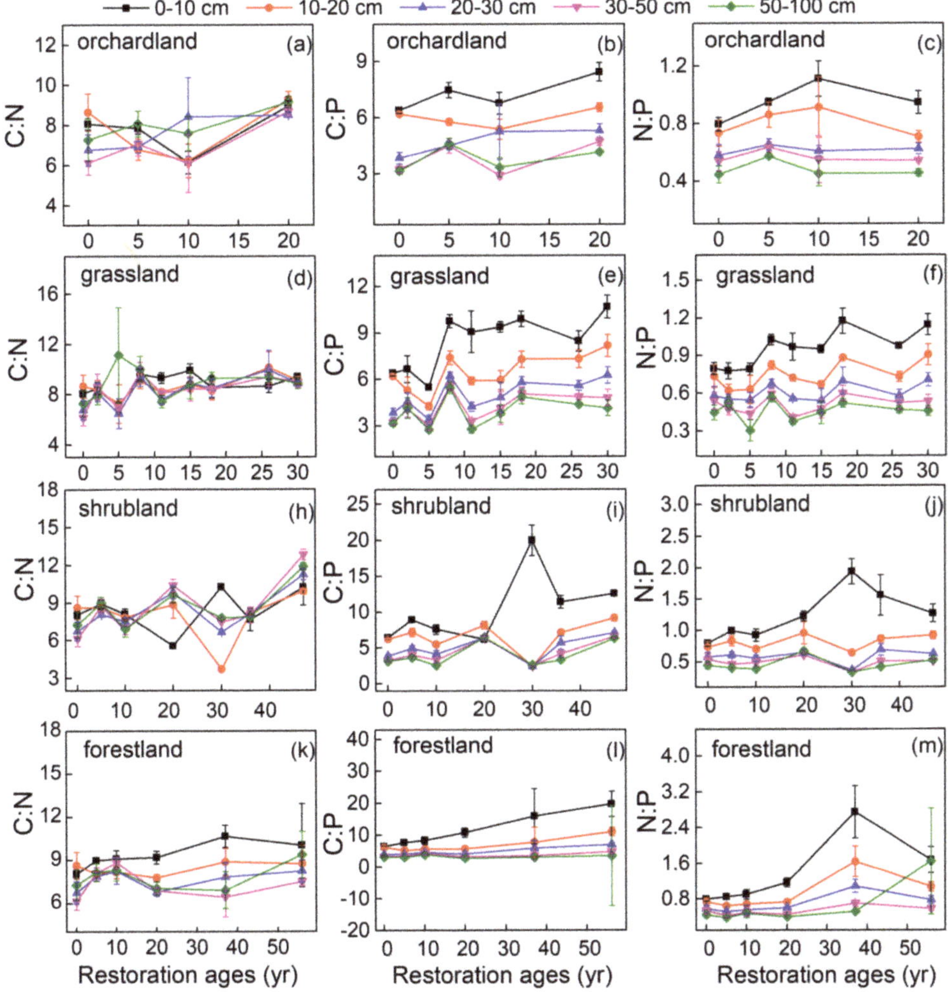

Figure 4. Stoichiometric characteristics of soil organic carbon (SOC), soil total nitrogen (TN), and soil total phosphorus (TP) with restoration age. Note: values are mean ± standard error.

The C:P and N:P ratios at 0–30 cm in grassland30 were significantly lower than those in natural grassland (Figure 5b,c), and the C:P and N:P ratios at 0–30 cm depth in shrubland47 and forestland56 were significantly lower than those in natural forest (Figure 5e,f). In addition, the C:N ratio decreased with soil depth in forestland (Figure 4k). Overall, the C:P and N:P ratios gradually decreased with soil depth in the three land types (Figure 4b,c,e,f,i,j,l,m).

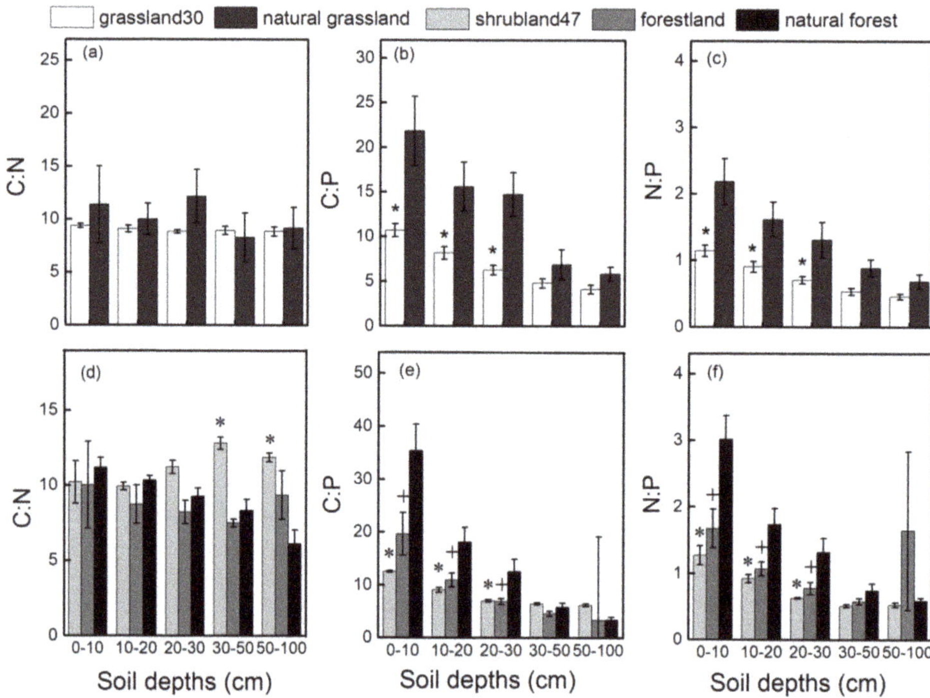

Figure 5. Stoichiometry of soil organic carbon (SOC), soil total nitrogen (TN), and soil total phosphorus (TP) at different soil depths in grassland at 30 years compared to that of natural grassland, and those of shrubland at 47 years and forestland at 56 years compared to that of natural forestland. Note: values are mean ± standard error. (**a–c**) means the C:N, C:P and N:P at different soil depths in grassland30 compared to those of natural grassland. (**d–f**) means the C:N, C:P and N:P at different soil depths in shrubland47 and forestland56 compared to those of natural forest * denotes significant differences between the natural grassland and grassland at 30 years after vegetation restoration ($p < 0.05$). * denotes significant differences between the natural forest and shrubland at 47 years after vegetation restoration ($p < 0.05$). + denotes significant differences between the natural forest and forestland at 56 years after vegetation restoration ($p < 0.05$).

4. Discussion

4.1. Restoration Ages Altered SOC, Soil TN, and Soil TP Stocks and Stoichiometry

Our results showed that the SOC stocks at 0–30 cm soil depth in orchardland, grassland, and shrubland first increased and then stabilized with restoration age, while the SOC stocks at 0–50 cm in forestland gradually increased with restoration age. SOC is mainly derived from surface litter, root secretions, and animal residues [21,22]. After vegetation restoration, a large input of litter and organic matter can enhance the accumulation of SOC [23], and with an improvement in soil structure, the levels of surface runoff, soil erosion, and soil nutrient loss may be reduced [24]. In addition, soil microbial activity is strengthened as restoration age increases [25], and soil nutrient conversion and storage are further enhanced. However, vegetation restoration involves the coordinated development of the plant community and soil environment, and the plant community structure, soil structure, and microbial diversity reach a stable level as restoration age increases [26–28]. However, in forestland, SOC stocks had not reached a steady level at 57 years, and when these stocks may stabilize needs to be evaluated. In our study, the soil TN stocks at 0–30 cm depth decreased in the later stages of orchardland, shrubland, and forestland restoration. Plants may enter a senescence phase, and the soil nitrogen nutrients absorbed during the plant growth process are then fed back to the soil by litter; thus, the nitrogen absorption rate is lower than the release rate, resulting in the soil TN stock decreasing [29]. We also found that soil TP stocks did not significantly change with restoration age, in contrast to the results of some previous studies [30]. Soil TP is mainly influenced by parent material, land use, and biogeochemical processes in the soil [31,32]. As the parent material and climate were similar for all vegetation types, the variability and migration rate of TP were not as obvious as those of SOC and soil TN, and restoration age had little effect on soil TP. Within the range of restoration ages that we studied, the SOC and soil TN stocks in grassland, shrubland, and forestland at 0–30 cm soil depth failed to reach the stock levels found in natural grassland or natural forest. Severe soil erosion and lack of water are typical characteristics of the Loess Plateau, which may lead to the loss of soil nutrients [24,33]. Additionally, soil tillage can also cause soil nutrient loss by affecting the soil physical structure and microbial activity [34].

The C:N ratio increased with restoration age after vegetation restoration. The main factors affecting the C:N ratio are changes in the SOC and soil TN contents [35]. Both SOC and soil TN contents increased overall with restoration age (Table S3), and the rate of increase of SOC was greater than that of soil TN; consequently, the C:N ratio increased (Table S3). A previous study showed that the C:N ratio was negatively correlated with the rate of decomposition of organic matter [36], so the decomposition rate of organic matter increases with restoration age. The C:P ratio indicates the availability of soil TP in the soil [37], We found that the rate of increase of SOC was also greater than that of soil TP, resulting in an increase in the C:P ratio with restoration age (Table S3). In addition, the rate of increase of soil TN was greater than that of soil TP, resulting in an increase in the N:P ratio with restoration age (Table S3). The N:P ratio was reduced in the later stages of restoration, which may be related to the lower soil TN content (Figure S1). Soil N and soil TP are essential mineral nutrients for plant growth and common limiting elements in ecosystems, and the N:P ratio is a predictor of nutrient limitation [38]. In the grassland, shrubland, and forestland types, the N:P ratio at 0–30 cm soil depth in the later stages of recovery was lower than those of natural grassland and natural forest, which may result from the more alkaline soil and lower soil TN content in the Loess Plateau Region; the soil TP content did not differ significantly between the restoration and control sites. In the grassland, shrubland, and forestland restoration types, the C:P ratio in the 0–30 cm soil layer was lower than those of natural grassland and natural forest at the later stages of recovery, which may be related to the lower SOC content in restoration soils than in natural soils.

4.2. Vertical Distribution of Stocks and Stoichiometry of SOC, Soil TN, and Soil TP

Soil depth is an important factor influencing SOC, soil TN, and soil TP distribution [39]. In our study, the SOC and soil TN stocks decreased with soil depth at all restoration ages in orchardland, grassland, shrubland, and forestland (Figure 6a,b,d,e,h,I,k,l), which is consistent with the results of previous studies [40,41]. Our study also revealed that the overall rates of SOC and soil TN content change decreased with soil depth in orchardland, grassland, shrubland, and forestland (Table S3), which indicated that the SOC and soil TN sequestration rates gradually decreased with soil depth. Meanwhile, SOC, soil TN, and soil TP were most sensitive to change in the surface soil (0–30 cm). The SOC and soil TN content at 0–30 cm represented more than 65% of the total SOC and soil TN stocks from 0–100 cm depth for all restoration types; the soil TP content at 0–30 cm represented more than 60% of the total soil TP stocks from 0–100 cm (Figure S1). Such differences in the SOC, soil TN, and soil TP profiles can be explained partly by root distribution. The surface soil is affected by external environmental factors, soil microorganisms, and the return of nutrients from surface litter, resulting in a concentration of nutrients in the surface soil [42]. With increasing soil depth, the input of organic matter is limited by the permeability of the soil, microbial decomposition activity, and root absorption [21,39]. Moreover, SOC and soil TN stocks are not only affected by soil parent material, but also by the decomposition of litter and absorption and utilization by plants [40], resulting in large spatial variability. While the soil TP content of orchardland and forestland decreased, it showed little variation with soil depth in grassland and shrubland (Figure 6c,f,j,m). Soil TP is mainly affected by the soil parent material, which, in this case, is a sedimentary mineral with low mobility in soil; therefore, there was little vertical variation in soil TP [43].

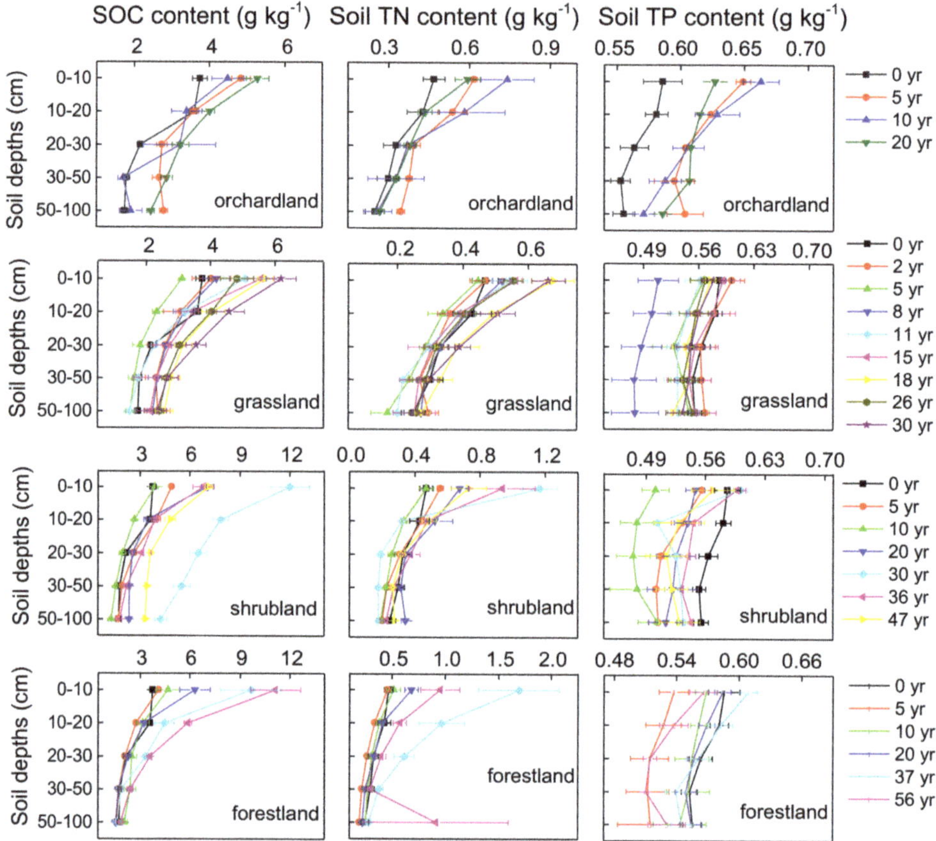

Figure 6. Vertical distributions of soil organic carbon (SOC), soil total nitrogen (TN), and soil total phosphorus (TP) contents for different vegetation types. Note: values are mean ± standard error.

Our study found that the overall C:N ratio gradually decreased with soil depth. It may be that the surface SOC and soil TN contents were higher, but as the soil depth increased, the SOC content change was larger than that of the soil TN content. When the decomposition process occurs, easily decomposed material vanishes, and soil TN is immobilized in decayed products, leaving behind more durable material with slower decomposition rates in the deeper layers [44]. This results in a relatively lower C:N ratio in the deeper soil layers. There was a significant difference in the C:P and N:P ratios at different soil depths. It may be that the soil TP content is relatively stable at different soil depths, and the C:P ratio and N:P ratio are mainly affected by SOC and soil TN content, so they showed greater variation.

4.3. Effect of Restoration Type on SOC, Soil TN, and Soil TP Stocks and Stoichiometry.

Our study demonstrated that the SOC and soil TN stocks in the 0–20 cm soil layer at 5 years were highest in orchardland, which may be related to the use of fertilizer. In other years, SOC, soil TN, and soil TP stocks showed no difference between orchardland, grassland, shrubland, and forestland

(Figure S2). However, the rate of SOC and soil TN change at 0–10 cm soil depth was the highest in forestland, while the rates of SOC and soil TN change at other depths varied among different restoration types (Table S3). The rapid increases in surface SOC and soil TN are closely related to the input of litter [45,46]. Guo et al. [46] showed that forest litter was 19 times greater in mass than that of shrubs in the Loess Plateau Region. At the same time, the presence of the higher amount of organic carbon over a long period of time indicated that forestland likely had developed a higher degree of humification. The long-term development of soil organic carbon stocks suggested that the afforestation improved the humification process. It has been shown that the process of humification in the soil is critical for the ecosystem, due to its strong contributions to the improvement of fertility and, hence, the storage of both carbon and nitrogen [47,48]. This also explains why the increase in SOC and soil TN in the topsoil of forests was higher than that of shrubs. As soil depth increases, root secretions and soil microorganisms are the main sources of soil nutrition [21,22]. There are significant differences in the effects of different plant roots and litters on the community composition of microorganisms [49], which may explain the large differences in the rates of SOC and soil TN change between different restoration types.

The C:N ratios of orchardland5 and grassland5 at 0–20 cm soil depth were lower than those of shrubland5 and forestland5, and the C:N ratio of orchardland10 was significantly lower than those of grassland10, shrubland10, and forestland10 (Figure S2). The C:N ratio of orchardland was lower than those of grassland, shrubland, and forestland, which may be related to anthropogenic N deposition in orchardland (Figure S2). The lower C:N ratio in grassland5 may occur because grassland retains more organic matter content and greater nutrient absorption takes place through plant roots [8]. There were no significant differences in the C:N ratio at other soil depths of orchardland, grassland, shrubland, and forestland as there were no significant differences in the stocks of SOC and soil TN (Figure S2). This finding is related to the nutrient conditions of the soil in the study area and the feedback between plant and soil, so the soil stoichiometric changes in different restoration types showed the same characteristics in the same environmental context. In addition, the overall C:P and N:P ratios of orchardland5 and shrubland5 at 0–20 cm soil depth were higher than those of grassland5 and forestland5, which is related to the relatively higher SOC and soil TN content in orchardland5 and shrubland5 (Figure S2). However, there were no significant differences in the C:P and N:P ratios at 30–100 cm soil depth between the four restoration types because there were no significant differences in the SOC, soil TN, and soil TP stocks of orchardland, grassland, shrubland, and forestland (Figure S2).

5. Conclusions

We examined the changes in SOC, soil TN, and soil TP stocks and stoichiometry at depths of 0–100 cm following vegetation restoration in the Loess Hilly Region. Our results revealed that the SOC stocks appeared to increase and reach stable levels. The soil TN stocks first increased and then decreased with restoration age, but they did not reach the levels seen in natural grassland or natural forest in the Loess Hilly Region in the absence of appropriate management. Soil TP stocks failed to improve substantially with restoration age. The C:N, C:P, and N:P ratios gradually increased with restoration age. At the same time, the SOC, soil TN, and soil TP stocks and the C:N, C:P, and N:P ratios decreased with soil depth. Reforested land had the highest sequestration rate of SOC and TN at 0–10 cm soil depth. The results of this study provide data for the assessment of the long-term SOC, soil TN, and soil TP stocks and stoichiometry after vegetation restoration under different restoration types in the Loess Hilly Region.

Supplementary Materials: The following are available online at http://www.mdpi.com/1999-4907/10/1/27/s1, Figure S1: Changes in soil organic carbon (SOC), soil total nitrogen (TN), and soil total phosphorus (TP) content with restoration age. Figure S2: Changes in content and stoichiometry of soil organic carbon (SOC), soil total nitrogen (TN), and soil total phosphorus (TP) for different vegetation types at 5, 10, and 20 years. Table S1: The details of the sample sites selected for the study. Table S2: F and *sig* values for independent factors (soil depth, restoration age) and their interactions. Table S3: The slope parameters of the linear regression models for soil organic carbon (SOC), soil total nitrogen (TN), and soil total phosphorus (TP) contents and C:N, C:P, and N:P ratios with recovery year at different soil depths under four restoration types.

Author Contributions: The research design was completed by H.X. and S.X. The manuscript was written by H.X. and Q.Q. The collection and analysis of soil samples were performed by H.X., P.L., Q.Q., Z.G. and E.W.

Funding: This research was funded by National Science and Technology Support Program (2015BAC01B03), Natural Science Foundation of China (41771557) and Foundation for Western Young Scholars, Chinese Academy of Sciences (XAB2015A05).

Acknowledgments: Many thanks to Juan Li, Dou Zhang and Zhao Fang for their technical assistance in the laboratory work. We would like to thank Mancai Guo and Minggang Wang for providing statistics assistance.

Conflicts of Interest: All the authors declare no conflicts of interest.

References

1. Reed, S.C.; Yang, X.; Thornton, P.E. Incorporating phosphorus cycling into global modeling efforts: A worthwhile, tractable endeavor. *New Phytol.* **2015**, *208*, 324–329. [CrossRef] [PubMed]
2. Lv, J.L.; Yan, M.J.; Song, B.L. Ecological stoichiometry characteristics of soil carbon, nitrogen, and phosphorus in an oak forest and a black locust plantation in the Loess hilly region. *Acta Ecol. Sin.* **2017**, *10*, 3385–3393, (In Chinese with English Abstract). [CrossRef]
3. Rodríguez, E.S.; Lozano, Y.M.; Bardgett, R.D. Influence of soil microbiota in nurse plant systems. *Funct. Ecol.* **2016**, *30*, 30–40. [CrossRef]
4. Rudel, T.K.; Coomes, O.T.; Moran, E.; Achard, F.; Angelsen, A.; Xu, J.; Lambin, E. Forest transitions: Towards a global understanding of land use change. *Glob. Environ.* **2005**, *15*, 23–31. [CrossRef]
5. Winans, K.; Whalen, J.; Cogliastro, A.; Rivest, D.; Ribaudo, L. Soil Carbon Stocks in Two Hybrid Poplar-Hay Crop Systems in Southern Quebec, Canada. *Forests* **2014**, *5*, 1952–1966. [CrossRef]
6. Jiang, P.; Chen, Y.; Cao, Y. C:N:P Stoichiometry and Carbon Storage in a Naturally-Regenerated Secondary Quercus variabilis Forest Age Sequence in the Qinling Mountains, China. *Forests* **2017**, *8*, 281. [CrossRef]
7. Wang, W.; Sardans, J.; Zeng, C.; Zhong, C.; Li, Y.; Peñuelas, J. Responses of soil nutrient concentrations and stoichiometry to different human land uses in a subtropical tidal wetland. *Geoderma* **2014**, *232–234*, 459–470. [CrossRef]
8. Gao, Y.; He, N.; Yu, G.; Chen, W.; Wang, Q. Long-term effects of different land use types on C, N, and P stoichiometry and storage in subtropical ecosystems: A case study in China. *Ecol. Eng.* **2014**, *67*, 171–181. [CrossRef]
9. Zhao, F.; Sun, J.; Ren, C.; Kang, D.; Deng, J.; Han, X.; Yang, G.; Feng, Y.; Ren, G. Land use change influences soil C, N and P stoichiometry under 'Grain-to-Green Program' in China. *Sci. Rep.* **2015**, *5*, 10195. [CrossRef]
10. Fang, J.; Chen, A.; Peng, C.; Zhao, S.; Ci, L. Changes in Forest Biomass Carbon Storage in China Between 1949 and 1998. *Science* **2001**, *292*, 2320–2322. [CrossRef]
11. Post, W.M.; Kwon, K.C. Soil carbon sequestratrion and land-use change: Processes and potential. *Glob. Environ.* **2010**, *6*, 317–327. [CrossRef]
12. Xiao, H.; Li, Z.; Dong, Y.; Chang, X.; Deng, L.; Huang, J.; Nie, X.; Liu, C.; Liu, L.; Wang, D.; et al. Changes in microbial communities and respiration following the revegetation of eroded soil. *Agric. Ecosyst. Environ.* **2017**, *246*, 30–37. [CrossRef]
13. Yang, W.; Cheng, H.; Hao, F.; Ouyang, W.; Liu, S.; Lin, C. The influence of land-use change on the forms of phosphorus in soil profiles from the Sanjiang Plain of China. *Geoderma* **2012**, *189–190*, 207–214. [CrossRef]
14. Liu, X.; Ma, J.; Ma, Z.; Li, L. Soil nutrient contents and stoichiometry as affected by land-use in an agro-pastoral region of northwest China. *Catena* **2017**, *150*, 146–153. [CrossRef]
15. Li, C.; Zhao, L.; Sun, P.; Zhao, F.Z.; Kang, D.; Yang, G.H. Deep Soil C, N, and P Stocks and Stoichiometry in Response to Land Use Patterns in the Loess Hilly Region of China. *PLoS ONE* **2016**, *11*, e0159075. [CrossRef] [PubMed]
16. Fei, W.; Rui, L.I.; Zhong, W. Problems and Proposals on Policy of Converting Cropland Into Forest and Grassland: A Case-based Study. *Nat. Sci. Ed.* **2003**, *3*, 60–65, (In Chinese with English Abstract). [CrossRef]
17. Xin, Z.; Ran, L.; Lu, X.X. Soil Erosion Control and Sediment Load Reduction in the Loess Plateau: Policy Perspectives. *Int. J. Water Resour. Dev.* **2012**, *28*, 325–341. [CrossRef]

18. Nelson, D.W.; Sommers, L.E. Total carbon, organic carbon and organic matter. In Methods of Soil Analysis Part 3—Chemical Methods. *Chem. Microbiol. Prop.* **1982**. [CrossRef]

19. Bremner, J.; Mulvaney, C. Nitrogen—total. *Methods of Soil Analysis Chemical Methods Part.* **1996**, *72*, 532–535. [CrossRef]

20. Schimel, J.P.; Weintraub, M.N. The implications of exoenzyme activity on microbial carbon and nitrogen limitation in soil: A theoretical model. *Soil Biol. Biochem.* **2003**, *35*, 549–563. [CrossRef]

21. Clemmensen, K.E.; Bahr, A.; Ovaskainen, O.; Dahlberg, A.; Ekblad, A.; Wallander, H.; Stenlid, J.; Finlay, R.D.; Wardle, D.A.; Lindahl, B.D. Roots and Associated Fungi Drive Long-Term Carbon Sequestration in Boreal Forest. *Science* **2013**, *339*, 1615–1618. [CrossRef] [PubMed]

22. Fontaine, S.; Barot, S.; Barré, P.; Bdioui, N.; Mary, B.; Rumpel, C. Stability of organic carbon in deep soil layers controlled by fresh carbon supply. *Nature* **2007**, *450*, 277–280. [CrossRef] [PubMed]

23. Yang, Y.; Luo, Y. Carbon:nitrogen stoichiometry in forest ecosystems during stand development. *Glob. Ecol. Biogeogr.* **2011**, *20*, 354–361. [CrossRef]

24. Li, Z.; Liu, C.; Dong, Y.; Chang, X.; Nie, X.; Liu, L.; Xiao, H.; Lu, Y.; Zeng, G. Response of soil organic carbon and nitrogen stocks to soil erosion and land use types in the Loess hilly–gully region of China. *Soil Tillage Res.* **2017**, *166*, 1–9. [CrossRef]

25. Leon, D.G.D.; Moora, M.; Öpik, M.; Neuenkamp, L.; Gerz, M.; Jairus, T.; Vasar, M.; Bueno, C.G.; Davison, J.; Zobel, M. Symbiont dynamics during ecosystem succession: Co-occurring plant and arbuscular mycorrhizal fungal communities. *FEMS Microbiol. Ecol.* **2016**, *92*, 1–9. [CrossRef]

26. Hao, W.F.; Liang, Z.S.; Chen, C.G.; Tang, L. Study of species diversity evolvement process during vegetation restoration of abandoned farmland in the hilly loess plateau. *Pratac. Sci.* **2005**, *22*, 1–8, (In Chinese with English Abstract). [CrossRef]

27. Novara, A.; Mantia, T.L.; Rühl, J.; Badalucco, L.; Kuzyakov, Y.; Gristina, L.; Laudicina, V.A. Dynamics of soil organic carbon pools after agricultural abandonment. *Geoderma* **2014**, *235–236*, 191–198. [CrossRef]

28. O'Brien, S.L.; Jastrow, J.D. Physical and chemical protection in hierarchical soil aggregates regulates soil carbon and nitrogen recovery in restored perennial grasslands. *Soil Biol. Biochem.* **2013**, *61*, 1–13. [CrossRef]

29. Jia, H.E.; Yu-Fu, H.U.; Shu, X.Y.; Wang, Q.; Jia, A.D.; Yan, X.; University, S.A. Effect of Salix cupularis plantations on soil stoichiometry and stocks in the alpine-cold desert of northwestern. *Sin. Acta Agrestia Sin.* **2018**. (In Chinese with English Abstract). [CrossRef]

30. Lane, P.N.J.; Noske, P.J.; Sheridan, G.J. Phosphorus enrichment from point to catchment scale following fire in eucalypt forests. *Catena* **2011**, *87*, 157–162. [CrossRef]

31. Cheng, Y.; Li, P.; Xu, G.; Gao, H. Spatial distribution of soil total phosphorus in Yingwugou watershed of the Dan River, China. *Catena* **2016**, *136*, 175–181. [CrossRef]

32. Kooijman, A.M.; Jongejans, J.; Sevink, J. Parent material effects on Mediterranean woodland ecosystems in NE Spain. *Catena* **2005**, *59*, 55–68. [CrossRef]

33. Comino, J.R.; Sinoga, J.D.R.; González, J.M.S.; Guerra-Merchán, A.; Seeger, M.; Ries, J.B. High variability of soil erosion and hydrological processes in Mediterranean hillslope vineyards (Montes de Málaga, Spain). *Catena* **2016**, *145*, 274–284. [CrossRef]

34. Duan, X.N.; Wang, X.K.; Lu, F.; Ouyang, Z.Y. Soil carbon sequestration and its potential by grassland ecosystems in China. *Acta Ecol. Sin.* **2008**. (In Chinese with English Abstract). [CrossRef]

35. Reich, P.B.; Oleksyn, J. Global patterns of plant leaf N and P in relation to temperature and latitude. *Proc. Natl. Acad. Sci. USA* **2004**, *101*, 11001–11006. [CrossRef] [PubMed]

36. An, H.; Li, G.Q. Effects of grazing on carbon and nitrogen in plants and soils in a semiarid desert grassland, China. *J. Arid Land* **2015**, *7*, 341–349. [CrossRef]

37. Yu, Y.F.; Peng, W.X.; Song, T.Q. Stoichiometric characteristics of plant and soil C, N and P in different forest types in depressions between karst hills, southwest China. *Chin. J. Appl. Ecol.* **2014**, *25*, 947–954, (In Chinese with English Abstract). [CrossRef]

38. Chen, L.; Li, C.L. Research advances in ecological stoichiometry of marine plankton. *Chin. J. Appl. Ecol.* **2014**, *25*, 3047–3055, (In Chinese with English Abstract). [CrossRef]

39. Berger, T.W.; Neubauer, C.; Glatzel, G. Factors controlling soil carbon and nitrogen stores in pure stands of Norway spruce (*Picea abies*) and mixed species stands in Austria. *For. Ecol. Manag.* **2002**, *159*, 3–14. [CrossRef]

40. Liu, X.Z.; Zhou, G.Y.; Zhang, D.Q.; Liu, S.Z.; Chu, G.W.; Yan, J.H. N and P stoichiometry of plant and soil in lower subtropical forest successional series in southern China. *J. Plant Ecol.* **2010**, *34*, 64–71, (In Chinese with English Abstract). [CrossRef]

41. Zeng, Q.; Liu, Y.; Fang, Y.; Ma, R.; Lal, R.; An, S.; Huang, Y. Impact of vegetation restoration on plants and soil C:N:P stoichiometry on the Yunwu Mountain Reserve of China. *Ecol. Eng.* **2017**, *109*, 92–100. [CrossRef]

42. Barreto, P.A.B.; Gama-Rodrigues, E.F.; Fontes, A.G.; Polidoro, J.C.; Moço, M.K.S.; Machado, R.C.R.; Baligar, V.C. Distribution of oxidizable organic C fractions in soils under cacao agroforestry systems in Southern Bahia, Brazil. *Agrofor. Syst.* **2011**, *81*, 213–220. [CrossRef]

43. Zhang, X. Ecological stoichiometry characteristics of robinia pseudoacacia forest soil in different latitudes of loess plateau. *Acta Ecol. Sin.* **2013**, *50*, 818–825. [CrossRef]

44. Han, L.; Li, Z.; Ceng, Y. Carbon, nitrogen, and phosphorous stoichiometry of herbaceous plant leaf and soil in riparian zone of Taihu Lake basin, East China under effects of different land use types. *Chin. J. Ecol.* **2013**, *32*, 3281–3288, (In Chinese with English Abstract). [CrossRef]

45. Vesterdal, L.; Ritter, E.; Gundersen, P. Change in soil organic carbon following afforestation of former arable land. *For. Ecol. Manag.* **2002**, *169*, 137–147. [CrossRef]

46. Guo, S.L.; Ma, Y.H.; Che, S.G.; Sun, W.Y. Effects of artificial and natural vegetations on litter production and soil organic carbon change in loess hilly areas. *Sci. Silvae Sin.* **2009**, *45*, 14–18, (In Chinese with English Abstract). [CrossRef]

47. Pizzeghello, D.; Francioso, O.; Concheri, G.; Muscolo, A.; Nardi, S. Land Use Affects the Soil C Sequestration in Alpine Environment, NE Italy. *Forests* **2017**, *8*, 197. [CrossRef]

48. De Feudis, M.; Cardelli, V.; Massaccesi, L.; Lagomarsino, A.; Fornasier, F.; Westphalen, D.J.; Coccco, S.; Corti, G.; Agnelli, A. Influence of Altitude on Biochemical Properties of European Beech (*Fagus sylvatica* L.) Forest Soils. *Forests* **2017**, *8*, 213. [CrossRef]

49. Cesarz, S.; Fender, A.C.; Beyer, F.; Valtanen, K.; Pfeiffer, B.; Gansert, D.; Hertel, D.; Leuschner, C. Roots from beech (*Fagus sylvatica* L.) and ash (*Fraxinus excelsior* L.) differentially affect soil microorganisms and carbon dynamics. *Soil Biol. Biochem.* **2013**, *61*, 23–32. [CrossRef]

Article

Mixed Broadleaved Tree Species Increases Soil Phosphorus Availability but Decreases the Coniferous Tree Nutrient Concentration in Subtropical China

Wen-Sheng Bu [1,2], Han-Jiao Gu [1,2], Can-can Zhang [1,2], Yang Zhang [1], Anand Narain Singh [3], Xiang-Min Fang [1,2], Jing Fan [1,2], Hui-Min Wang [4] and Fu-Sheng Chen [1,2,*,†]

[1] Jiangxi Provincial Key Laboratory of Silviculture, College of Forestry, Jiangxi Agricultural University, Nanchang 330045, China; bws2007@163.com (W.-S.B.); guhanjiao@163.com (H.-J.G.); brightzcc@163.com (C.-c.Z.) zhangyang0558@163.com (Y.Z.); xmfang2013@126.com (X.-M.F.); fxh2022@163.com (J.F.)

[2] Jiulianshan National Observation and Research Station of Chinese Forest Ecosystem, 2011 Collaborative Innovation Center of Jiangxi Typical Trees Cultivation and Utilization, Jiangxi Agricultural University, Nanchang 330045, China

[3] Department of Botany, Panjab University, Chandigarh 160014, India; dranand1212@gmail.com

[4] Qianyanzhou Ecological Station, Key Laboratory of Ecosystem Network Observation and Modeling, Institute of Geographic Sciences and Natural Resources Research, Chinese Academy of Sciences, Beijing 100101, China; wanghm@igsnrr.ac.cn

* Correspondence: chenfush@yahoo.com; Tel./Fax: +86-791-83813243

† Current address: No. 1101 Zhiminda Road, Economic & Technological Development Zone, Nanchang 330045, China.

Received: 10 March 2020; Accepted: 17 April 2020; Published: 19 April 2020

Abstract: Phosphorus (P) is a key limiting nutrient in subtropical forests and mixed forests with broadleaved species have been expected to stimulate P cycling, compared to pure conifer plantations. However, the mixture effect of Chinese fir (*Cunninghamia lanceolata* (Lamb.) Hook.) and broadleaved species on rhizosphere soil and coniferous tree P dynamics is unclear. In our study, eight plots of a single species of a Chinese fir plantation (pure plantation, PP) and eight mixed plantations (mixed plantation, MP) with broadleaved tree species (*Michelia macclurei* Dandy in Hunan Province or *Schima superba* Gardn. et Champ. in Fujian Province) were selected in subtropical China. Six P fractions in the rhizosphere and bulk soils were analyzed by a modified Hedley P fractionation method. Phosphorus fractions and nitrogen (N) concentrations in different root orders, different age fresh needles and twigs, and needle and twig litter of Chinese fir were measured. Our results showed that available P, slowly released P, occluded P, and the total extractable P in rhizosphere soil were significantly higher in MP than PP ($p < 0.05$). In contrast, P and N concentrations in the transportive roots and two-year old needles were generally higher in PP than MP. Meanwhile, the slowly released P, occluded P, total extractable P, and residual P in rhizosphere soil were negatively correlated with P concentrations in young (absorptive and transportive roots, one- and two-year old needles) but not old tissues (storative roots, three-year old needles and litters). In conclusion, mixture may increase soil P availability through the rhizosphere effect, but can decrease P and N concentration of Chinese fir tissues by competition between Chinese fir and broadleaved species. Clearly, the mixture effect may differ in soil and plant nutrients, and this issue needs be taken into consideration when converting a pure conifer plantation into a mixed-species forest.

Keywords: *Cunninghamia lanceolata*; mixture effect; nutrient cycling; rhizosphere effect; species competition

---

1. Introduction

Forest plantations in China have an area of 69 million hectares, occupy 31.8% of the total area of Chinese forests, and play important social, economic, and environmental roles [1]. However, most of these plantations were planted as monoculture conifers, such as Chinese fir (*Cunninghamia lanceolate* (Lamb.) Hook.) and *Pinus massoniana* Lamb., which might lead to nutrient imbalance, soil degradation, and the reduction of ecosystem stability [2]. In Southern China, Chinese fir has been widely planted and provides excellent commercial woods with easy processing because it is a native, fast-growing woody species with a 20–25 year rotation period and more than 1000 years of management practice history [3]. The removal of harvested wood and accumulated litter could alter the biogeochemistry in Chinese fir plantations, which may result in the decline of forest productivity due to multiple rotations [4]. Therefore, converting Chinese fir monoculture into mixed with broadleaved trees has become a common trend of forest management, since the mixed plantations seemed to have many advantages over pure stands, such as higher rates of litter decomposition and maintaining soil nutrient cycling [5], i.e., there were hardly any, or even negative effects of mixture on ecosystem functions, such as tree biomass accumulation and useful wood yield [6].

Phosphorus (P) is a key limiting nutrient in terrestrial ecosystems especially in the tropics and subtropics, where P is limited in the highly weathered soil and is found in pools with low amount available for plant [7]. Moreover, soil P utilization percentage is only 10%–25% in subtropical China [8] in comparison to P-rich soil from other tropics of the world. Shenoy et al. found that suboptimal levels of P could decrease crop growth and resulted in 5%–15% yield losses [9]. Many Chinese fir plantations have been established in these P-deficient sites, therefore, the availability of soil P is one of the most important limiting factors causing a decline in forest productivity because Chinese fir has a high use efficiency of which P is required for its optimal growth [8]. As mentioned above, the mixed-species forests have many advantages over pure plantations, and can improve root distribution patterns, plant residue compositions, and soil microbial diversity [10]. Theoretically, mixed forest plantations with Chinese fir and broadleaved trees would stimulate P cycling by mobilizing soil fixed P and increasing plant uptake, compared to pure stands of Chinese fir. However, the effect of mixed species on soil–tree P cycling in the Chinese fir forest ecosystem is little studied [8] and future P scenario cannot be effectively forecasted with the increasing conversion from pure plantations to mixed forests.

The nutrient availability is usually a major constraint to tree growth in most terrestrial ecosystems, while trees take up most of the mineral nutrients through the rhizosphere. The rhizosphere is a vital region of ecosystem biogeochemistry and governs the nutrient availability through plant roots to stimulate soil microbial activity [11]. Nutrients in the rhizosphere can rapidly transform from unavailable to available forms through the interaction of root exudates and microorganisms [12]. Plants generally secrete 10%–30% root exudates into rhizosphere soil through the root system. In order to increase the carbon availability in rhizosphere, microorganisms utilized those root exudates as food and energy [13]. Therefore, the microorganism population density might be much larger in the rhizosphere than in bulk soil. Plants absorb most mineral nutrients with effective nutrient transformation and mineralization dominated by rich microorganisms in rhizosphere soil [11]. Unfortunately, the difference of nutrient transformation and availability between the rhizosphere and bulk soil is less understood especially in forest ecosystems [14]. The modified P functional fractionation method in acidic soils provides an effective tool to explore the P supply process in the rhizosphere soil of subtropical forests [15].

On the other hand, the effects of species interactions on ecological processes such as nutrient cycling and tree growth in mixed stands have raised significant interest in recent years [2,16]. However, little is known about the influence of mixture on tree internal nutrient cycling [17], for example, the nutrient concentrations in various components of fresh tissues found in trees and their litters [18]. This result suggests that there is a change in plant physiology based on what is around the plant, or is it simply a function of competition? The understanding of internal nutrient cycling is of great importance because it not only depends on site nutrient availability but also reflects the competition among the different tree species in the mixed forest [19]. By now, mixed-species plantations containing a leguminous species have shown the potential to improve nutrient cycling [20]. However, there is a lack of evidence to validate the potential of mixed-species plantations without leguminous species, particularly in the tropics and subtopics [16]. There is no significant difference between mixtures and monocultures, when all of the empirical studies are combined [21]. The lack of a general trend, and the limited number of studies on the impact of mixed stands on tree nutrients, switches further studies. In addition, nutrient dynamics generally vary with the age of plant tissue. For example, more nutrients were resorbed to young leaves from old leaves when the acquisition of nutrients from the soil became more difficult [22]. Thus, the differentiation among the different ages of plant tissues might be helpful to explore nutrient internal cycles in response to the changing environment.

In the present study, eight plots of single-species Chinese fir plantation (pure plantation, PP) and eight plots of mixed plantation with a broadleaved tree species (mixed plantation, MP) with similar standing density in mature forest were selected to assess the effect of mixed species on soil and tree P cycling including belowground rhizosphere processes and nutrient concentrations of different functional root orders, and nutrient concentrations in aboveground components (fresh twigs, needles, and their litters) of different ages in Hunan and Fujian provinces of China. Both are the center distribution zones of Chinese fir production in China [23]. Our main hypotheses are: (1) the mixture with broadleaved trees improves soil P availability, and then increase tree nutrient levels; (2) the mixture effect (the difference between PP and MP) on rhizosphere soil P supply varies with P fraction, and depends on its rhizosphere effect (the increasing percentage in rhizosphere soil compared to bulk soils); and (3) the response of Chinese fir nutrients to the mixed with broadleaved tree varies with root orders, needle and twig ages due to combined effects of soil nutrient supplies, tree species competition, and plant physiological requirement. In addition, we explored the relationships among the nutrients of belowground component (root–soil), and aboveground nutrients (fresh twigs, needles, and their litters) in these ecosystems. Our results will help assess the potential effect of stand transformation from the monoculture plantation to mixed forest, and will provide theoretical support for forest management.

2. Materials and Methods

2.1. Study Sites

This study was conducted at two selected sites in subtropical China. One is located at Huitong Forest Ecological Station in Hunan Province (26°52′N, 109°42′E), and the other at Datian Experimental Forest Farm in Fujian Province (25°45′N, 117°33′E) (Figure S1). Both sites are approximately 1000 km away from each other but have a similar humid, mid-subtropical, monsoonal climate with the average annual temperatures of 15~18 °C and average annual precipitation of 1350~1450 mm. The zonal vegetation is evergreen broadleaved forest; but establishing forests by plantation is more pronounced, and this has occupied more than 90% due to logging. The altitudes of both sites are 500–600 m ASL, soil types belong to red soil, but are classified as Oxisol in Hunan Province and Humic Planosol in Fujian Province, respectively, according to the FAO classification system. Chinese fir has become the most important afforestation tree due to its fast-growing, easy processing, and wide utilization in both provinces [24].

2.2. Experimental Design and Investigation

At each selected site from both provinces, we randomly established total eight plots (circles with 200-m^2 size); At each site of both provinces, there are four plots in pure Chinese fir plantations and four plots in mixed plantations with a broadleaved tree species (*Michelia macclurei* in Hunan Province or *Schima superba* in Fujian Province), respectively, which are located on the mid-slope with the slopes ranging from 25° to 30°. Both *M. macclurei* and *S. superba* are native evergreen tree species, and widely afforested as pure or mixed forest due to rapid growth characteristic. In Hunan Province, both stand types were established in 1983 and common management practices were used in the early stages of the two stand types, including weeding and thinning. In Fujian Province, both stands were established in 1991 and similarly managed, such as weed-controlling during the first three years and thinning during the first 10 years (for more details see [23]). All sixteen selected plots were independently distributed in different hills, and generally far away, with more than 500 m between two plots of the same stand type.

In May 2013, the diameter at breast-height (DBH) and species of trees with a DBH > 2 cm were measured and recorded in all plots of both sites. The stand basal area was obtained at plot level. Meanwhile, general stand characteristics, including stem density, shrub density, and herb cover were investigated (See Table 1).

Table 1. Stand characteristics (means and standard errors, $n = 4$) of pure Chinese fir plantations (PP) and its mixed plantations with a broadleaved tree species (MP) at two sites located in Hunan and Fujian provinces of China.

Parameters	Hunan		Fujian	
	PP	MP	PP	MP
Basal area (m^2 ha^{-1})				
Chinese fir	35.31 ± 2.20	23.47 ± 2.26	69.75 ± 4.16	42.43 ± 2.02
Broadleaved tree	—	11.04 ± 1.21	—	11.97 ± 1.49
Total	35.31 ± 2.20	34.51 ± 2.23	69.75 ± 4.16	54.40 ± 2.96
Stem density (trees ha^{-1})				
Chinese fir	1410 ± 102	620 ± 86	2510 ± 129	2010 ± 76
Broadleaved tree	—	730 ± 68	—	710 ± 29
Total	1410 ± 102	1350 ± 47	2510 ± 129	2720 ± 85
Shrub density (stems ha^{-1})	8160 ± 1063	6320 ± 1038	14000 ± 2449	8500 ± 1000
Herb cover (%)	27.4 ± 12.4	14.1 ± 4.1	8.6 ± 4.0	2.6 ± 2.6
Understory plant biomass (g m^{-2})	154 ± 79	145 ± 42	214 ± 76	157 ± 63

Note: The broadleaved tree species are *Michelia macclurei* and *Schima superba* in Hunan and Fujian provinces, respectively.

2.3. Sampling

Within each circular plot in both PP and MP, three individuals of Chinese fir tree were chosen as reference plants and the rhizosphere and bulk soils, twig and needle litters under these tree canopies, roots of various orders, stems, fresh twigs, and needles varying with age in the trees were sampled. Rhizosphere soils were sampled using a hand shaking method and defined as the soils <4 mm away from the fine roots distributed 0–20 cm in the surface soil layer (>50% of the total Chinese fir fine roots [23]). Bulk soil was collected using soil cores (10 cm diameter) at 0–20 cm depth in the middle locations of forest gaps. Roots (living roots with <4 mm diameter) were extracted by shovel, the Chinese fir roots carefully collected by hand, and divided as three functional orders of absorptive roots (AR, the first three orders), transportive roots (TR, the 4th–5th orders), and storative roots (SR, >5th orders) in the laboratory [25]. Stems were sampled at breast height of the reference trees using an increment borer. Fresh needles and twigs from one first-order branch were collected from all three reference trees. Needles and twigs were divided into the first, second, and third orders of branching based on their ages (one-year old, two-year old, and three-year old needles or twigs) (see [18]). Meanwhile, twig and needle litters were directly cut from the trees using the combined method of people climbing and a tree pruner, since part of new branch litters (including the twig and needle litters) generally remain on the Chinese fir trunks. In our study, the concentrations of nutrients remaining in twig and needle litters were defined as nutrient resorption proficiencies in twig and needle organs, respectively [26]. All same samples were mixed as a sample within a plot. Soil and plant samples were air-dried at room temperature and dried at 60 °C for more than 72 h in an oven before nutrient measurement.

2.4. Chemical Analyses

Rhizosphere and bulk soils were cleared of roots and all organic debris, and ground to pass through a 0.25-mm sieve before analyses of P fractions, total P, organic carbon (C), and total N. In this study, total P concentrations of soil were determined by the molybdenum-antimony colorimetric method after samples through digestion with 1.84 M H_2SO_4 [27]. For the determination of extractable P, we employed improved Hedely P fractionation methods to quantify soil P functional fractions [28,29]. Air-dried soil samples were processed to follow the soil P fractionation sequential procedure [15]. The corresponding supernatants sequentially exacted with anion exchange resin (weak base), 0.5 M $NaHCO_3$, 0.1 M NaOH, 0.1 M NaOH with sonication, 1.0 M HCl were collected by centrifuge at 1.7×10^4 m·s^{-1} (3200 rpm) for five minutes, followed by filtering samples through a 0.45-μm micropore filter. Phosphorus concentration in each supernatant was determined by the phosphomolybdic acid blue color method. The extractable P, including Resin-P, $NaHCO_3$-P, NaOH-P, sonication-P, and HCl-P, are defined as available P, soluble P, slowly released P, occluded P, and weathered mineral P, respectively, based on their functions in soils [15]. The residual P is the difference between total P and extractable P.

Soil pH was determined using a pH Meter with a soil:water ratio of 1:2.5. Soil organic carbon (SOC) was determined by dichromate oxidation and titration with ferrous ammonium sulfate [29]. Total N was measured using the microkjeldahl method after digestion with 1.84 M H_2SO_4. Plant samples were washed with dematerialized water to remove dust, oven-dried, and ground and screened with a 0.25 mm sieve. Total N and total P concentrations in plant tissues were determined using the micro-Kjeldahl method and molybdenum-antimony colorimetric method, respectively [29].

2.5. Statistical Analyses

The data were tested for homogeneity of variances (Brown and Forsythe's variation of Levene's test) before statistical analysis. In order to assess the Chinese fir fine root function on soil P supply, the rhizosphere effect (%) of P availability was defined as the difference between rhizosphere and bulk soils. Multiple-way analysis of variance (ANOVA) was used to determine the effect of stand type, site location, soil sources (rhizosphere vs. bulk soil), or tissue ages, and the interactions between stand type and site location on nutrient variables. One-way ANOVA and least significant difference (LSD) was used to compare the differences of mean values among the three orders of roots, or three age classes of needles and twigs. Pearson's tests were used for comparing the significance of correlations among soil P fractions and tree nutrient concentrations. SPSS 16.0 software (SPSS, Inc., Chicago, IL, USA) [30] was used to perform all analyses. The standard 0.05 level was used throughout as a cutoff for statistical significance.

3. Results

3.1. Soil General Properties

Organic C and total N concentrations in rhizosphere and bulk soils were generally higher in MP than PP in both sites. However, the C/N ratio of rhizosphere soil in PP at the Hunan site was higher than that in MP, while it showed the opposite trend at the Fujian site. The C/N ratio in bulk soil was not significantly different between PP and MP in both sites. Organic C and total N were generally higher in rhizosphere than bulk soil under all the treatments. Meanwhile, organic C and C/N in rhizosphere and bulk soils were generally higher in the Fujian than in the Hunan sites, but the total N was not significantly different between both sites. In addition, there is not a significant difference in pH between rhizosphere and bulk soils and among the four treatments (Table 2).

3.2. Phosphorus Fractions in Rhizosphere and Bulk Soils

Average concentrations of soil available P, slowly released P, and occluded P (not soluble P), mineral P and residual P reflected significantly higher concentrations in MP than PP whether in rhizosphere or bulk soil (Figure 1 and Table S1). Average concentrations of soil soluble P, slowly released P, and residual P were higher in rhizosphere than bulk soil, and these three P fractions showed positive rhizosphere effects in all treatments. In contrast, the rhizosphere effects of available P, occluded P, and mineral P varied with forest type and site location (Figure 1 and Table S1). Soil P fractions including slowly released P, occluded P, mineral P, and residual P, but not available P, and soluble P was generally higher in the Fujian than the Hunan sites (Figure 1 and Table S1).

Table 2. General chemical properties of pure Chinese fir plantations (PP) and mixed plantations with a broadleaved tree species (MP) in Hunan and Fujian provinces of China. Values are means and standard errors of four replicates).

Parameter	Hunan		Fujian	
	PP	MP	PP	MP
pH				
Rhizosphere soil	4.36 ± 0.20 a	4.39 ± 0.09 a	4.35 ± 0.03 a	4.29 ± 0.03 a
Bulk soil	4.36 ± 0.21 a	4.32 ± 0.08 a	4.33 ± 0.01 a	4.20 ± 0.03 a
Organic carbon (g kg^{-1})				
Rhizosphere soil	16.0 ± 0.1 c	18.1 ± 0.8b c	21.7 ± 1.1 b	34.4 ± 2.7 a
Bulk soil	11.3 ± 2.0 d	15.6 ± 1.6 c	18.9 ± 2.0 b	23.3 ± 1.1 a
Total nitrogen (g kg^{-1})				
Rhizosphere soil	0.6 ± 0.0 b	1.0 ± 0.0 a	0.8 ± 0.1 ab	1.0 ± 0.1 a
Bulk soil	0.5 ± 0.00 b	0.8 ± 0.1 a	0.6 ± 0.1b	0.7 ± 0.1 ab
C/N				
Rhizosphere soil	27.6 ± 1.0 c	18.0 ± 0.9 d	30.7 ± 5.3 b	36.0 ± 1.2 a
Bulk soil	19.1 ± 1.5 b	20.9 ± 2.6 b	33.7 ± 2.0 a	35.3 ± 4.4 a

Note: Different letters indicate significant differences among four treatments at probability level of $p < 0.05$.

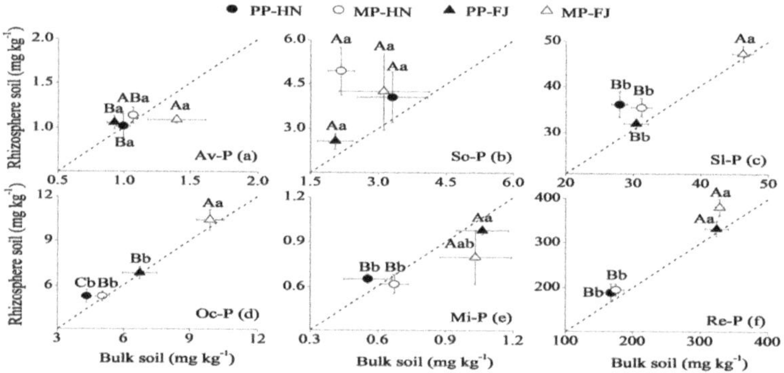

Figure 1. Variations in rhizosphere and bulk soil phosphorus fractions in pure Chinese fir plantations (PP) and mixed plantations with broadleaved tree species (MP) in Hunan and Fujian provinces of China. Note: The bar represents standard error and meaning of abbreviated words are as follow: PP-HN indicates the pure plantation in Hunan province, MP-HN indicates the mixed plantation in Hunan province, PP-FJ indicates the pure plantation in Fujian province, MP-FJ indicates the mixed plantation in Fujian province. Av-P indicates available P, So-P indicates soluble P, Sl-P indicates slowly released P, Oc-P indicates occluded P, Mi-P indicates mineral P, Re-P indicates residual P. The symbols located above and below the dotted line indicate the positive and negative rhizosphere effect, respectively. Among treatments, different small and capital letters indicate the significant differences in rhizosphere and bulk soils, respectively, probability level $p < 0.05$.

When the data from Hunan and Fujian were pooled, total P values were not significantly different between PP and MP either in rhizosphere or bulk soil, but the total extractable P was higher in MP than PP in both rhizosphere and bulk soils (Figure 2). The percentage of total extractable P to total P increased from 14.8% to 16.3% in bulk soil and from 13.8% to 16.8% in rhizosphere soil with the broadleaved tree species mixture. Among the five extractable fractions, slowly released P was the dominant form and its percentage to total extractable P ranged from 73.6% to 77.7%, followed by occluded P (11.2%~14.1%) in our study (Figure 2).

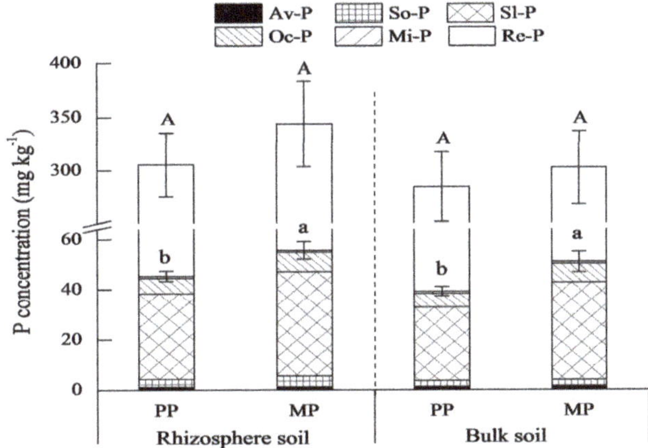

Figure 2. The components of soil phosphorus fractions in pure Chinese fir plantations (PP) and mixed plantations with broadleaved tree species (MP) in Hunan and Fujian provinces of China. Note: The bar represents standard error. Av-P indicates available P, So-P indicates soluble P, Sl-P indicates slowly released P, Oc-P indicates occluded P, Mi-P indicates mineral P, Re-P indicates residual P. Different small and capital letters indicate the significant differences of total extractable P (the sum of Av-P, So-P, Sl-P, Oc-P, and Mi-P) and total P, respectively, among the four soil sources (two forest types in rhizosphere or bulk soils, probability level $p < 0.05$.

3.3. Nutrient Levels in Different Root Orders

Total P concentration in absorptive and transportive roots is higher in PP than MP in the Fujian site, whereas it was not significantly different in roots between PP and MP in the Hunan site. Total N concentrations in absorptive and transportive roots were higher in PP than MP in each site of Hunan and Fujian provinces. Thus, the average N concentration in roots was also higher in PP than MP, although there were not statistical differences between both forest types in storative roots. The N/P ratio in all orders of roots seemed to be higher in PP than MP in the Hunan site (Figure 3 and Table S2). Meanwhile, total P and N concentrations were generally higher in absorptive than transportive and storative roots, and the N/P ratio was highest in storative roots, followed by transportive roots, and lowest in absorptive roots. In addition, the average P concentration was significantly higher in the Fujian than the Hunan site, while the average N concentration and N/P ratio showed the opposite trends (Figure 3).

3.4. Nutrient Distributions in Aboveground Tissues

Average P concentrations in the two-year old fresh needles were generally higher in PP than MP, although P concentrations in other aged needles were not significantly different between the two forest types (Figure 4a,b). In contrast, average N concentrations of all ages' needles and in each ages' needles were not significantly different between two forest types except higher N concentration in PP than MP in the two-year old fresh needles (Figure 4a,b and Table S3). Average P concentration in needles across all plots was highest in fresh needle of one-year old, followed by two- and three-year old needles and lowest in needle litter, while average N concentration was higher in fresh needles of one- and two-year old than three-year old needles and needle litter (Figure 4a,b). The ratio of N/P varied slightly between the two forest types and between the two site locations, but was higher in needle litter than fresh needles of various ages (Figure 4c).

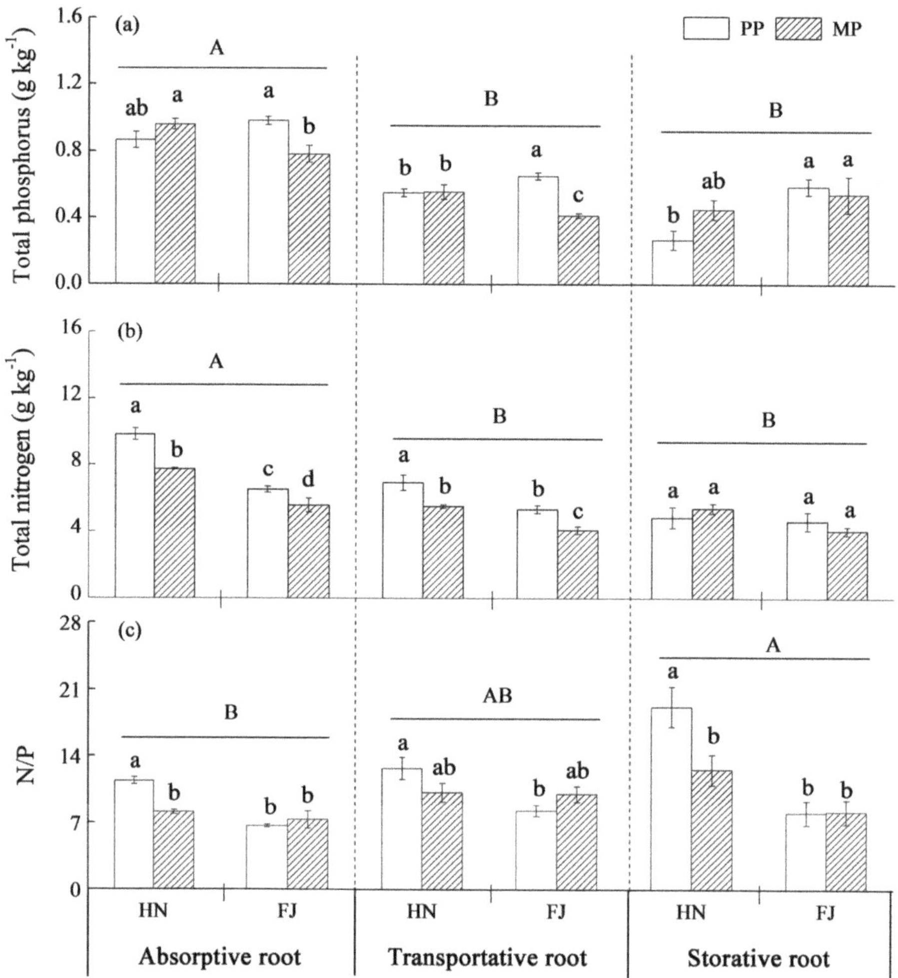

Figure 3. Phosphorus and nitrogen concentrations and N/P ratios of Chinese fir various functional roots in pure Chinese fir plantations (PP) and mixed plantations with broadleaved tree species (MP) in Hunan and Fujian provinces of China. (**a**) Total Phosphorus (g kg^{-1}), (**b**) Total nitrogen (g kg^{-1}), (**c**) N/P. Note: The bar represents standard error. Different small and capital letters indicate the significant differences among the four treatments within the root order, and among the three root orders, respectively, probability level $p < 0.05$.

Both P and N concentrations of twig were not significantly different between PP and MP, although N concentrations of the two- and three-year old twigs were significantly larger in PP than those in MP at the Hunan site. Additionally, those nutrient concentrations were highest in fresh twigs of one-year old, and lowest in twig litters, and significantly higher in Hunan than Fujian sites when just considered the effect of site location (Figure 5a,b and Table S4). The average N/P ratio was not significantly different between two forest types and between two site locations, but highest in twig litters and lowest in fresh twig of one-year old (Figure 5c and Table S4). In addition, the stem N and P concentrations and N/P ratio were not significantly different between both forest types except for a higher N/P ratio in PP than MP in the Fujian site (Figure 5).

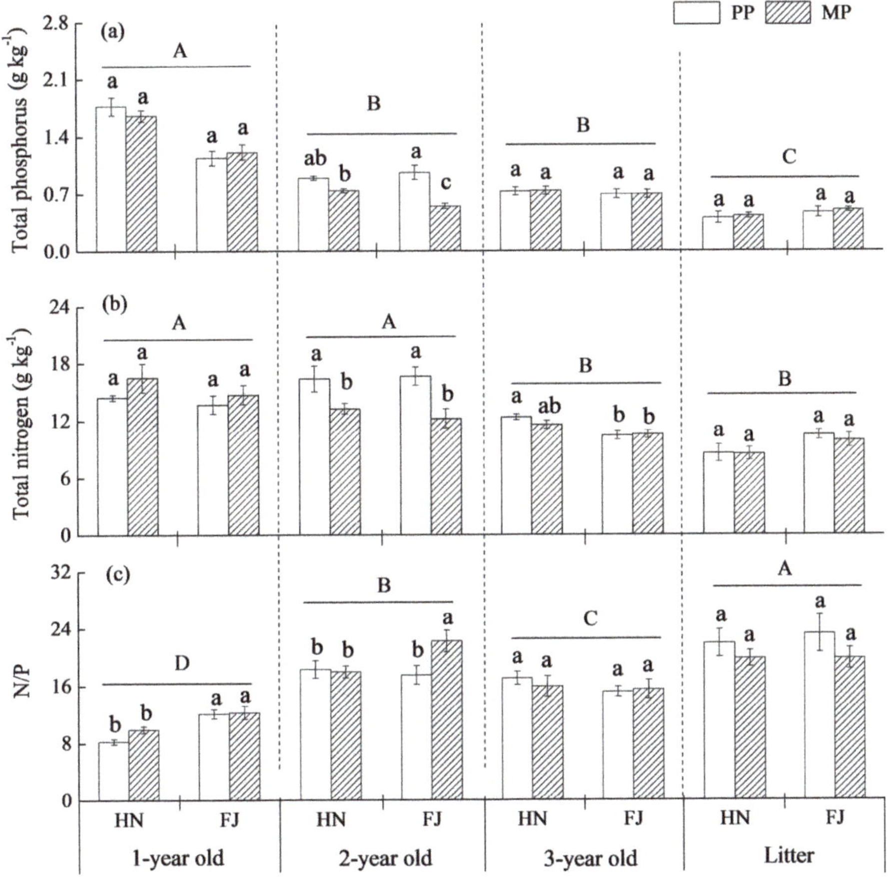

Figure 4. Phosphorus and nitrogen concentrations and N/P ratios of Chinese fir different-aged fresh needles and needle litters between pure Chinese fir plantations (PP) and mixed plantations with broadleaved tree species (MP) in Hunan and Fujian provinces of China. (**a**) Total Phosphorus (g kg^{-1}), (**b**) Total nitrogen (g kg^{-1}), (**c**) N/P. Note: The bar represents standard error. Different small and capital letters indicate the significant differences among the four treatments within a same age needle, and among these four age needles, respectively, probability level $p < 0.05$.

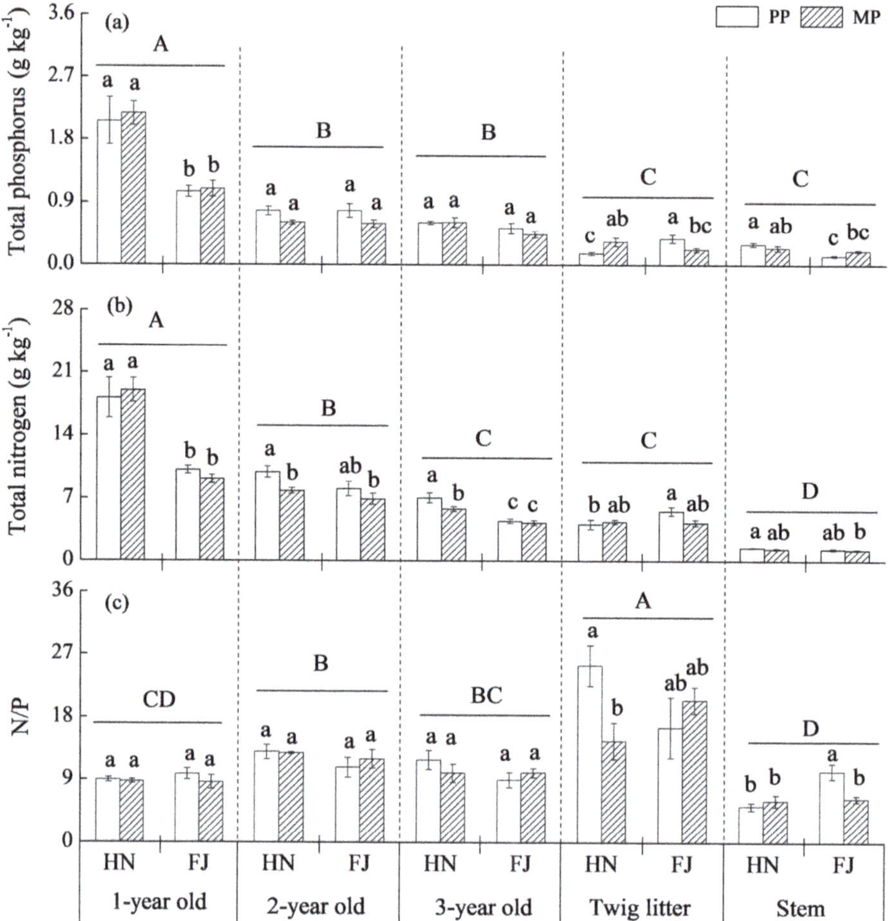

Figure 5. Phosphorus and nitrogen concentrations and N/P ratio of Chinese fir different-aged fresh twigs, twig litters, and stems between pure Chinese fir plantations (PP) and mixed plantations with broadleaved tree species (MP) in Hunan and Fujian provinces of China. (**a**) Total Phosphorus (g kg^{-1}), (**b**) Total nitrogen (g kg^{-1}), (**c**) N/P. Note: The bar represents standard error. Different small and capital letters indicate the significant differences among the four treatments within a same-aged twig, and among the four ages of twigs and stems, respectively, probability level $p < 0.05$.

3.5. Linkages Among Soil, Root and Needle Nutrients

Unlike our expectations, we did not find any positive relationships between all P fractions in rhizosphere soils and P concentrations in roots of different functions and needles of various ages. In contrast, slowly released P, occluded P, and residual P negatively correlated with the P concentrations in some functional roots and fresh needles of some ages, and the extractable P negatively correlated with P concentration in one-year old needles (Table 3). Meanwhile, the rhizosphere effect of some P fractions negatively correlated with P concentration in some roots, fresh needles, and needle litters except for a positive correlation between soluble P and P concentration in absorptive roots (Table 3).

Likewise, slowly released P, occluded P, mineral P, extractable P, and residual P in rhizosphere soils negatively correlated with N concentrations in transportive or storative roots, and in two- or three-year old needles. However, rhizosphere soil soluble P concentration was positively correlated with absorptive root N concentration, and both mineral P and extractable P were positively correlated with N concentration in one-year old needle (Table 4). Meanwhile, the rhizosphere effects of slowly released P and residual P were positively correlated with N concentrations in two- and three-year old needles, but negatively correlated with N concentration in one-year old needle. In addition, N concentration in needle litter was not significantly correlated with any P fractions and their rhizosphere effects (Table 4).

Table 3. Correlation efficiencies ($n = 16$) between rhizosphere soil phosphorus supplies (rhizosphere soil P fractions and their rhizosphere effects) and tree tissue phosphorus concentrations (total P of various Chinese fir tissues) under all studied stands in Hunan and Fujian provinces, China.

Variables	Total P of Roots			Total P of Fresh Needles			Total P of Needle Litter
	Absorp-tive	Transpor-tive	Storative	1-yr old	2-yr old	3-yr old	
P fractions in rhizosphere soil							
Av-P	-0.04^{ns}	-0.16^{ns}	-0.11^{ns}	-0.34^{ns}	-0.12^{ns}	0.05^{ns}	-0.21^{ns}
So-P	0.08^{ns}	-0.26^{ns}	-0.50^{ns}	0.19^{ns}	-0.26^{ns}	-0.16^{ns}	-0.35^{ns}
Sl-P	$\mathbf{-0.58}$ *	$\mathbf{-0.77}$ **	-0.11^{ns}	-0.21^{ns}	$\mathbf{-0.78}$ **	-0.02^{ns}	-0.04^{ns}
Oc-P	-0.39^{ns}	$\mathbf{-0.63}$ **	0.25^{ns}	$\mathbf{-0.64}$ **	$\mathbf{-0.62}$*	-0.10^{ns}	0.21^{ns}
Mi-P	0.34^{ns}	0.25^{ns}	0.25^{ns}	-0.38^{ns}	0.16^{ns}	0.17^{ns}	0.24^{ns}
Ex-P	-0.18^{ns}	-0.31^{ns}	0.47^{ns}	$\mathbf{-0.86}$ **	-0.40^{ns}	-0.06^{ns}	0.39^{ns}
Re-P	-0.49^{ns}	$\mathbf{-0.76}$ **	-0.11^{ns}	-0.29^{ns}	$\mathbf{-0.76}$ **	-0.06^{ns}	-0.04^{ns}
Rhizosphere effect of phosphorus fractions							
Av-P	0.43^{ns}	0.36^{ns}	-0.27^{ns}	-0.12^{ns}	0.43^{ns}	0.21^{ns}	-0.34^{ns}
So-P	$\mathbf{0.51}$ *	0.00^{ns}	-0.17^{ns}	0.08^{ns}	-0.08^{ns}	0.10^{ns}	-0.27^{ns}
Sl-P	-0.02^{ns}	0.02^{ns}	$\mathbf{-0.51}$ *	0.42^{ns}	0.21^{ns}	-0.03^{ns}	$\mathbf{-0.60}$ *
Oc-P	0.07^{ns}	-0.12^{ns}	-0.39^{ns}	-0.04^{ns}	0.05^{ns}	-0.18^{ns}	$\mathbf{-0.54}$ *
Mi-P	-0.02^{ns}	0.27^{ns}	0.01^{ns}	0.48^{ns}	0.35^{ns}	-0.08^{ns}	0.03^{ns}
Ex-P	0.15^{ns}	$\mathbf{-0.5\,2}$ *	-0.31^{ns}	0.06^{ns}	-0.39^{ns}	0.38^{ns}	-0.03^{ns}
Re-P	0.18^{ns}	0.03^{ns}	$\mathbf{-0.50}$ *	0.34^{ns}	0.19^{ns}	-0.06^{ns}	$\mathbf{-0.69}$ **

Note: Av-P indicates available P, So-P indicates soluble P, Sl-P indicates slowly released P, Oc-P indicates occluded P, Mi-P indicates mineral P, Re-P indicates residual P. ns indicates not significant, * $p < 0.05$, ** $p < 0.01$.

Table 4. Correlation efficiencies ($n = 16$) between rhizosphere soil phosphorus supplies (rhizosphere soil P fractions and their rhizosphere effects) and tree tissue nitrogen concentrations (total N of various Chinese fir tissues) under all studied stands in Hunan and Fujian provinces of China.

Variables	Total N of Roots			Total N of Fresh Needles			Total N of Needle Litter
	Absorp-tive	Transpor-tive	Storative	one-year old	two-year old	3-year old	
Phosphorus fractions in rhizosphere soil							
Av–P	-0.18^{ns}	-0.31^{ns}	-0.14^{ns}	-0.24^{ns}	-0.09^{ns}	-0.12^{ns}	-0.14^{ns}
So–P	$\mathbf{0.51}$ *	-0.33^{ns}	0.03^{ns}	-0.49^{ns}	0.11^{ns}	-0.10^{ns}	0.01^{ns}
Sl–P	-0.05^{ns}	$\mathbf{-0.55}$ *	-0.13^{ns}	-0.02^{ns}	-0.38^{ns}	-0.46^{ns}	-0.32^{ns}
Oc–P	-0.11^{ns}	-0.46^{ns}	$\mathbf{-0.60}$ *	0.34^{ns}	$\mathbf{-0.71}$ **	$\mathbf{-0.73}$ **	-0.39^{ns}
Mi–P	-0.07^{ns}	0.16^{ns}	-0.33^{ns}	$\mathbf{0.69}$ **	$\mathbf{-0.51}$ *	-0.28^{ns}	-0.33^{ns}
Ex–P	-0.31^{ns}	-0.36^{ns}	$\mathbf{-0.70}$ **	$\mathbf{0.50}$ *	$\mathbf{-0.76}$ **	$\mathbf{-0.72}$ **	-0.43^{ns}
Re–P	0.03^{ns}	$\mathbf{-0.57}$ *	-0.25^{ns}	-0.01^{ns}	-0.44^{ns}	$\mathbf{-0.54}$ *	-0.34^{ns}
Rhizosphere effect of phosphorus fractions							
Av–P	-0.12^{ns}	0.27^{ns}	0.11^{ns}	-0.12^{ns}	0.12^{ns}	0.17^{ns}	-0.14^{ns}
So–P	0.46^{ns}	-0.14^{ns}	-0.03^{ns}	-0.35^{ns}	-0.02^{ns}	-0.03^{ns}	0.11^{ns}
Sl–P	-0.09^{ns}	0.27^{ns}	$\mathbf{0.55}$ *	$\mathbf{-0.58}$*	$\mathbf{0.61}$ *	$\mathbf{0.72}$ **	0.24^{ns}
Oc–P	-0.06^{ns}	0.02^{ns}	-0.10^{ns}	-0.49^{ns}	0.31^{ns}	0.28^{ns}	0.01^{ns}
Mi–P	0.04^{ns}	0.47^{ns}	0.25^{ns}	0.07^{ns}	0.30^{ns}	$\mathbf{0.56}$ *	0.26^{ns}
Ex–P	0.38^{ns}	-0.47^{ns}	0.14^{ns}	-0.21^{ns}	0.12^{ns}	-0.24^{ns}	-0.28^{ns}
Re–P	0.05^{ns}	0.20^{ns}	0.40^{ns}	$\mathbf{-0.68}$ **	$\mathbf{0.54}$ *	$\mathbf{0.62}$ *	0.23^{ns}

Note: Av–P indicates available P, So–P indicates soluble P, Sl–P indicates slowly released P, Oc–P indicates occluded P, Mi–P indicates mineral P, Re–P indicates residual P. ns indicates not significant, * $p < 0.05$, ** $p < 0.01$.

4. Discussion

Overall, there is a trend of converting pure plantations into mixed plantations for increasing stability and sustainability due to, in theory, complementary resource use, environmental benefits, and soil improvement in mixed forests [5]. However, the practical effect does not always match with our expectation. Until now, there have been few reports on the negative effect of mixture with broadleaved trees and coniferous trees [31].

4.1. Mixture Effect on Nutrient Concentrations in Rhizosphere and Bulk Soils

Our results showed that organic C, total N, and total extractable P in both rhizosphere and bulk soils were higher in MP than PP. In other words, mixed Chinese fir plantations with broadleaved species increased nutrient concentrations in both rhizosphere and bulk soils. These results were in line with the hypothesis that the mixture with broadleaved trees improves soil P availability in Chinese fir plantations. Greater litter production or changes in the timing of litter inputs or differences of leaf nutrient concentration in mixed-species plantations could increase the soil nutrient supply, relative to monocultures, if decomposition rates were constant or faster [30]. Generally, the decomposition rate of litter and fine roots for coniferous species are slower than broadleaf species [32]. Additionally, some studies support that the decomposition rate of leaves and roost for *M. macclurei* and *S. superba* were faster than those of Chinese fir [33,34]. Moreover, compared with a pure Chinese fir plantation, the soil fertility and nutrient return were raised significantly in the mixed Chinese fir plantation with *M. macclurei*, due to substantial amount of litter and higher turnover rate of fine root [24]. Therefore, broadleaved trees with higher nutrient concentrations in the mixture could increase soil nutrient availability by altering the amount and quality of litter input as well as the amount and chemical composition of the root mass and exudates through fine root decomposition.

4.2. Mixture Effect Varies with Soil P Fraction

The average concentrations of soil available P, slowly released P, and occluded P, but not soluble P, mineral P, and residual P, were significantly higher in MP than PP and three of these showed positive rhizosphere effects. Since P is an element of depositional cycle, P absorption by plants depends on its concentration gradient and diffusivity in the soil near the roots. Both available P and soluble P are labile and are considered as the available fractions for plant growth [35]. Some studies have found that amending phosphates increases the immediate phosphorus availability and the rate of available P dissolution can be enhanced by the rhizosphere effect [36]. Moreover, the litter decomposition experiment of *S. superba* on a Chinese fir plantation shows that P concentration in the *S. superba* litter dropped rapidly, which could release 32% of the initial P into the soil during the first three months in Hunan Province [37].

Furthermore, the total extractable P was higher in MP than PP in both rhizosphere and bulk soils through the introduction of a broadleaf species in the mixed plantation. Recent studies indicated that the inactive P fractions could be converted into plant available forms with the help of the necessary manipulation of the rhizosphere environment [38]. The release of root exudates, such as organic ligands, is an activity of the root that can alter the concentration of P in the soil solution [12]. Some studies showed that cyclic dipeptides, which caused autoinhibition of Chinese fir, may be released into the soil through litter decomposition and root exudation. Moreover, root exudates provided more contributions to soil cyclic dipeptide levels than litter in Chinese fir plantations [39]. Thus, the introduction of broadleaf species to a pure Chinese fir plantation may relieve the autoinhibition of Chinese fir and alter the concentration of P in the soil solution. Furthermore, *S. superba* is a Mn-accumulating subtropical tree species and Mn hyperaccumulation is associated with Mn mobilization in the rhizosphere, most likely due to the release of protons. The carboxylates generated to produce the protons released into the rhizosphere are used internally in the plant to mobilize soil inorganic and organic P [40]. Therefore, the overall combination of changes in the P fractions in soil demonstrates that a mixture effect may induce an increase in soil P availability through the solubility of the unavailable form of soil P and the rhizosphere effect. After all, the mixture effect on P availability in rhizosphere soil varies with the P fraction, and may depend on the litter decomposition and root exudates of introduced broadleaf species.

4.3. The Response of Chinese Fir to Mixture Effect Varies with Root Orders, Needle, and Twig Ages

Our results showed that P and N concentrations in one-year old fresh needles and one-year old twigs were not significantly different between PP and MP in both study sites, although the mixture effect may induce an increase in soil P availability. Generally, evergreen trees tend to maintain a relatively favorable nutrient status in active young leaves for positive carbon (C) gain and high N use efficiency [41]. Trees can transport nutrients from old and senescing leaves to new organs to support new growth, which has been verified as a key mechanism of nutrient conservation and reuse in plants [42]. Thus, old leaves might be more sensitive to environmental variation, such as those mixed with a broadleaf species than young leaves. Generally, P and N concentrations in absorptive and transportive roots, two-year old fresh needles and two-year old twigs were significantly higher in PP than MP, although these concentrations in the other tissues were not different between both stand types. These results suggest that the introduction of a broadleaf species to pure Chinese fir plantation might decrease the nutrient concentration of Chinese fir. In contrast, most of studies suggested that mixed species stands have higher nutrient availability than monoculture, even in the absence of N-fixing species [43], though few studies have contrasted nutrient concentrations for species in mixtures [21]. Therefore, foliar nutrient concentrations did not show a general trend in the mixtures studied so far. Moreover, since the Chinese fir biomass in the mixed plantation was lower than that in the pure Chinese fir stand in our study, we deduced that lower P concentration in Chinese fir tissue may be attributing to the stronger competition with the broadleaved tree. Obviously, our result did not support the hypothesis that the mixture can increase tree tissue nutrient availability but, rather, supported the hypothesis that the response of Chinese fir nutrients to the mixed with broadleaved tree varies with root orders, and needle and twig ages. These results may be caused by the lower competition of Chinese fir than broadleaved species for nutrient uptake.

4.4. Linkages between Soil, Root, and Needle Nutrients

For the linkages among soil, root, and needle nutrients, we did not find any positive relationships between all P fractions in rhizosphere soil and P concentration in roots of different orders and in leaves of various ages. In contrast, slowly released P, occluded P, and residual P negatively correlated with the P concentrations in some root orders and fresh leaves of some ages, and the extractable P negatively correlated with P concentration in one-year old needle. Many studies suggest that any shift in plant species composition, which is able to alter soil nutrient stoichiometry, can influence rhizosphere microbial and soil enzyme activities, which can further induce plant community species shifts and alter ecosystem function [44]. Two adjacent plants can simultaneously participate in competition and facilitation processes, and the direction and intensity of plant interactions are determined by the sum of the co-occurring negative and positive effects of one to another. These negative linkages may cause by the negative plant–plant interactions between Chinese fir and the broadleaf species. In general, plants take up most mineral nutrients through the rhizosphere where microorganisms interact with root exudates. A study reported that the seedling survivorship of *S. superba* was significantly inhibited by *eucalyptus* robusta Smith litter addition alone, meanwhile the seedling height of *S. superba* and *M. macclurei* was significantly suppressed when eucalyptus roots were present [45]. This negative nutrient feedback (homeostasis) supports plant species coexistence with lower proportional changes in consumer stoichiometry compared with resource stoichiometry [46]. Evaluating homeostatic relationships can provide valuable insight into assessing plant competition or plant coexistence [47]. Therefore, negative relationships between P fractions in rhizosphere soil and tree tissue nutrients of Chinese fir may be caused by plant competition between Chinese fir and the broadleaved species.

5. Conclusions

We found that, compared with the Chinese fir plantation, the mixed Chinese fir plantation with a broadleaved species increased nutrient availability in both rhizosphere and bulk soils. Meanwhile, the introduction of broadleaf species to pure Chinese fir plantation generally decreased the tree nutrient availability of the transportive root and two-year old needles of Chinese fir, possibly due to the lower competition of Chinese fir than broadleaved species for nutrient uptake. Therefore, the effect of mixture may bring about a negative nutrient feedback between P availability in rhizosphere soil and P concentration in plant tissues attributed to root exudates and plant competition between Chinese fir and broadleaf species. In conclusion, replacing monoculture plantations of Chinese fir into mixed-species forest promotes soil nutrient availability, and it might be an option for multi-purpose forest management in China, from where the largest areas of Chinese fir plantations alter the functioning of the soil ecosystem. However, the mixed forest may affect P and N stocks of Chinese fir tissues. Thus, the mixture effect may differ in soil and plant nutrients, and when soil nutrient availability is expected to be improved, converting a pure conifer plantation into a mixed-species forest can be considered.

Supplementary Materials: The following are available online at http://www.mdpi.com/1999-4907/11/4/461/s1, Figure S1: The study sites in Huitong Forest Ecological Station of Hunan Province and Datian Experimental Forest Farm of Fujian Province, Table S1: Summary of ANOVA about effects of forest type, soil type, site location and their interaction on phosphorus fractions under pure Chinese fir plantation (PP) and its mixed plantation with broadleaved tree species (MP) at Hunan and Fujian provinces, China, Table S2: ANOVA of effect of forest type, root functional order, site location and their interaction on root nitrogen, phosphorus concentrations and N/P between pure Chinese fir plantation (PP) and mixed plantation with broadleaved tree species (MP) at Hunan and Fujian province, China.. Table S3: ANOVA of effect of forest type, leaf age, site location and their interaction on leaf nitrogen, phosphorus concentrations and N/P ratio between pure Chinese fir plantation (PP) and mixed plantation with broadleaved tree species (MP) at Hunan and Fujian province, China. Table S4: ANOVA of effect of forest type, twig age, site location and their interaction on twig nitrogen, phosphorus concentrations and N/P ratio between pure Chinese fir plantation (PP) and mixed plantation with broadleaved tree species (MP) at Hunan and Fujian province, China.

Author Contributions: F.-S.C. contributed to the study conception and design. Material preparation, data collection and analysis were performed by W.-S.B., H.-J.G., C.-c.Z., Y.Z., A.N.S., X.-M.F., J.F., H.-M.W. and F.-S.C. The first draft of the manuscript was written by W.-S.B. and all authors commented on previous versions of the manuscript. All authors have read and agreed to the published version of the manuscript

Funding: This study was supported by grants from the National Natural Science Foundation of China (31730014, 31,870,427, and 31760134), Jiangxi Provincial Department of Science and Technology (20181ACH80006), the China Scholarship Council (201908360227), and the Jiangxi Province Science Foundation for Youths (20181BAB214014). We greatly appreciate Xiaoli Fu and Shebao Yu for their help in field sampling.

Conflicts of Interest: The authors declare no conflict of interest.

References

1. Xu, J. The 8th forest resources inventory results and analysis in China. *For. Econ.* **2004**, *24*, 6–8.
2. Wen, L.; Lei, P.; Xiang, W.; Yan, W.; Liu, S. Soil microbial biomass carbon and nitrogen in pure and mixed stands of Pinus massoniana and Cinnamomum camphora differing in stand age. *Ecol. Manag.* **2014**, *328*, 150–158. [CrossRef]
3. Chen, H. Phosphatase activity and P fractions in soils of an 18−year−old Chinese fir (*Cunninghamia lanceolata*) plantation. *For. Ecol. Manag.* **2003**, *178*, 301–310. [CrossRef]
4. Ma, X.; Heal, K.V.; Liu, A.; Jarvis, P.G. Nutrient cycling and distribution in different−aged plantations of Chinese fir in southern China. *For. Ecol. Manag.* **2007**, *243*, 61–74. [CrossRef]
5. Meng, J.; Lu, Y.; Zeng, J. Transformation of a degraded *Pinus massoniana* plantation into a mixed-species irregular forest: Impacts on stand structure and growth in Southern China. *Forests* **2014**, *5*, 3199–3221. [CrossRef]
6. Nunes, L.; Coutinho, J.; Nunes, L.F.; Rego, F.C.; Lopes, D. Growth, soil properties and foliage chemical analysis comparison between pure and mixed stands of *Castanea saliva* Mill.; *Pseudotsuga menziesii* (Mirb.) Franco, in Northern Portugal. *System* **2011**, *20*, 496–507.

7.   Yang, X.; Thornton, P.E.; Ricciuto, D.M.; Post, W.M. The role of phosphorus dynamics in tropical forests—A modeling study using CLM–CNP. *Biogeosciences* **2014**, *11*, 1667–1681. [CrossRef]

8.   Zou, X.; Wu, P.; Chen, N.; Wang, P.; Ma, X. Chinese fir root response to spatial and temporal heterogeneity of phosphorus availability in the soil. *Can. J. For. Res.* **2015**, *45*, 402–410. [CrossRef]

9.   Shenoy, V.; Kalagudi, G. Enhancing plant phosphorus use efficiency for sustainable cropping. *Biotechnol. Adv.* **2005**, *23*, 501–513. [CrossRef]

10.   Hasegawa, M.; Ota, A.T.; Kabeya, D.; Okamoto, T.; Saitoh, T.; Nishiyama, Y. The effects of mixed broad–leaved trees on the collembolan community in larch plantations of central Japan. *Appl. Soil. Ecol.* **2014**, *83*, 125–132. [CrossRef]

11.   Pii, Y.; Mimmo, T.; Tomasi, N.; Terzano, R.; Cesco, S.; Crecchio, C. Microbial interactions in the rhizosphere: Beneficial influences of plant growth–promoting rhizobacteria on nutrient acquisition process. A review. *Biol. Fert. Soils* **2015**, *51*, 403–415. [CrossRef]

12.   Hinsinger, P. Bioavailability of soil inorganic P in the rhizosphere as affected by root–induced chemical changes: A review. *Plant Soil* **2001**, *237*, 173–195. [CrossRef]

13.   Brzostek, E.R.; Greco, A.; Drake, J.E.; Finzi, A.C. Root carbon inputs to the rhizosphere stimulate extracellular enzyme activity and increase nitrogen availability in temperate forest soils. *Biogeochemistry* **2013**, *115*, 65–76. [CrossRef]

14.   Dotaniya, M.L.; Meena, V.D. Rhizosphere effect on nutrient availability in soil and Its uptake by plants: A Review. *Proc. Nat. Acad. Sci. India Sect. B Biol. Sci.* **2015**, *85*, 1–12. [CrossRef]

15.   Hu, X.F.; Chen, F.S.; Nagle, G.; Fang, Y.T.; Yu, M.Q. Soil phosphorus fractions and tree phosphorus resorption in pine forests along an urban–to–rural gradient in Nanchang, China. *Plant Soil* **2011**, *346*, 97–106. [CrossRef]

16.   Redondo-Brenes, A.; Montagnini, F. Growth, productivity, aboveground biomass, and carbon sequestration of pure and mixed native tree plantations in the Caribbean lowlands of Costa Rica. *For. Ecol. Manag.* **2006**, *232*, 168–178.

17.   Lemma, B. Soil chemical properties and nutritional status of trees in pure and mixed-species stands in south Ethiopia. *J. Plant Nutr. Soil Sci.* **2012**, *175*, 769–774. [CrossRef]

18.   Chen, F.S.; Niklas, K.J.; Liu, Y.; Fang, X.M.; Wan, S.Z.; Wang, H. Nitrogen and phosphorus additions alter nutrient dynamics but not resorption efficiencies of Chinese fir leaves and twigs differing in age. *Tree Physiol.* **2015**, *35*, 1106–1117. [CrossRef]

19.   Pretzsch, H. Canopy space filling and tree crown morphology in mixed-species stands compared with monocultures. *For. Ecol. Manag.* **2014**, *327*, 251–264. [CrossRef]

20.   Montagnini, F. Accumulation in above–ground biomass and soil storage of mineral nutrients in pure and mixed plantations in a humid tropical lowland. *For. Ecol. Manag.* **2000**, *134*, 257–270. [CrossRef]

21.   Rothe, A.; Binkley, D. Nutritional interactions in mixed species forests: A synthesis. *Can. J. For. Res.* **2001**, *31*, 1855–1870. [CrossRef]

22.   Estiarte, M.; Peñuelas, J. Alteration of the phenology of leaf senescence and fall in winter deciduous species by climate change: effects on nutrient proficiency. *Glob. Chang. Biol.* **2015**, *21*, 1005–1017. [CrossRef] [PubMed]

23.   Fu, X.; Wang, J.; Di, Y.; Wang, H. Differences in Fine–Root Biomass of Trees and Understory Vegetation among Stand Types in Subtropical Forests. *PLoS ONE* **2015**, *10*, e0128894. [CrossRef] [PubMed]

24.   Wang, Q.; Wang, S. Soil microbial properties and nutrients in pure and mixed Chinese fir plantations. *J. Res. (Harbin)* **2008**, *19*, 131–135. [CrossRef]

25.   Guo, D.; Xia, M.; Wei, X.; Chang, W.; Liu, Y.; Wang, Z. Anatomical traits associated with absorption and mycorrhizal colonization are linked to root branch order in twenty–three Chinese temperate tree species. *New Phytol.* **2008**, *180*, 673–683. [CrossRef]

26.   Killingbeck, K.T. Nutrients in senesced leaves: Keys to the search for potential resorption and resorption proficiency. *Ecology* **1996**, *77*, 1716–1727. [CrossRef]

27.   Allen, S.E.; Grimshaw, H.M.; Parkinson, J.A.; Quarmby, C. *Chemical Analysis of Ecological Materials*; Blackwell Scientific Publications: Oxford, UK, 1974.

28.   Chen, F.S.; Li, X.; Nagle, G.; Zhan, S.X. Topsoil phosphorus signature in five forest types along an urban–suburban–rural gradient in Nanchang, southern China. *J. Res.* **2010**, *21*, 39–44. [CrossRef]

29.   Hedley, M.J.; Stewart, J.W.B.; Chauhan, B.S. Changes in inorganic and organic soil phosphorus fractions induced by cultivation practices and by laboratory incubations. *Soil Sci. Soc. Am. J.* **1982**, *46*, 970–976. [CrossRef]

30. Spss, I. *SPSS for Windows (16.0)*; SPSS Inc: Chicago, IL, USA, 2007.
31. Richards, A.E.; Forrester, D.I.; Bauhus, J.; Scherer-Lorenzen, M. The influence of mixed tree plantations on the nutrition of individual species: a review. *Tree Physiol.* **2010**, *30*, 1192–1208. [CrossRef]
32. Eshel, A.; Beeckman, T. *Plant Roots: The Hidden Half*; CRC Press: Bocaraton, FL, USA, 2013.
33. Wang, Q.; Wang, S.; Fan, B.; Yu, X. Litter production, leaf litter decomposition and nutrient return in *Cunninghamia lanceolata* plantations in south China: effect of planting conifers with broadleaved species. *Plant Soil* **2007**, *297*, 201–211. [CrossRef]
34. Wang, Q.; Wang, S.; Huang, Y. Leaf litter decomposition in the pure and mixed plantations of *Cunninghamia lanceolata* and *Michelia macclurei* in subtropical China. *Biol. Fert. Soils* **2009**, *45*, 371–377. [CrossRef]
35. Li, G.; Li, H.; Leffelaar, P.A.; Shen, J.; Zhang, F. Dynamics of phosphorus fractions in the rhizosphere of fababean (*Vicia faba* L.) and maize (*Zea mays* L.) grown in calcareous and acid soils. *Crop Pasture Sci.* **2017**, *66*, 1151. [CrossRef]
36. Phillips, R.P.; Fahey, T.J. Tree species and mycorrhizal associations influence the magnitude of rhizosphere effects. *Ecology* **2006**, *87*, 1302–1313. [CrossRef]
37. Zhong, M.; Wang, Q.; Gao, H.; Yu, X. Decomposition and nitrogen and phosphorus release of leaf litters from main tree species in a mid–subtropical forest. *Chin. J. Ecol.* **2013**, *32*, 1653–1659.
38. Gerke, J. The acquisition of phosphate by higher plants: Effect of carboxylate release by the roots. A critical review. *J. Plant Nutr. Soil Sci.* **2015**, *178*, 351–364. [CrossRef]
39. Chen, L.; Kong, C. Autoinhibition and soil allelochemical (cyclic dipeptide) levels in replanted Chinese fir (*Cunninghamia lanceolata*) plantations. *Plant Soil* **2014**, *374*, 793–801. [CrossRef]
40. Lambers, H.; Hayes, P.E.; Laliberte, E.; Oliveira, R.S.; Turner, B.L. Leaf manganese accumulation and phosphorus–acquisition efficiency. *Trends Plant Sci.* **2015**, *20*, 83–90. [CrossRef]
41. Di, Y.; Fu, X.; Wang, H.; Li, W.; Wang, S. N concentration of old leaves and twigs is more sensitive to stand density than that of young ones in Chinese fir plantations: A case study in subtropical China. *J. For. Res.* **2018**, *29*, 163–169. [CrossRef]
42. Wang, M.; Murphy, M.T.; Moore, T.R. Nutrient resorption of two evergreen shrubs in response to long–term fertilization in a bog. *Oecologia* **2014**, *174*, 365–377. [CrossRef]
43. Zeugin, F.; Potvin, C.; Jansa, J.; Scherer-Lorenzen, M. Is tree diversity an important driver for phosphorus and nitrogen acquisition of a young tropical plantation? *For. Ecol. Manag.* **2010**, *260*, 1424–1433. [CrossRef]
44. De Graaff, M.A.; Classen, A.T.; Castro, H.F.; Schadt, C.W. Labile soil carbon inputs mediate the soil microbial community composition and plant residue decomposition rates. *New Phytol.* **2010**, *188*, 1055–1064. [CrossRef] [PubMed]
45. Zhang, C.; Fu, S. Allelopathic effects of eucalyptus and the establishment of mixed stands of eucalyptus and native species. *For. Ecol. Manag.* **2009**, *258*, 1391–1396. [CrossRef]
46. Bell, C.; Carrillo, Y.; Boot, C.M.; Rocca, J.D.; Pendall, E.; Wallenstein, M.D. Rhizosphere stoichiometry: are C: N: P ratios of plants, soils, and enzymes conserved at the plant species-level? *New Phytol.* **2014**, *201*, 505–517. [CrossRef] [PubMed]
47. Elgersma, K.J.; Yu, S.; Vor, T.; Ehrenfeld, J.G. Microbial–mediated feedbacks of leaf litter on invasive plant growth and interspecific competition. *Plant Soil* **2012**, *35*, 341–355. [CrossRef]

Article

Understory Plant Functional Types Alter Stoichiometry Correlations between Litter and Soil in Chinese Fir Plantations with N and P Addition

Junyi Xie, Haifu Fang, Qiang Zhang, Mengyun Chen, Xintong Xu, Jun Pan, Yu Gao, Xiangmin Fang, Xiaomin Guo and Ling Zhang *

Jiangxi Provincial Key Laboratory of Silviculture, Jiangxi Agricultural University, Nanchang 330045, China
* Correspondence: lingzhang@jxau.edu.cn or lingzhang09@126.com; Tel.: +86-79183813243

Received: 25 July 2019; Accepted: 23 August 2019; Published: 28 August 2019

Abstract: Research Highlights: This study identifies the effect of nitrogen (N) and phosphorus (P) addition on stoichiometry correlations between understory plants and soil in subtropical Chinese fir plantations. Background and Objectives: Nitrogen and P are two nutrients limiting forest ecosystem production. To obtain more wood production, N and P are usually applied in plantation management. Changes in soil N and P will generally alter the stoichiometric characteristics of understory plants, which control carbon (C) and nutrient cycles between plants and soil. However, different correlations between plant and soil stoichiometry among functional groups of understory plants have not been investigated, which also impacted element cycling between plants and soil. Materials and Methods: Subtropical Chinese fir plantations were selected for N (100 kg ha^{-1} year^{-1}) and P (50 kg ha^{-1} year^{-1}) addition study. We collected fresh litter and the corresponding soil of four understory plants (*Lophatherum gracile* Brongn., *Woodwardia japonica* (L.f.) Sm., *Dryopteris atrata* (Kunze) Ching and *Dicranopteris dichotoma* (Thunb.) Berhn.) for study of C, N, and P stoichiometric ratios. Results: Nitrogen and P addition affected C, N, and P concentrations and stoichiometric ratios in litter and soil as well as correlations between litter and soil stoichiometric ratios. Understory plant species with different functional types impacted the correlations between plants and soil in C, N, and P stoichiometric ratios, especially correlations between litter C and soil C and N. Conclusions: Changes in soil N and P affect the stoichiometric ratios of understory plants. Functional groups impacted the correlation in C, N, and P stoichiometric ratios between plants and soil, indicating functional groups varied in their impacts on element cycling between plants and soil in plantations with exogenous nutrient addition, which should be considered in future management of plantations with intensive fertilization practice.

Keywords: nitrogen and phosphorus addition; understory plants; stoichiometric ratio; litter decomposition

---

1. Introduction

Combining stoichiometry with biology, chemistry, and physics, ecological stoichiometry addresses the balance of carbon (C), nitrogen (N), and phosphorus (P) [1]. Carbon, N, and P, which act on nutrient cycling and utilization, play a very important role in the nutrient cycling process of the ecosystem and ecological stoichiometry [2]. Thus, numerous scholars focus on ecological stoichiometry [3]. Studying the stoichiometric characteristics of C, N, and P of the forest ecosystem can provide insight into the stability of the forest ecosystem in a certain region; in addition, it can clarify the proportion of the demand for environmental nutrients by plants and the drivers of change. Moreover, studying the stoichiometric characteristics in forest ecosystems can uncover the amount of nutrients returned to the soil from plant litter, as well as the nutrient supply of soil to plants in the ecosystem [4,5]. The C:N:P ratio of forest ecosystems has been shown to be significantly different with different vegetation types, climatic factors, and environmental factors [3,6,7]. The input of N and P directly changes the environmental factors of the forest ecosystem, and the addition of N and P affects the structure, function, and nutrient cycles of forest ecosystem [8]. Nitrogen inputs affect forest ecosystems in different ways. For example, N addition may cause an increase in leaf biomass and photosynthetic efficiency, and reduce the number of thin roots [9,10]. Nitrogen and P addition acts directly on leaves. Nitrogen and P addition leads to changes in the correlation of the stoichiometric ratio between plants and soil, affecting plant productivity and species diversity in communities, finally resulting in a change in the forest ecosystem [11–14]. Nitrogen addition reduced the C:N ratio and increased the N:P ratio of leaves, thus affecting the N and P balance in plants, resulting in the restriction of plant growth by P or N and P, thus reducing the productivity of the ecosystem [14,15]. The addition of P to the soil generally increases the P content in plant tissues and reduces the N:P ratio [16,17]. Nitrogen addition promotes P absorption in subtropical plants and accelerates plant growth, but the results are different from those in some temperate forests [18–20]. Therefore, to a large extent, the nutrient status of leaves can reflect a plant's environmental status; there is a close relationship between leaf nutrient status and soil stoichiometric ratio [21].

Research has shown that the effect of N and P addition on vegetation type was related to ecosystem type, time, and dose of N and P addition [22–24]. At present, research is increasing on the nutrient circulation of forest ecosystems with a focus on ecological stoichiometry. Due to the improvement of forest quality and the advantage of high canopy productivity, there are many studies on the C, N, and P ecological stoichiometry of the plant-litter-soil system [25,26]. However, most previously published studies have focused on overstory plant species, largely ignoring understory plant species, which have a faster C and N turnover rate than the overstory plants. Moreover, understory species tend to be more sensitive to environmental changes, which limits our understanding of the nutrient cycles of ecosystems [27]. Under the current state of global climate change, understory vegetation plays a very important role in the C and N cycle of forest ecosystems and helps to maintain forest biodiversity and ecosystem stability [28]. Understory vegetation plays an important ecological role in nutrient cycling, soil physical and chemical properties, and community succession [29–31]. Previous research has shown that the addition of N and P to soil changed the species richness of understory vegetation and reduced plant diversity, highlighting that more research is needed on common understory plant species [32–34].

Chinese fir (*Cunninghamia lanceolata* (Lamb.) Hook.) is widely distributed in 14 provinces in China, including Zhejiang, Jiangxi, and Hunan [35–37]. Chinese fir has a history of more than 1000 years of planting because it is one of the most important fast-growing timber species in southern China; therefore, it plays an irreplaceable role in economic and ecological benefits [38]. In recent years, the management of understory vegetation in plantations and the application of fertilizer have significantly influenced the productivity of forest ecosystems [39,40]. However, the stoichiometric responses of subtropical fir plantation litter and soil to persistent exogenous N and P input are still not clear. Therefore, under the global climate change, in order to improve our understanding for the interaction between plant and soil in plantation forests ecosystem, we studied the response of C, N, and P concentrations and stoichiometric characteristics of four representative plants (two different functional types) in the lower layer of Chinese fir plantation to the addition of N and P in the subtropical red land area. We try to solve the following problems: (1) What is the reaction of C, N, P concentration and stoichiometric ratio in plant litter and soil to N and P addition; (2) How does N or P addition affect the stoichiometric ratios of different functional plants litter and soil?

2. Materials and Methods

2.1. Study Sites

This study was conducted at the Qianyanzhou Ecological Research station in Taihe County, Ji'an City, Jiangxi Province, China (115°03′29.2″ E, 26°44′29.1″ N). The research site is a typical red hilly region with subtropical monsoonal climate. The annual average precipitation is up to 1600 mm and the annual average temperature is approximately 18 °C [41]. The experimental site is within the native distribution of Chinese fir. The study was conducted in an experimental Chinese fir plantation that was established approximately 15 years ago. The coverage of the sample plot was between 75 and 85%, and the slope was between 20 and 30°. The natural atmospheric N deposition in this area was 49 kg N ha^{-1} year^{-1} [42].

2.2. Experimental Design

The experiment began in November 2011, and four treatments (control: CK, N addition, P addition, and N and P addition together) were established. These four treatments were arranged in a random complete block design, with a total of four blocks. There were 16 of 20 × 20 m plots. To prevent influence between the plots, the spacing between plots was greater than 20 m. Nitrogen and P were applied first in December 2011 and then four times each year (March, June, September, and December). Nitrogen (100 kg ha^{-1} year^{-1}) was added in the form of NH_4NO_3, and P (50 kg ha^{-1} year^{-1}) was added in the form of NaH_2PO_4 [43]. To more evenly distribute N and P addition plots in the study site, we added sand (8 kg plot^{-1}) in the N plots, phosphate plots, and control (CK) plots. For fertilization we chose a date where the forecast predicted no rain for four consecutive days [27].

2.3. Sample Collection

In November 2018, we investigated the presence or absence of understory species in the Chinese fir plantation in each plot. From this preliminary data, we selected four dominant species (*Lophatherum gracile* Brongn., *Woodwardia japonica* (L.f.) Sm., *Dryopteris atrata* (Kunze) Ching and *Dicranopteris dichotoma* (Thunb.) Berhn.), which appeared in all 16 plots and formed communities, as sampling targets. Four communities were randomly selected in each plot. To collect plant samples, five individual plants and the corresponding soils were collected from each plot. These samples were immediately brought back to the laboratory in cooler. After sampling, the four collected plants were divided into different functional types (Gramineae and ferns) for nutritional analysis.

The leaves of each plant from all plots were collected and combined to produce one leaf sample. Leaf samples were air-dried, and then ground to pass through 0.149 mm sieve for determination of C, N, and P. Sub-sample of air-dried leaves were oven dried at 60 °C to obtain dry mass of measured leaves without residual water content. Soil samples were also processed by removing visible plant debris and rocks, passed through a 0.149 mm sieve to determine elements, or passed through a 2 mm sieve for determination of available nutrients. Sub-samples of soils were oven dried to a constant weight to obtain water content in the air-dried soil.

2.4. Litter and Soil Nutrient Measurement

Fresh soil samples through a 2 mm sieve were used for the determination of soil dissolved organic C (DOC) (VarioTOC, Elementar, Germany) and available N (AN, including NH_4^+ and NO_3^-) [44]. Soil/water (2:5) was used to determine the pH value of soil [44]. Air-dried soil samples that had been passed through a 2 mm sieve were used for the determination of soil available P (AP) [44]. We measured AN and AP with a flow injection auto analyzer (Smartchem 200 Alliance Corp. France). Before analyzing total organic C (TOC), total N (TN), and total P (TP) in litter and soil samples, we ground the litter and air-dried soil samples and passed them through a 0.149 mm sieve. Then, we used a digester (FOSS. Tecator Digestor 20) to digest plant litter and soil and used a flow injection auto analyzer to analyze the TN and TP of the digested products. The stoichiometric ratio of C, N, and P of litter and soil was reported as mass ratio.

2.5. Data Analysis

Using the One-way analysis of variance (ANOVA) for four kinds of plants soil DOC, NH_4^+, NO_3^-, AN, AP, and pH difference analysis, when the difference is significant, using LSD (least significant difference) comparison ($p = 0.05$). Nested ANOVAs were used to analyze the effects of N addition (N), phosphorous addition (P), function type plant (f), and understory species (S, nested within Function type plant) on litter and soil C, N, and P and their stoichiometry. The variation trends of C, N, and P concentrations in the litter of two functional plants and the soil were fitted by one-dimensional regression equation. All statistical analyses were conducted by JMP 9.0 (SAS Institute, Cary, NC, USA).

3. Results

3.1. Availability of Soil Nutrients and Carbon, Nitrogen, and Phosphorus (CNP) Concentration of Plant Litter and Soil

Seven years after the annual addition of N and P to the plantation, we found that the soil of *L. gracile* has a higher NO_3^--N content compared with the other three understory species. Moreover, the AP content in the soil of *Dicranopteris dichotoma* is significantly lower than that of the other plants ($p < 0.01$). DOC, $NH4^+-N$, AN, and pH do not change among the four understory plant species (Table 1). Among the four species, *L. gracile* has higher concentrations of C and N in litter and N and P in associated soil (Table A1). The litter C, N, and P and soil C concentrations of *W. japonica* are all low (Table A1). The litter and soil C concentrations of *Dryopteris atrata* are all higher than the other species ($p < 0.01$). The litter and soil P concentrations of *Dicranopteris dichotoma* are all low (Table 1).

Table 1. Litter and soil nutrients of four understory plant species (means ± SE). Superscript letters in the same line indicate a significant difference at $p = 0.05$ based on LSD (least significant difference) comparison. Lines without letters indicate that the difference is not significant. LC: litter total organic carbon (C); LN: litter total nitrogen (N); LP: litter total phosphorus (P); SC: soil total organic C; SN: soil total N; SP: soil total P.

Factors	*Lophatherum gracile* Brongn. (Gramineae)	*Woodwardia japonica* (L.f.) Sm. (Fern)	*Dryopteris atrata* (Kunze) Ching (Fern)	*Dicranopteris dichotoma* (Thunb.) Berhn.) (Fern)
DOC (mg kg^{-1})	88.05 ± 6.88	100.79 ± 8.23	103.53 ± 10.59	85.67 ± 4.64
NH_4^+-N (mg kg^{-1})	5.26 ± 0.40	5.01 ± 0.40	5.23 ± 0.32	5.45 ± 0.51
NO_3^--N (mg kg^{-1})	1.92 ± 0.19 ^{A}	1.41 ± 0.09 ^{B}	1.40 ± 0.09 ^{B}	1.42 ± 0.15 ^{B}
AN (mg kg^{-1})	7.18 ± 0.50	6.42 ± 0.44	6.63 ± 0.37	6.87 ± 0.64
AP (mg kg^{-1})	1.17 ± 0.15 ^{A}	1.15 ± 0.18 ^{A}	1.34 ± 0.20 ^{A}	0.56 ± 0.09 ^{B}
pH	3.32 ± 0.05	3.31 ± 0.04	3.37 ± 0.04	3.33 ± 0.04
LC (g kg^{-1})	465.09 ± 8.61 ^{A}	419.44 ± 7.18 ^{B}	459.67 ± 6.32 ^{A}	475.82 ± 6.62 ^{A}
LN (g kg^{-1})	24.39 ± 0.69 ^{A}	17.07 ± 0.57 ^{C}	20.10 ± 0.77 ^{B}	15.05 ± 0.92 ^{C}
LP (g kg^{-1})	1.39 ± 0.12 ^{B}	1.18 ± 0.06 ^{BC}	1.94 ± 0.18 ^{A}	1.02 ± 0.07 ^{C}
SC (g kg^{-1})	31.57 ± 2.10 ^{BC}	27.87 ± 0.93 ^{C}	37.78 ± 1.39 ^{A}	35.88 ± 2.00 ^{AB}
SN (g kg^{-1})	1.30 ± 0.04 ^{A}	1.10 ± 0.05 ^{B}	0.62 ± 0.05 ^{C}	1.22 ± 0.10 ^{AB}
SP (g kg^{-1})	0.43 ± 0.02 ^{A}	0.39 ± 0.02 ^{A}	0.29 ± 0.02 ^{B}	0.32 ± 0.01 ^{B}

3.2. Effects of N and P Addition on Carbon, Nitrogen, and Phosphorus (C, N, and P) Concentrations and Stoichiometric Ratios of Different Functional Groups

Litter C, N, and the C:N ratio were significantly affected by N addition, the plant functional group, understory species, and the randomized complete block design. Litter P and the C:P ratio depended on P addition, understory species, and the randomized complete block design (Table 2). Litter P also depended on understory species × P addition ("S × P[f]"). The N:P ratio of litter depended on P, N × P, and the plant functional group. However, the results varied with blocks except L N/P (Table 2).

Table 2. ANOVAs on the effects of nitrogen (N) and phosphorus (P) deposition, plant functional group (F), understory species (Sp; nested within plant functional group, f), and their interactions on litter (raw decomposed leaves) C, N, P, and C/N/P. B&R indicates randomized complete block design. The six indicators include litter total organic C (LC), litter total N (LN), litter total P (LP), litter C to N ratio (L C/N), litter C to P ratio (L C/P), and litter N to P ratio (L N/P) along the first row of the table. df is the degrees of freedom. *: $p < 0.05$; **: $p < 0.01$; ***: $p < 0.001$.

Factor	df	LC (g kg^{-1})		LN (g kg^{-1})		LP (g kg^{-1})		L C/N		L C/P		L N/P	
		MS	F	MS	F	MS	F	MS	F	MS	F	MS	F
N	2	8256	5.92 ***	113	10.10 ***	0.70	2.30	376	5.19 **	323	0.02	90.33	2.47
P	1	26.78	0.02	17.18	1.54	10.37	34.22 ***	56.65	0.78	618,936	37.03 ***	982	26.84 ***
N*P	2	2440	26.71	28.54	2.56	0.81	2.66	70.56	0.98	27,073	1.62	156	4.26 *
f	1	23,734	17.00 ***	2066	185.37 ***	0.01	0.04	2777	38.41 ***	2941	0.18	1472	40.26 ***
N*f	2	10,323	7.40 ***	21.97	1.97	0.12	0.40	49.92	0.69	28,014	1.68	58.89	1.61
P*f	1	497.30	0.36	31.06	2.79	0.11	0.38	64.55	0.89	2928	0.18	71.35	1.95
N*P*f	2	3603	2.58	2.88	0.26	0.02	0.05	36.05	0.50	884	0.05	23.27	0.64
Sp[f]	2	47,217	33.83 ***	398	35.69 ***	13.89	45.85 ***	2517	34.81 ***	660,744	39.53 ***	174	4.76 ***
Sp*N[f]	4	2545	1.82	11.58	1.04	0.26	0.86	68.92	0.95	4379	0.26	1.59	0.04
Sp*P[f]	2	2714	1.94	19.89	1.79	3.35	11.04 ***	7.05	0.10	38,456	2.30	80.97	2.21
B&R	3	5465	3.92 ***	201	18.01 ***	1.16	3.82 **	447	6.19 ***	59,442	3.56*	43.52	1.19

3.3. Correlation between the Litter of Different Functional Groups and Soil Carbon, Nitrogen, and Phosphorus (C, N, and P) with the Addition of N and P

Based on the unitary regression analysis, the litter N of the fern functional group was positively correlated with soil P ($p = 0.051$, Figure A1f) compared with the CK. In addition, for the fern functional group, the litter P was positively correlated with soil C ($p = 0.002$, Figure A1g) and negatively correlated with soil P ($p = 0.019$, Figure A1i). With the addition of N, the litter C of Gramineae was positively correlated with soil P ($p = 0.029$, Figure A2c); the litter N was positively correlated with soil N ($p = 0.015$, Figure A2e). The litter P and soil P were positively correlated in the Gramineae group but negatively correlated in the fern group (Figure A2i). With the addition of P, there was a negative correlation between the C and soil P of the litter of the fern group ($p = 0.051$, Figure A3c); the litter N and soil N were negatively correlated ($p < 0.001$, Figure A3e) in ferns but positively correlated ($p = 0.008$, Figure A3f) in Gramineae. Litter P is positively correlated with soil P in both the Gramineae and fern groups (Figure A3i). With the application of both N and P, the change of the litter C concentration in the fern group was affected by the change of soil C, soil N, and soil P concentration. The litter C concentration increased with the increase of soil C but decreased with the increase of soil N and soil P concentration (Figure 1a–c). The litter C concentration was negatively correlated with soil N ($p = 0.059$, Figure 1b) in the fern group, whereas soil N was positively correlated with litter C ($p = 0.003$, Figure 1b) in the Gramineae group. The litter N concentration of the fern group decreased with the increase of the soil N concentration ($p < 0.001$, Figure 1e). There was no correlation between litter N, soil C, and soil P of the two different plant functional groups (Figure 1d, f). Moreover, the litter P concentration was positively correlated with both soil N and soil P (Figure 1h, i) in the fern functional group.

3.4. Effects of N and P Addition on Carbon, Nitrogen, and Phosphorus (C, N, and P) Concentrations and Stoichiometric Ratios of the Soils Associated with Different Plant Functional Groups

A nested ANOVA showed that soil C was significantly affected by the understory species. Soil N and soil P were significantly affected by N addition, plant functional group, understory species, and randomized complete block design. The soil N also depended on P addition*plant functional group (Table 3). The soil C/N ratio and C:P ratio depended on the plant functional group, understory species, understory species*P addition, and randomized complete block design (Table 3; Figure 2). The soil C/P ratio depended on P addition*plant functional group. The soil N:P ratio depended on N addition, understory species, and randomized complete block design (Figure 2). Similarly, results of SN, SP, S C/N, SC/P, and SN/P varied with blocks (Table 3).

Table 3. ANOVAs on the effects of N, P, and Species (Sp, nested within f) and their interactions with soil carbon, nitrogen, and phosphorus (C, N, P), soil C/N, soil C/P, and soil N/P. Terms are the same as in Table 2. B&R is randomized complete block design. df is the degrees of freedom. *: $p < 0.05$; **: $p < 0.01$; ***: $p < 0.001$.

Factor	df	SC (g kg^{-1})		SN (g kg^{-1})		SP (g kg^{-1})		S C/N		S C/P		S N/P	
		MS	F	MS	F	MS	F	MS	F	MS	F	MS	F
N	2	182	1.41	0.90	5.59 **	0.05	3.49 *	914	0.24	5964	2.92	18.93	7.39 ***
P	1	109	0.84	0.12	0.73	0.05	3.29	1178	0.31	1865	0.91	3.38	1.32
N*P	2	16.63	0.13	0.33	2.03	0.01	0.64	1479	0.39	13.27	0.01	5.32	2.08
f	1	186	1.44	3.75	23.39 ***	0.30	21.79 ***	31,687	8.31 **	41,992	20.54 ***	5.85	2.28
N*f	2	278	2.15	0.05	0.31	0.02	1.78	821	0.22	4381	2.14	2.97	1.16
P*f	1	38.80	0.30	0.62	3.86 *	0.05	3.58	7680	2.01	11,111	5.43 *	0.55	0.22
N*P*f	2	356	2.76	0.02	0.12	0.01	0.92	1074	0.28	2776	1.36	2.39	0.93
Sp[f]	2	1327	10.28 ***	4.77	29.77 ***	0.13	9.57 ***	90,281	23.68 ***	68,901	33.70 ***	20.70	8.08 ***
Sp*N[f]	4	176	1.36	0.12	0.78	0.01	1.07	5364	1.41	3802	1.86	2.05	0.80
Sp*P[f]	2	22.99	0.18	0.12	0.76	0.04	2.91	15,940	4.18 *	8820	4.31 *	5.02	1.96
B&R	3	325	2.52	1.41	8.83 ***	0.17	12.44 ***	26,816	7.04 ***	6256	3.06 *	33.72	13.16 ***

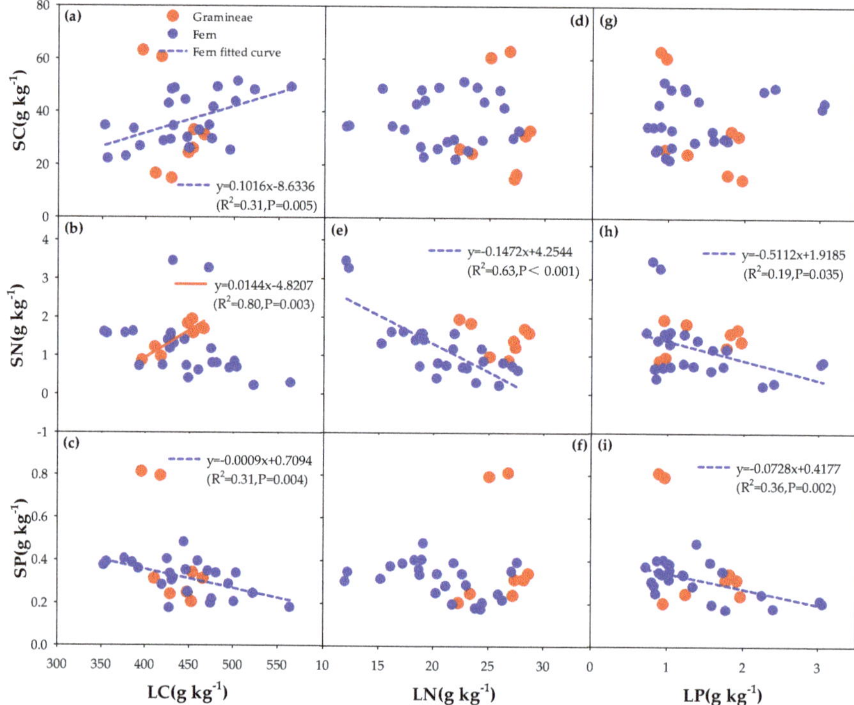

Figure 1. Correlations between litter and soil carbon, nitrogen, and phosphorus (C, N, and P) of four dominate species with two different functional types (Gramineae *Lophatherum gracile* and fern *Woodwardia japonica*, *Dryopteris atrata* and *Dicranopteris dichotoma*) with N (100 kg ha^{-1} year^{-1}) and P (50 kg ha^{-1} year^{-1}) addition.

4. Discussion

Ecological stoichiometry plays an important role in the study of plants and soils within an entire ecosystem. Especially under global climate change, it is valuable to study the understory plant community and the nutrient cycle [45,46]. In the Chinese fir plantation, this study identified the differences in N and P nutrients availability in the soil associated with four understory species. Soil pH was lower than that of other study [43], and the soil of understory plants has a very low pH (3.31–3.37, soil/water, 2:5), indicating that these four species were acid-tolerant. In addition, there were significant differences in the concentrations of C, N, and P in litter and soil among the four species, indicating that species may be the factors contributing to this difference [2,47]. After the continuous addition of N and P, the C, N, and P concentration and stoichiometric ratio of the litter of different plant functional groups showed different degrees of variation compared to the CK. This change may be caused by changes in the C, N, and P cycles in the understory ecosystem after the addition of N and P [12,48,49]. With the addition of N and P, the C, N, and P concentrations of different plant functional groups and soil showed different degrees of correlation, and the correlation would change with different rates of N and P addition. However, most results were significantly affected by blocks, indicating potential variations in both litter and soil C and nutrients with spatial distribution of plots, which should be further considered in future studies.

4.1. Effects of Different Understory Vegetation Types on Stoichiometric Ratio

Understory vegetation plays an important role in maintaining biodiversity in forest ecosystems and soil nutrient cycling [50,51]. Most ecological indicators are correlated with different understory vegetation types [52]. Understory vegetation often participates in soil C and N nutrient cycling processes via multiple methods [53]. There are different correlations between water-soluble organic C and microbial biomass C of different understory vegetation types [54]. In this study, litter C, N, and C:N and soil N, P, C:N, and C:P in species of different functional groups are significantly correlated (Tables 2 and 3). The litter from different species is significantly correlated with soil C, N, and P and the stoichiometric ratio (Tables 2 and 3). This is consistent with Güsewell's conclusion that plants have different C, N, and P concentrations in leaves due to the species and nutritional conditions [2]. These findings are consistent with Cheng et al.'s conclusions that indicated that the soil N utilization efficiency of Gramineae is higher than ferns [55]. In addition, litter from the Gramineae group has a lower C:N ratio, which is similar to the results of a study on the understory species of a *Pinus massoniana* plantation. Gramineae may be more conducive to improving the nutrient cycling rate in its dominant area [54].

4.2. Effect of Nitrogen (N) and Phosphorus (P) Addition on Stoichiometric Ratio

Nitrogen and P are the main limiting elements of plant growth in terrestrial ecosystems. It is generally believed that a N:P mass ratio lower than 14 indicates that plant growth is limited by N, while a N:P mass ratio higher than 16 indicates that plant growth is limited by P [47]. An unbalanced input of N and P will seriously affect the ecological stoichiometry, ultimately affecting the function of the ecosystem [15,56,57]. We found that N:P ratio of Gramineae litter (>16) was higher than ferns, After the addition of P, the N:P ratio of Gramineae decreased, but it was not significantly different from the CK (Figure 2i). Plants regulate their growth rate by adjusting the C:N:P ratio [58–60]. Because of the growth dilution effect, the available plant P concentration would decrease with increasing N [60]. However, our study did not find that N addition significantly reduced the P concentration of plant litter (Figure 2c, Table 2), which was in agreement with a study on the effect of long-term application of N and P fertilizer on plant N:P in a Tibetan alpine meadow [23]. It is possible that this effect is species specific or depends on the N application rate [61,62]. In comparison, the addition of N increased the N:P of litter and soil in the fern group (Figure 2i,l). The addition of P reduced the C:P and N:P of litter and soil in the fern group (Figure 2h,i,k,l). The separate addition of N and P had no significant effect on the CNP stoichiometric ratio of Gramineae, which indicates that the fern group was affected by the changes of a single environmental factor. This suggests that the fern population could rapidly expand if the environment was conducive to their favored growth conditions. Gramineae species have a good adaptability with the single change of N and P [53]. The co-addition of N and P reduced the C:N and C:P of litter (Figure 2g,h) of two of the different functional groups, increased the soil N:P (Figure 2i) of the fern group, and reduced the C:N ratio of understory species, which indicated that the addition of N and P together increased the N concentration in plants.

Coupling between N and P was observed, which plays an important role in the regulation of nutrient limitations and the strategies for plants to obtain nutrients in the changing environment [63,64]. These results show that the addition of N and P affects the correlation between litter and soil C, N, and P, and has different degrees of influence in different plant functional groups.

With the increasing deposition of N in southern China, the availability of soil P is limited, impacting the balance of C, N, and P in the ecosystem. In the context of global climate change, long-term experiments adding N and P are beneficial to understanding the nutrient cycle of understory vegetation in forest ecosystems and to cope with environmental changes. Future studies on understory vegetation management should focus on the effects of N and P addition and understory vegetation types on soil C and N cycles in plantations.

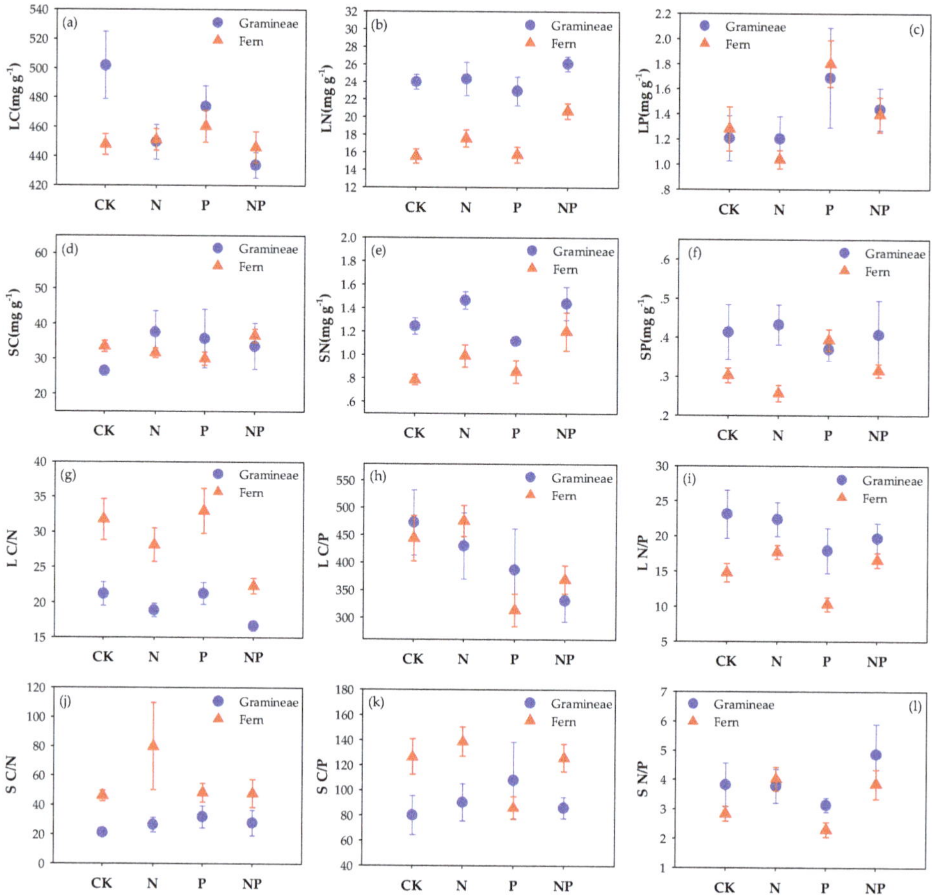

Figure 2. Litter and soil carbon, nitrogen, and phosphorus (C, N, P), C/N, C/P and N/P of four dominate species with two different functional types (Gramineae *Lophatherum gracile* and fern *Woodwardia japonica*, *Dryopteris atrata* and *Dicranopteris dichotoma*) under four nutrient addition modes (CK, N, P and NP). Terms are the same as in Table 2.

5. Conclusions

The C, N, and P concentrations of plants and soil vary greatly among different species, and these differences may be species dependent. The input of exogenous N and P changes the stoichiometric characteristics of understory plants to different degrees; thus, we believe that different species have different response mechanisms to changes of environmental factors. Different functional plant groups show different changes with the addition of N and P. The addition of N and P had an interactive effect on the N:P of litter, indicating that the growth of understory plants was dually restricted by N and P. Over the long-term, the addition of N and P may change the competitiveness of different functional plant groups in the forest, leading to a change in the community of understory plant species. Therefore, the response of the stoichiometric characteristics of understory plants to different N and P inputs should be studied in future research on understory vegetation.

Author Contributions: Conceptualization, L.Z. and J.X.; methodology, L.Z. and J.X.; software, J.X. and M.C.; investigation, J.X., H.F., Q.Z. and Y.G.; resources, L.Z.; data curation, J.P. and J.X.; writing—original draft preparation, J.X.; writing—review and editing, L.Z., X.X., X.F. and J.X.; supervision, X.G. and L.Z.; funding acquisition, X.F. and L.Z.

Funding: This research was funded by the National Natural Science Foundation of China (31770749) and the Research Project of Jiangxi Provincial Department of Forestry (No.201806).

Acknowledgments: The authors thank the following people for their help in this research. Li-Li Ma, Shu-Li Wang, Xi Yuan provided field assistance. Jia-Shun Peng assisted with lab analyses. Jiao Wu assisted with syntax modification.

Conflicts of Interest: The authors have declared that no competing interests exist.

Abbreviations

C	carbon
N	nitrogen
P	phosphorus
DOC	dissolved organic carbon
TOC	total organic carbon
TN	total nitrogen
TP	total phosphorus
AN	available nitrogen
AP	available phosphorus
LC	total litter organic carbon
LN	total litter nitrogen
LP	total litter phosphorus
SC	total soil organic carbon
SN	total soil nitrogen
SP	total soil phosphorus
Sp	understory species
F	plant functional group
B&R	block as random effect
L C/N	litter carbon to nitrogen ratio
L C/P	litter carbon to phosphorus ratio
L N/P	litter nitrogen to phosphorus ratio
S C/N	soil carbon to nitrogen ratio
S C/P	soil carbon to phosphorus ratio
S N/P	soil nitrogen to phosphorus ratio

Appendix A

Table A1. The values of organic C (C), total N (N) and total P (P) in litter and soil of four plants under the addition of N and P (means ± SE).

Species			No P Addition		P Addition	
			CK	N	CK	N
Lophatherum gracile (Gramineae)	Litter	C	502.10 ± 22.99	449.88 ± 11.72	474.09 ± 13.99	434.28 ± 8.70
		N	24.04 ± 0.85	24.38 ± 1.89	23.00 ± 1.63	26.14 ± 0.82
		P	1.21 ± 0.18	1.21 ± 0.18	1.69 ± 0.40	1.44 ± 0.17
	soil	C	26.61 ± 1.37	37.68 ± 6.08	35.88 ± 8.32	33.71 ± 6.54
		N	1.25 ± 0.07	1.47 ± 0.07	1.12 ± 0.03	1.44 ± 0.14
		P	0.41 ± 0.07	0.43 ± 0.05	0.37 ± 0.03	0.41 ± 0.09
Woodwardia japonica (Fern)	Litter	C	420.27 ± 12.05	421.31 ± 9.00	442.22 ± 19.48	393.95 ± 11.78
		N	15.36 ± 1.39	17.56 ± 1.23	16.43 ± 0.93	18.95 ± 0.67
		P	1.23 ± 0.08	1.13 ± 0.18	1.33 ± 0.13	1.05 ± 0.08
	soil	C	29.38 ± 2.08	29.88 ± 2.15	25.22 ± 1.75	32.07 ± 2.97
		N	0.96 ± 0.07	1.18 ± 0.10	1.00 ± 0.13	1.34 ± 0.14
		P	0.38 ± 0.03	0.29 ± 0.05	0.40 ± 0.03	0.39 ± 0.02

Table A1. *Cont.*

Species			No P Addition		P Addition	
			CK	N	CK	N
Dryopteris atrata (Fern)	Litter	C	443.67 ± 6.65	457.00 ± 6.40	454.05 ± 15.97	483.93 ± 15.52
		N	16.62 ± 0.68	19.82 ± 1.65	18.77 ± 1.14	25.17 ± 0.69
		P	1.88 ± 0.45	1.16 ± 0.09	2.57 ± 0.42	2.18 ± 0.22
	soil	C	40.89 ± 1.94	35.70 ± 3.45	31.08 ± 1.68	38.16 ± 3.02
		N	0.64 ± 0.05	0.60 ± 0.17	0.40 ± 0.03	0.74 ± 0.13
		P	0.24 ± 0.03	0.19 ± 0.04	0.40 ± 0.08	0.25 ± 0.03
Dicranopteris dichotoma (Fern)	Litter	C	480.50 ± 6.65	476.03 ± 13.68	485.45 ± 19.59	461.29 ± 10.67
		N	14.74 ± 1.94	15.37 ± 1.90	12.01 ± 1.54	18.08 ± 1.55
		P	0.75 ± 0.04	0.84 ± 0.04	1.53 ± 0.17	0.97 ± 0.06
	soil	C	30.64 ± 2.27	29.51 ± 0.70	34.01 ± 4.97	39.89 ± 3.83
		N	0.77 ± 0.07	1.19 ± 0.144	1.18 ± 0.15	1.53 ± 0.42
		P	0.30 ± 0.02	0.29 ± 0.02	0.38 ± 0.03	0.32 ± 0.01

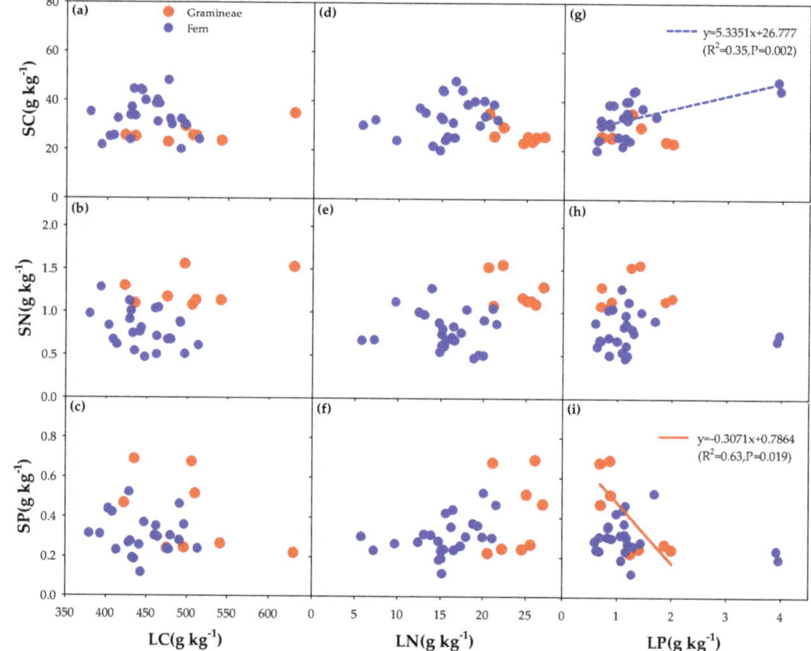

Figure A1. Correlations between litter and soil C, N and P of four dominate species with two different functional types (Gramineae *Lophatherum gracile* and fern *Woodwardia japonica*, *Dryopteris atrata* and *Dicranopteris dichotoma*) in CK treatment.

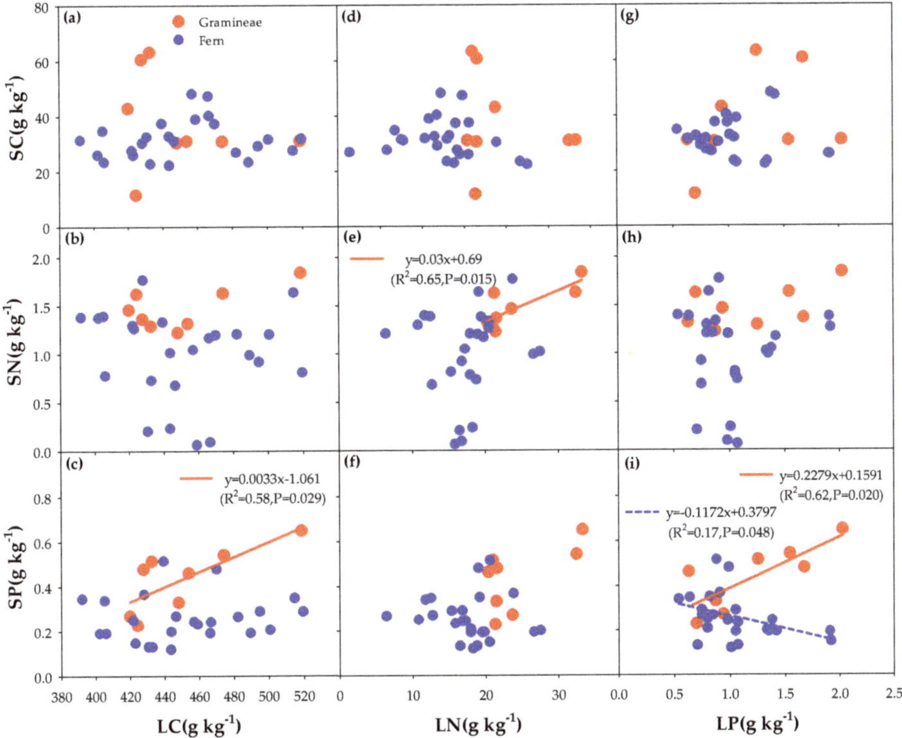

Figure A2. Correlations between litter and soil C, N and P of four dominate species with two different functional types (Gramineae *Lophatherum gracile* and fern *Woodwardia japonica*, *Dryopteris atrata* and *Dicranopteris dichotoma*) in N addition treatment.

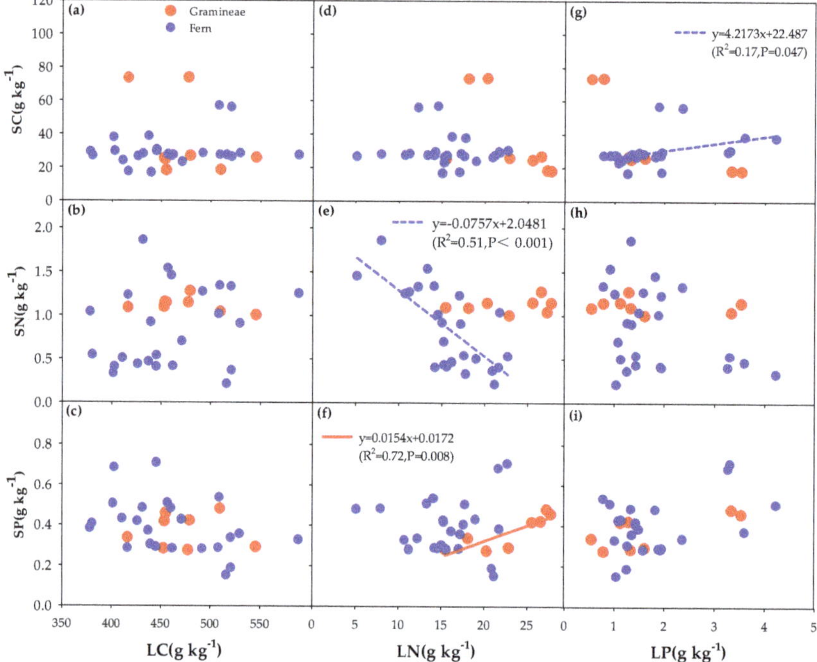

Figure A3. Correlations between litter and soil C, N and P of four dominate species with two different functional types (Gramineae *Lophatherum gracile* and fern *Woodwardia japonica*, *Dryopteris atrata* and *Dicranopteris dichotoma*) in P addition treatment.

References

1. Elser, J.; Sterner, R.; Fagan, W.; Markow, T.; Cotner, J.; Hobbie, S.; Odell, G.; Weider, L.; Gorokhova, E.; Harrison, J. Biological stoichiometry from genes to ecosystems. *Ecol. Lett.* **2000**, *3*, 540–550. [CrossRef]
2. Güsewell, S. N:P ratios in terrestrial plants: Variation and functional significance. *New Phytol.* **2004**, *164*, 243–266. [CrossRef]
3. Han, W.; Fang, J.; Guo, D.; Zhang, Y. Leaf nitrogen and phosphorus stoichiometry across 753 terrestrial plant species in China. *New Phytol.* **2005**, *168*, 377–385. [CrossRef] [PubMed]
4. Reich, P.B.; Oleksyn, J. Global patterns of plant leaf N and P in relation to temperature and latitude. *Proc. Natl. Acad. Sci. USA* **2004**, *101*, 11001–11006. [CrossRef] [PubMed]
5. Ladanai, S.; Ågren, G.I.; Olsson, B.A. Relationships between tree and soil properties in *picea abies* and *pinus sylvestris* forests in sweden. *Ecosystems* **2010**, *13*, 302–316. [CrossRef]
6. Bui, E.N.; Henderson, B.L. C:N:P stoichiometry in Australian soils with respect to vegetation and environmental factors. *Plant Soil* **2013**, *373*, 553–568. [CrossRef]
7. Tian, H.; Chen, G.; Zhang, C.; Melillo, J.M.; Hall, C.A.S. Pattern and variation of C: N:P ratios in China's soils: A synthesis of observational data. *Biogeochemistry* **2010**, *98*, 139–151. [CrossRef]
8. Siddique, I.; Vieira, I.C.G.; Schmidt, S.; Lamb, D.; Carvalho, C.J.R.; Figueiredo, R.D.O.; Blomberg, S.; Davidson, E.A. Nitrogen and phosphorus additions negatively affect tree species diversity in tropical forest regrowth trajectories. *Ecology* **2010**, *91*, 2121–2131. [CrossRef]
9. Wei, X.; Blanco, J.A.; Jiang, H.; Kimmins, J.H. Effects of nitrogen deposition on carbon sequestration in Chinese fir forest ecosystems. *Sci. Total Environ.* **2012**, *416*, 351–361. [CrossRef]
10. Litton, C.M.; Raich, J.W.; Ryan, M.G. Carbon allocation in forest ecosystems. *Glob. Chang. Biol.* **2007**, *13*, 2089–2109. [CrossRef]
11. Clark, C.M.; Tilman, D. Loss of plant species after chronic low-level nitrogen deposition to prairie grasslands. *Nature* **2008**, *451*, 712–715. [CrossRef] [PubMed]

12. Peñuelas, J.; Sardans, J.; Rivas-Ubach, A.; Janssens, I.A. The human-induced imbalance between C, N and P in Earth's life system. *Glob. Chang. Biol.* **2012**, *18*, 3–6. [CrossRef]

13. Strengbom, J.; Walheim, M.; Näsholm, T.; Ericson, L. Regional differences in the occurrence of understorey species reflect nitrogen deposition in Swedish forests. *Ambio* **2003**, *32*, 91–97. [CrossRef] [PubMed]

14. Menge, D.N.L.; Field, C.B. Simulated global changes alter phosphorus demand in annual grassland. *Glob. Chang. Biol.* **2007**, *13*, 2582–2591. [CrossRef]

15. Sardans, J.; Rivas-Ubach, A.; Peñuelas, J. The elemental stoichiometry of aquatic and terrestrial ecosystems and its relationships with organismic lifestyle and ecosystem structure and function: A review and perspectives. *Biogeochemistry* **2012**, *111*, 1–39. [CrossRef]

16. Mayor, J.R.; Wright, S.J.; Turner, B.L. Species-specific responses of foliar nutrients to long-term N and P additions in a lowland tropical forest. *J. Ecol.* **2014**, *102*, 36–44. [CrossRef]

17. Huang, W.; Zhou, G.; Liu, J.; Zhang, D.; Xu, Z.; Liu, S. Effects of elevated carbon dioxide and nitrogen addition on foliar stoichiometry of nitrogen and phosphorus of five tree species in subtropical model forest ecosystems. *Environ. Pollut.* **2012**, *168*, 113–120. [CrossRef] [PubMed]

18. Koller, E.K.; Press, M.C.; Callaghan, T.V.; Phoenix, G.K. Tight coupling between shoot level foliar N and P, leaf area, and shoot growth in Arctic dwarf shrubs under simulated climate change. *Ecosystems* **2016**, *19*, 326–338. [CrossRef]

19. Zhang, Q.; Xie, J.; Lyu, M.; Xiong, D.; Wang, J.; Chen, Y.; Li, Y.; Wang, M.; Yang, Y. Short-term effects of soil warming and nitrogen addition on the N:P stoichiometry of *Cunninghamia lanceolata* in subtropical regions. *Plant Soil* **2016**, *411*, 395–407. [CrossRef]

20. Braun, S.; Thomas, V.F.; Quiring, R.; Flückiger, W. Does nitrogen deposition increase forest production? The role of phosphorus. *Environ. Pollut.* **2010**, *158*, 2043–2052. [CrossRef]

21. Chen, F.S.; Niklas, K.J.; Zeng, D.H. Important foliar traits depend on species-grouping: Analysis of a remnant temperate forest at the Keerqin sandy lands, China. *Plant Soil* **2011**, *340*, 337–345. [CrossRef]

22. Heuck, C.; Smolka, G.; Whalen, E.D.; Frey, S.; Gundersen, P.; Moldan, F.; Fernandez, I.J.; Spohn, M. Effects of long-term nitrogen addition on phosphorus cycling in organic soil horizons of temperate forests. *Biogeochemistry* **2018**, *141*, 167–181. [CrossRef]

23. Zhang, J.; Yan, X.; Su, F.; Li, Z.; Wang, Y.; Wei, Y.; Ji, Y.; Yang, Y.; Zhou, X.; Guo, H.; et al. Long-term N and P additions alter the scaling of plant nitrogen to phosphorus in a Tibetan alpine meadow. *Sci. Total Environ.* **2018**, *625*, 440–448. [CrossRef] [PubMed]

24. Craft, C.B.; Vymazal, J.; Richardson, C.J. Response of everglades plant communities to nitrogen and phosphorus additions. *Wetlands* **1995**, *15*, 258–271. [CrossRef]

25. Yang, Y.; Liu, B.R.; An, S.S. Ecological stoichiometry in leaves, roots, litters and soil among different plant communities in a desertified region of Northern China. *Catena* **2018**, *166*, 328–338. [CrossRef]

26. Wang, W.Q.; Wang, C.; Sardans, J.; Zeng, C.S.; Tong, C.; Peñuelas, J. Plant invasive success associated with higher N-use efficiency and stoichiometric shifts in the soil–plant system in the Minjiang River tidal estuarine wetlands of China. *Wetl. Ecol. Manag.* **2015**, *23*, 865–880. [CrossRef]

27. Wang, F.; Chen, F.; Wang, G.G.; Mao, R.; Fang, X.; Wang, H.; Bu, W. Effects of experimental nitrogen addition on nutrients and nonstructural carbohydrates of dominant understory plants in a Chinese fir plantation. *Forests* **2019**, *10*, 155. [CrossRef]

28. Takafumi, H.; Hiura, T. Effects of disturbance history and environmental factors on the diversity and productivity of understory vegetation in a cool-temperate forest in Japan. *For. Ecol. Manag.* **2009**, *257*, 843–857. [CrossRef]

29. Dobson, A.M.; Blossey, B.; Richardson, J.B. Invasive earthworms change nutrient availability and uptake by forest understory plants. *Plant Soil* **2017**, *421*, 175–190. [CrossRef]

30. Wang, F.; Zou, B.; Li, H.; Li, Z. The effect of understory removal on microclimate and soil properties in two subtropical lumber plantations. *J. For. Res.* **2014**, *19*, 238–243. [CrossRef]

31. Jules, M.J.; Sawyer, J.O.; Jules, E.S. Assessing the relationships between stand development and understory vegetation using a 420-year chronosequence. *For. Ecol. Manag.* **2008**, *255*, 2384–2393. [CrossRef]

32. Lu, X.; Mo, J.; Gilliam, F.S.; Zhou, G.; Fang, Y. Effects of experimental nitrogen additions on plant diversity in an old-growth tropical forest. *Glob. Chang. Biol.* **2010**, *16*, 2688–2700. [CrossRef]

33. Hurd, T.M.; Brach, A.R.; Raynal, D.J. Response of understory vegetation of Adirondack forests to nitrogen additions. *Can. J. For. Res.* **1998**, *28*, 799–807. [CrossRef]

34. Rainey, S.M.; Nadelhoffer, K.J.; Silver, W.L.; Downs, M.R. Effects of chronic nitrogen additions on understory species in a red pine plantation. *Ecol. Appl.* **1999**, *9*, 949–957. [CrossRef]

35. Bi, J.; Blanco, J.A.; Seely, B.; Kimmins, J.P.; Ding, Y.; Welham, C. Yield decline in Chinese-fir plantations: A simulation investigation with implications for model complexity. *Can. J. For. Res.* **2007**, *37*, 1615–1630. [CrossRef]

36. Lu, Z.H.; Wu, G.; Ma, X.; Bai, G.X. Current situation of Chinese forestry tactics and strategy of sustainable development. *J. For. Res.* **2002**, *13*, 319–322.

37. Chen, G.S.; Yang, Z.J.; Gao, R.; Xie, J.S.; Guo, J.F.; Huang, Z.Q.; Yang, Y.S. Carbon storage in a chronosequence of Chinese fir plantations in southern China. *For. Ecol. Manag.* **2013**, *300*, 68–76. [CrossRef]

38. Tang, X.; Pérez-Cruzado, C.; Fehrmann, L.; Álvarez-González, J.G.; Lu, Y.; Kleinn, C. Development of a compatible taper function and stand-level merchantable volume model for Chinese fir plantations. *PLoS ONE* **2016**, *11*, e0147610. [CrossRef]

39. Peng, Y.; Tian, D.; Thomas, S.C. Forest management and soil respiration: Implications for carbon sequestration. *Environ. Rev.* **2008**, *16*, 93–111. [CrossRef]

40. Powers, R.F.; Busse, M.D.; McFarlane, K.J.; Zhang, J.W.; Young, D.H. Long-term effects of silviculture on soil carbon storage: Does vegetation control make a difference? *Forestry* **2013**, *86*, 47–58. [CrossRef]

41. Chen, F.S.; Niklas, K.J.; Liu, Y.; Fang, X.M.; Wan, S.Z.; Wang, H. Nitrogen and phosphorus additions alter nutrient dynamics but not resorption efficiencies of Chinese fir leaves and twigs differing in age. *Tree Physiol.* **2015**, *35*, 1106–1117. [CrossRef] [PubMed]

42. Lu, C.; Tian, H. Spatial and temporal patterns of nitrogen deposition in China: Synthesis of observational data. *J. Geophys. Res. Space Phys.* **2007**, *112*, 10–15. [CrossRef]

43. Tang, Y.; Zhang, X.; Li, D.; Wang, H.; Chen, F.; Fu, X.; Fang, X.; Sun, X.; Yu, G. Impacts of nitrogen and phosphorus additions on the abundance and community structure of ammonia oxidizers and denitrifying bacteria in Chinese fir plantations. *Soil Biol. Biochem.* **2016**, *103*, 284–293. [CrossRef]

44. Bao, S.D. *Soil and Agricultural Chemistry Analysis*, 3rd ed.; Agriculture Press: Beijing, China, 2008. (In Chinese)

45. Zechmeister-Boltenstern, S.; Keiblinger, K.M.; Mooshammer, M.; Peñuelas, J.; Richter, A.; Sardans, J.; Wanek, W. The application of ecological stoichiometry to plant–microbial–soil organic matter transformations. *Ecol. Monogr.* **2015**, *85*, 133–155. [CrossRef]

46. Metcalfe, D.B.; Fisher, R.A.; Wardle, D.A. Plant communities as drivers of soil respiration: Pathways, mechanisms, and significance for global change. *Biogeosciences* **2011**, *8*, 2047–2061. [CrossRef]

47. Meuleman, A.F.M.; Koerselman, W. The Vegetation N:P Ratio: A new tool to detect the nature of nutrient limitation. *J. Appl. Ecol.* **1996**, *33*, 1441–1450.

48. Marklein, A.R.; Houlton, B.Z. Nitrogen inputs accelerate phosphorus cycling rates across a wide variety of terrestrial ecosystems. *New Phytol.* **2011**, *193*, 696–704. [CrossRef]

49. Lu, M.; Yang, Y.; Luo, Y.; Fang, C.; Zhou, X.; Chen, J.; Yang, X.; Li, B. Responses of ecosystem nitrogen cycle to nitrogen addition: A meta-analysis. *New Phytol.* **2011**, *189*, 1040–1050. [CrossRef]

50. Wu, J.; Liu, Z.; Chen, D.; Huang, G.; Zhou, L.; Fu, S. Understory plants can make substantial contributions to soil respiration: Evidence from two subtropical plantations. *Soil Biol. Biochem.* **2011**, *43*, 2355–2357. [CrossRef]

51. Wu, J.; Liu, Z.; Wang, X.; Sun, Y.X.; Zhou, L.; Lin, Y.; Fu, S. Effects of understory removal and tree girdling on soil microbial community composition and litter decomposition in two Eucalyptus plantations in South China. *Funct. Ecol.* **2011**, *25*, 921–931. [CrossRef]

52. Yimer, F.; Ledin, S.; Abdelkadir, A. Soil organic carbon and total nitrogen stocks as affected by topographic aspect and vegetation in the Bale Mountains, Ethiopia. *Geoderma* **2006**, *135*, 335–344. [CrossRef]

53. Zhao, J.; Wan, S.; Li, Z.; Shao, Y.; Xu, G.; Liu, Z.; Zhou, L.; Fu, S. Dicranopteris-dominated understory as major driver of intensive forest ecosystem in humid subtropical and tropical region. *Soil Biol. Biochem.* **2012**, *49*, 78–87. [CrossRef]

54. Pan, P.; Zhao, F.; Ning, J.; Zhang, L.; Ouyang, X.; Zang, H. Impact of understory vegetation on soil carbon and nitrogen dynamic in aerially seeded *Pinus massoniana* plantations. *PLoS ONE* **2018**, *13*, e0191952. [CrossRef] [PubMed]

55. Chen, J.; Stark, J.M. Plant species effects and carbon and nitrogen cycling in a sagebrush–crested wheatgrass soil. *Soil Biol. Biochem.* **2000**, *32*, 47–57. [CrossRef]

56. Finzi, A.C.; Austin, A.T.; Cleland, E.E.; Frey, S.D.; Houlton, B.Z.; Wallenstein, M.D. Responses and feedbacks of coupled biogeochemical cycles to climate change: Examples from terrestrial ecosystems. *Front. Ecol. Environ.* **2011**, *9*, 61–67. [CrossRef]

57. Penuelas, J.; Poulter, B.; Sardans, J.; Ciais, P.; Van Der Velde, M.; Bopp, L.; Boucher, O.; Godderis, Y.; Hinsinger, P.; Llusià, J.; et al. Human-induced nitrogen–phosphorus imbalances alter natural and managed ecosystems across the globe. *Nat. Commun.* **2013**, *4*, 2934. [CrossRef] [PubMed]

58. Klausmeier, C.A.; Litchman, E.; Daufresne, T.; Levin, S.A. Optimal nitrogen-to-phosphorus stoichiometry of phytoplankton. *Nature* **2004**, *429*, 171–174. [CrossRef] [PubMed]

59. Moe, S.J.; Stelzer, R.S.; Forman, M.R.; Harpole, W.S.; Daufresne, T.; Yoshida, T. Recent advances in ecological stoichiometry: Insights for population and community ecology. *Oikos* **2005**, *109*, 29–39. [CrossRef]

60. Van Heerwaarden, L.M.; Toet, S.; Aerts, R. Nitrogen and phosphorus resorption efficiency and proficiency in six sub-arctic bog species after 4 years of nitrogen fertilization. *J. Ecol.* **2003**, *91*, 1060–1070. [CrossRef]

61. Gusewell, S. Variation in nitrogen and phosphorus concentrations of wetland plants. *Perspect. Plant Ecol. Evol. Syst.* **2002**, *5*, 37–61. [CrossRef]

62. Bubier, J.L.; Smith, R.; Juutinen, S.; Moore, T.R.; Minocha, R.; Long, S.; Minocha, S. Effects of nutrient addition on leaf chemistry, morphology, and photosynthetic capacity of three bog shrubs. *Oecologia* **2011**, *167*, 355–368. [CrossRef] [PubMed]

63. Niklas, K.J.; Cobb, E.D. N, P, and C stoichiometry of *Eranthis hyemalis* (Ranunculaceae) and the allometry of plant growth. *Am. J. Bot.* **2005**, *92*, 1256–1263. [CrossRef] [PubMed]

64. Elser, J.J.; Bracken, M.E.; Cleland, E.E.; Gruner, D.S.; Harpole, W.S.; Hillebrand, H.; Ngai, J.T.; Seabloom, E.W.; Shurin, J.B.; Smith, J.E. Global analysis of nitrogen and phosphorus limitation of primary producers in freshwater, marine and terrestrial ecosystems. *Ecol. Lett.* **2007**, *10*, 1135–1142. [CrossRef] [PubMed]

 *forests*

Article

Effects of Experimental Nitrogen Addition on Nutrients and Nonstructural Carbohydrates of Dominant Understory Plants in a Chinese Fir Plantation

Fangchao Wang [1], Fusheng Chen [1,2,*], G. Geoff Wang [3], Rong Mao [1,2], Xiangmin Fang [1], Huimin Wang [4] and Wensheng Bu [1,2]

[1] Jiangxi Provincial Key Laboratory of Silviculture, Jiangxi Agricultural University, Nanchang 330045, China; wfangch@163.com (F.W.); maorong23@163.com (R.M.); xmin007@163.com (X.F.); bws2007@163.com (W.B.)

[2] Jiulianshan National Observation and Research Station of Chinese Forest Ecosystem, Key Laboratory of National Forestry and Grassland Administration on Forest Ecosystem Protection and Restoration of Poyang Lake Watershed, Jiangxi Agricultural University, Nanchang 330045, China

[3] Department Forestry and Environmental Conservation, Clemson University, Clemson, SC 29634, USA; gwang@g.clemson.edu

[4] Qianyanzhou Ecological Station, Key Laboratory of Ecosystem Network Observation and Modeling, Institute of Geographic Sciences and Natural Resources Research, Chinese Academy of Sciences, Beijing 100101, China; wanghm@igsnrr.ac.cn

[*] Correspondence: chenfusheng@jxau.edu.cn; Tel.: +86-791-838-13243

Received: 14 January 2019; Accepted: 11 February 2019; Published: 12 February 2019

Abstract: *Research Highlights:* This study identifies the nitrogen (N) deposition effect on understory plants by altering directly soil nutrients or indirectly altering environmental factors in subtropical plantation. *Background and Objectives:* N deposition is a major environmental issue and has altered forest ecosystem components and their functions. The response of understory vegetation to N deposition is often neglected due to a small proportion of stand productivity. However, compared to overstory trees, understory species usually have a higher nutrient cycle rate and are more sensitive to environmental change, so should be of greater concern. *Materials and Methods:* The changes in plant biomass, N, phosphorus (P), and nonstructural carbohydrates (NSCs) of three dominant understory species, namely *Dicranopteris dichotoma*, *Lophatherum gracile*, and *Melastoma dodecandrum*, were determined following four years of experimental N addition (100 kg hm^{-2} year^{-1} of N) in a Chinese fir plantation. *Results:* N addition increased the tissue N concentrations of all the understory plants by increasing soil mineral N, while N addition decreased the aboveground biomass of *D. dichotoma* and *L. gracile* significantly—by 82.1% and 67.2%, respectively. The biomass of *M. dodecandrum* did not respond to N addition. In contrast, N addition significantly increased the average girth growth rates and litterfall productivity of overstory trees—by 18.28% and 36.71%, respectively. NSCs, especially soluble sugar, representing immediate products of photosynthesis and main energy sources for plant growth, decreased after N addition in two of the three species. The plant NSC/N and NSC/P ratios showed decreasing tendencies, but the N/P ratio in aboveground tissue did not change with N addition. *Conclusions:* N addition might inhibit the growth of understory plants by decreasing the nonstructural carbohydrates and light availability indirectly rather than by changing nutrients and N/P stoichiometry directly, although species-specific responses to N deposition occurred in the Chinese fir plantation.

Keywords: experimental nitrogen addition; understory plant growth; plant nutrient; nonstructural carbohydrates

---

1. Introduction

Fossil fuel combustion and chemical fertilization are increasing atmospheric nitrogen (N) deposition throughout the world and altering regional and global N cycles [1]. The annual input of reactive N into the Earth's land surface has approximately doubled since 1970, and this trend will continue in the rapidly developing regions of the world [2]. Considering that N is an essential nutrient element limiting plant photosynthetic capacity and productivity, increased N deposition would produce a cascading effect on forest ecosystem structure, process, and function [3–5]. The understory plants' responses to N deposition in forest ecosystems need to be of greater concern in the future.

The vegetation of forest ecosystems includes both overstory (or canopy) and understory species. Because of the dominance of the forest canopy, understory species and their responses to environmental changes are often neglected, especially in plantation forests [6]. Understory vegetation, despite often accounting for a small proportion of stand productivity, is an important component of forest ecosystems and plays a key role in regulating ecosystem processes and functions. Moreover, compared with tree species in the canopy, understory species usually have faster nutrient turnover rates and thus are more sensitive to environmental change. Previous studies have found that N deposition exerted a substantial influence on soil nutrient supply and decreased understory plant biomass due to increasing soil acidification and phosphorus (P) limitation [7–9]. Furthermore, some studies reported that increased N availability exaggerated asymmetrical competition for other resources (e.g., light and water), favoring the growth of overstory species at the expense of understory species [6,10,11]. For example, Strengbom suggested that understory vegetation was mainly limited by light, since N addition increased shading from the canopy that decreased the light available to understory plants in a boreal forest [12]. However, these studies concentrated on boreal and temperate natural forests, the results from which may not be applicable to subtropical plantations given that species coexistence and ecosystem biogeochemical cycles vary across forest types and climatic regions [6,13,14].

Nitrogen and P are often recognized as limiting nutrients in forest ecosystems; thus, N and P concentrations and their stoichiometry have been widely used to determine plant adaptation and feedback in response to resource alteration [15]. Nonstructural carbohydrates (NSCs) are the immediate products of photosynthesis and are often used to indicate plant light environment and growth rate [16]. The proportional relationships among N, P, and NSCs to a large extent reflect the available C and energy utilization efficiency of plant growth [17]. For example, N deposition increases N supply and causes an imbalance between N and P in plant tissues [18]. Furthermore, N deposition alters plant photosynthetic processes and the associated accumulation and consumption of NSCs including starch (ST) and soluble sugar (SS) [19]. Therefore, identifying NSCs and their interactions with N and P in aboveground (i.e., leaves) and belowground tissues (i.e., roots) might provide a new perspective on the effect of N deposition on understory vegetation growth.

In this study, a chronic N addition experiment was conducted to simulate N deposition in a Chinese fir (*Cunninghamia lanceolata*) plantation of subtropical China. We measured soil available N, P and plant biomass and N, P, and NSC concentrations in major understory plants to assess the responses of understory vegetation to N deposition in subtropical plantations. We hypothesized that (1) N deposition leads to an imbalance between N and P in soils and understory plants due to elevated N availability; (2) N deposition causes a decline in understory plant biomass due to elevated P limitation or increased shading by stimulated overstory growth; and (3) the changes in NSCs and NSCs/nutrients in aboveground and belowground tissues help identify the potential resource (such as light) competition mechanisms of understory plants in response to N deposition [14]. Our results may have some implications for understory plant management in plantation forests of subtropical areas experiencing N deposition.

2. Materials and Methods

2.1. Study Region

This study was conducted in a 12-year-old Chinese fir planation in the Jian-Taihe Basin of Jiangxi Province, China (26°42′ N, 115°04′ E, 100 m asl). The area is a typical red soil hilly region with a subtropical moist monsoon climate. The soil belongs to the typical Hapludult Ultisols with 68% sand and 15% clay [20]. The month average temperatures range from 6.5 °C in January to 29.7 °C in July, and a mean annual temperature is 18 °C. The annual precipitation ranges from 945 to 2144 mm, with an average of 1500 mm [21]. The area belongs to the core distribution of Chinese fir and the center of N deposition in China with 49 kg N ha^{-1} year^{-1} [3,22].

2.2. Experimental Treatments and Sample Collection

The simulated N deposition experiment followed a paired design and was established in November 2011. Within each paired plot, two 20 m × 20 m plots were treated with four years of in situ N addition (100 kg hm^{-2} year^{-1} of N) or no N addition (control, CK), with a buffer zone of more than 20 m between the plots. Four replicates were established on four separate hilly slopes. Nitrogen mixed with sand was added four times each year (March, June, September, and December) in the form of NH$_4$NO$_3$. In order to evenly spread NH$_4$NO$_3$ in the N addition plots, we added N together with a small amount of sand (8 kg plot^{-1}) and also added sand in control plots. For fertilization we generally chose a date without rain in the two days before or two days after, combined with a local weather forecast. General properties in the Chinese fir forest plantation before the experiment were not significantly different between the experimental units that received the two treatments (Table A1).

In August 2015, eight 1 m × 1 m sample subplots were randomly established in each plot. The understory species in these subplots were measured and recorded to obtain their richness. We used the number of each understory species in subplot as understory richness. Furthermore, the harvested understory plants were divided into three representative species, namely *Dicranopteris dichotoma* (a fern belonging to Gleicheniaceae; a sun plant), *Lophatherum gracile* (a perennial grass belonging to Gramineae; a neutral plant), and *Melastoma dodecandrum* (a creeping small shrub belonging to Melastomataceae; a shade plant) (see Figure A1), and other understory plants. The average proportion of biomass contributed by the three major plants to the total understory vegetation biomass in our study plots was more than 90%. All three species are perennial understory plants. We collected samples of all three understory species in maturation stage. The aboveground tissues (leaves and stems) of these species in each subplot were brought back to the laboratory to measure their dry biomass. We did not harvest the belowground tissue (roots) in order to avoid damage to these permanent study plots.

In addition, we collected samples of aboveground and belowground tissues and from rhizosphere and bulk soils for each of the three understory species within a plot. Soil strongly adhering within 4 mm of roots was considered rhizosphere soil, and the remaining soil was considered bulk soil [23]. Rhizosphere soil samples were collected for each species by separating soil from roots through hand shaking, while the bulk soil samples were collected using a soil auger under/near the plant crown. Each soil sample was divided into two replicates: one used for available nutrient measurement within five days and another stored in a refrigerator at 4 °C and then used for determination of soil enzyme activities. All the plant samples were immediately microwaved for 90 s to stop all enzymatic activity [24], washed with distilled water and oven-dried to a constant mass at 60 °C. After being finely ground using a mixer mill and through a sieve (<0.2 mm), these plant samples were used for determining nutrient and NSC concentrations.

2.3. Soil Nutrient and Enzyme Measurement

Soil NH$_4{}^+$-N and NO$_3{}^-$-N were extracted with 2 M KCl for 30 min and then measured by spectrophotometry using the indophenol blue and cadmium reduction methods, respectively. Soil available P was extracted with 0.5 M NaHCO$_3$ for 30 min and determined using the molybdenum-antimony colorimetric method [25]. The activities of N-acetyl-β-D-glucosaminidase (NAG; EC 3.2.1.14) and acid phosphatase (AP; EC 3.1.3.2) in soils were determined by the fluorogenic microplate method [26].

2.4. Understory Plant Nutrient Measurement

Plant N and P concentrations were determined by the Kjeldahl method and the molybdenum blue spectrophotometric procedure, respectively, after the samples were digested with H_2SO_4 [25]. NSCs were measured by the anthrone colorimetry method [27]. Briefly, the powdered plant sample (0.5 g) was put into a 15 mL centrifuge tube, where 10 mL of 80% alcohol was added. The mixture was incubated in a 100 °C water bath for 20 min and then centrifuged at 4000 rpm for 10 min. The supernatants were retained for SS determination, and the residue was extracted two more times as described above. ST was extracted from the ethanol-insoluble pellet until ethanol was first removed by evaporation. The residue remaining after the SS extraction was extracted with 5 mL 1 M H_2SO_4, and the mixture was shaken for 15 min. The mixture was incubated in an 80 °C water bath for 40 min and then centrifuged at 4000 rpm for 10 min. The pellets were extracted two more times with 1 M H_2SO_4. SS and ST determinations were performed based on absorbance at 625 nm using the same anthrone reagent in a spectrophotometer [28,29]. NSC concentration was obtained by the sum of the total SS and ST.

2.5. Overstory Tree Growth and Litterfall Production Measurement

In each plot, 20 trees have been randomly selected to measure the girth growth rate at breast height using a self-made dendrometer (including a sheet steel, a wire spring two steel nails and a digital caliper) [3] in June and December since the establishment of the experiment. Meanwhile, five 75 cm × 75 cm litter traps were uniformly distributed under the stand canopies to measure litterfall biomass from November 2015 to April 2016. The growth rate and litterfall productivity of Chinese fir were calculated at a plot level in this study in order to assess the potential effect of overstory trees on light conditions of understory plants.

2.6. Data Analysis

The data were tested for homogeneity of variances (Brown and Forsythe's variation of Levene's test) before statistical analysis. A paired t-test was used to compare the differences in soil and plant variables between the control and N addition treatment. Multi-way analysis of variance (ANOVA) was used to determine the interactive effects of N addition, plant species and plant tissue on plant nutrient and NSC concentrations. Pearson's correlation analysis was performed to determine the relationship among soil available nutrients, plant nutrients and NSC parameters. All ANOVA and correlation analyses were conducted with a significance criterion of $p < 0.05$ using IBM SPSS 19 statistical software (SPSS, Chicago, IL, USA).

3. Results

3.1. Plant Growth and Soil Nutrients

After 4 years of N addition, the aboveground biomass of *D. dichotoma* and *L. gracile* significantly decreased by 82.1% and 67.2%, respectively ($p < 0.05$), while the biomass of *M. dodecandrum*, which was much lower than that of the other two species, did not respond to N addition (Table 1). Moreover, N addition did not alter the richness of understory plants (Table 1). In contrast, the average girth growth rates of overstory trees (Chinese fir) within four years after N addition treatment and litterfall productivity in the fifth year significantly increased by 18.28% and 36.71%, respectively (Figure 1, Table A2).

Table 1. The aboveground biomass and richness of the major understory plants in the Chinese fir forest plantation treated by nitrogen addition or in the control in 2015.

Variables	Control	Nitrogen Addition	*T*-test
Dicranopteris dichotoma			
Biomass (kg ha^{-1})	652.51 ± 83.71	117.37 ± 38.49	$p < 0.05$
Richness	4.00 ± 1.01	2.25 ± 0.63	ns
Lophatherum gracile			
Biomass (kg ha^{-1})	205.73 ± 16.15	67.84 ± 33.96	$p < 0.05$
Richness	7.67 ± 1.15	4.00 ± 2.45	ns
Melastoma dodecandrum			
Biomass (kg ha^{-1})	34.33 ± 16.40	14.80 ± 8.22	ns
Richness	3.57 ± 0.81	3.67 ± 1.20	ns

The values are the means $\pm$ SE (n = 4). ns = not significant at $p > 0.05$ level.

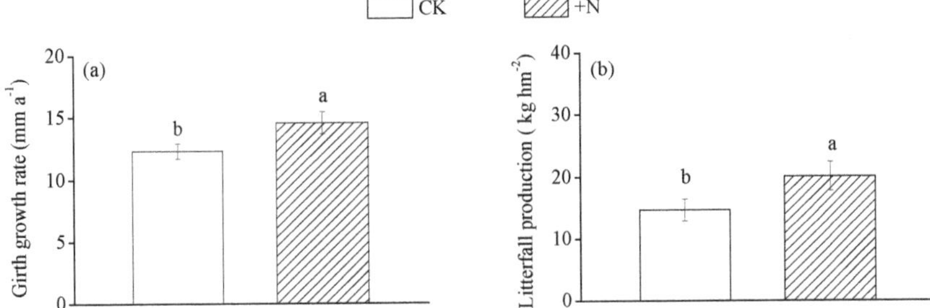

Figure 1. Girth growth from 2011 to 2015 (**a**) and litterfall production in 2016 (**b**) of the Chinese fir tree plantation treated by nitrogen addition (+N) or in the control (CK). Note: Values shown are the means $\pm$ SE (n = 4). Lowercase letters indicate significant differences at the $p < 0.05$ level between the control and N-treatment plots.

As expected, the rhizosphere soil NH_4^+-N and NO_3^--N concentrations of all three species were generally higher in the N addition treatment than in the control, while the available P concentration and AP and NAG activities were unaffected by N addition in both rhizosphere and bulk soils (except for NAG activity, which significantly increased in the rhizosphere soil of *L. gracile* in response to the N addition treatment) (Table 2).

Table 2. Available nutrients and enzyme activities in rhizosphere and bulk soils of the three understory plant species in the Chinese fir plantation treated by nitrogen addition or in the control.

Variables	Control	Nitrogen Addition	*T*-test
Rhizosphere Soil			
Dicranopteris dichotoma			
NH_4^+-N (mg kg^{-1})	16.03 ± 0.52	34.22 ± 0.89	$p < 0.05$
NO_3^--N (mg kg^{-1})	1.68 ± 0.11	2.93 ± 0.35	$p < 0.05$
Available P (mg kg^{-1})	3.81 ± 0.19	3.57 ± 0.57	ns
NAG activity (mmol g^{-1} h^{-1})	30.5 ± 8.47	14.3 ± 3.45	ns
AP activity (mmol g^{-1} h^{-1})	358.3 ± 57.33	419.8 ± 39.29	ns
Lophatherum gracile			
NH_4^+-N (mg kg^{-1})	9.95 ± 0.66	19.56 ± 0.23	$p < 0.05$
NO_3^--N (mg kg^{-1})	1.85 ± 0.31	3.04 ± 0.85	ns
Available P (mg kg^{-1})	3.82 ± 0.36	3.40 ± 0.15	ns
NAG activity (mmol g^{-1} h^{-1})	3 ± 2.38	20.8 ± 5.15	$p < 0.05$
AP activity (mmol g^{-1} h^{-1})	231.5 ± 52.34	303.5 ± 32.75	ns

Table 2. *Cont.*

Variables	Control	Nitrogen Addition	*T*-test
Melastoma dodecandrum			
NH_4^+-N (mg kg^{-1})	7.83 ± 0.71	12.49 ± 0.74	$p < 0.05$
NO_3^--N (mg kg^{-1})	1.16 ± 0.06	2.21 ± 0.37	$p < 0.05$
Available P (mg kg^{-1})	3.02 ± 0.42	3.77 ± 0.27	ns
NAG activity (mmol g^{-1} h^{-1})	22.5 ± 4.575	28.5 ± 7.89	ns
AP activity (mmol g^{-1} h^{-1})	272 ± 15.44	356.3 ± 56.79	ns
Bulk Soil			
Dicranopteris dichotoma			
NH_4^+-N (mg kg^{-1})	12.68 ± 3.09	31.99 ± 0.76	$p < 0.05$
NO_3^--N (mg kg^{-1})	1.55 ± 0.17	2.74 ± 0.84	ns
Available P (mg kg^{-1})	3.16 ± 0.14	3.46 ± 0.36	ns
NAG activity (mmol g^{-1} h^{-1})	26.01 ± 12.45	33.81 ± 3.04	ns
AP activity (mmol g^{-1} h^{-1})	344.82 ± 88.11	372.31 ± 29.19	ns
Lophatherum gracile			
NH_4^+-N (mg kg^{-1})	9.66 ± 0.81	18.66 ± 1.20	$p < 0.05$
NO_3^--N (mg kg^{-1})	1.39 ± 0.09	3.23 ± 0.26	$p < 0.05$
Available P (mg kg^{-1})	4.06 ± 0.45	3.06 ± 0.41	ns
NAG activity (mmol g^{-1} h^{-1})	12.80 ± 7.98	19.55 ± 6.88	ns
AP activity (mmol g^{-1} h^{-1})	237.8 ± 21.17	317.3 ± 20.07	ns
Melastoma dodecandrum			
NH_4^+-N (mg kg^{-1})	8.18 ± 1.47	14.88 ± 0.52	$p < 0.05$
NO_3^--N (mg kg^{-1})	0.92 ± 0.08	2.48 ± 0.37	$p < 0.05$
Available P (mg kg^{-1})	3.77 ± 0.27	3.54 ± 0.24	ns
NAG activity (mmol g^{-1} h^{-1})	13.34 ± 7.78	29.51 ± 10.51	ns
AP activity (mmol g^{-1} h^{-1})	361.3 ± 36.82	314.3 ± 19.26	ns

The values are the means $\pm$ SE ($n = 4$). Available P = Available phosphorus, NAG = N-acetyl-β-D-glucosaminidase, AP = acid phosphatase; ns = not significant at $p > 0.05$ level.

3.2. Nutrients and NSCs in Plant Tissues

Nitrogen addition, plant species, and plant tissue had significant interactive effects on the nutrients and NSC concentrations of the three understory plants (Figure 2). The average tissue N concentration increased, P and ST concentrations did not change, and average SS and NSC concentrations decreased in response to the N addition treatment (Figure 2). When analyzed by species and tissue type, N addition significantly increased the N concentration in both tissue types of all three species (Figure 2a); significantly increased the P concentration in the aboveground tissue of *M. dodecandrum* (Figure 2b); significantly decreased the ST concentration in the aboveground tissue of *M. dodecandrum* and belowground tissue of *D. dichotoma* (Figure 2d); significantly decreased the SS in the aboveground and belowground tissues of *M. dodecandrum* and the aboveground tissue of *D. dichotoma* (Figure 2c); and significantly decreased the NSC concentrations of *M. dodecandrum* while significantly increasing the NSC concentration in the aboveground tissue of *L. gracile* (Figure 2e).

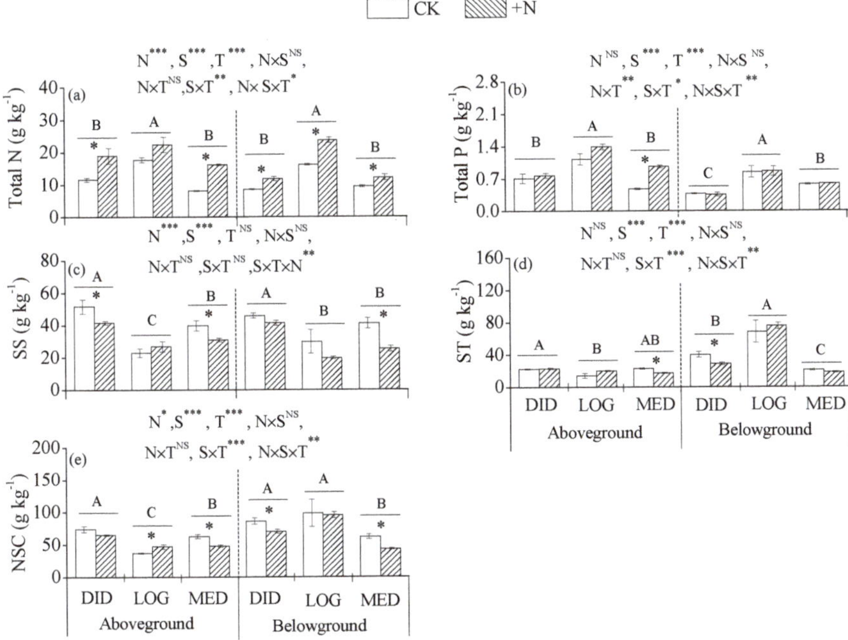

Figure 2. Total nitrogen (**a**), total phosphorus (**b**), soluble sugar (**c**), starch (**d**) and nonstructural ccarbohydrates (**e**) concentrations in aboveground and belowground tissues of the three understory plant species in the Chinese fir plantation treated by nitrogen addition (+N) or in the control (CK). Note: Values shown are the means ± SE (*n* = 4). The asterisks (*) indicate significant differences at the *p* < 0.05 level between the control and N-treatment plots within a single species. Different capital letters indicate significant differences (*p* < 0.05) among the three species within the same tissue based on Duncan's multiple range test. Total N = total nitrogen, Total P = total phosphorus, SS = soluble sugar, ST = starch, NSC = nonstructural carbohydrates; N = nitrogen addition, S = species, T = tissues. ^NS, not significant; * *p* < 0.05; ** *p* < 0.01; *** *p* < 0.001. DID: *D. dichotoma*, LOG: *L. gracile*, MED: *M. dodecandrum*.

3.3. The Ratios of N, P, and NSCs in Plant Tissues

Similarly, N addition, plant species, and plant tissue had significant interactive effects on the ratios of N, P, and the NSCs of the major understory plants (Figure 3). Nitrogen addition did not affect the N/P ratio in the aboveground tissue of any species, but it significantly increased the belowground N/P ratio in all plant tissues (Figure 3a). For *M. dodecandrum*, both the NSC/N and NSC/P ratios significantly decreased in the aboveground and belowground tissues due to N addition; for *D. dichotoma*, only the NSC/N ratio in the aboveground tissue significantly deceased with N addition; and for *L. gracile*, both the NSC/N and NSC/P ratios were unaffected by N addition (Figure 3b,c).

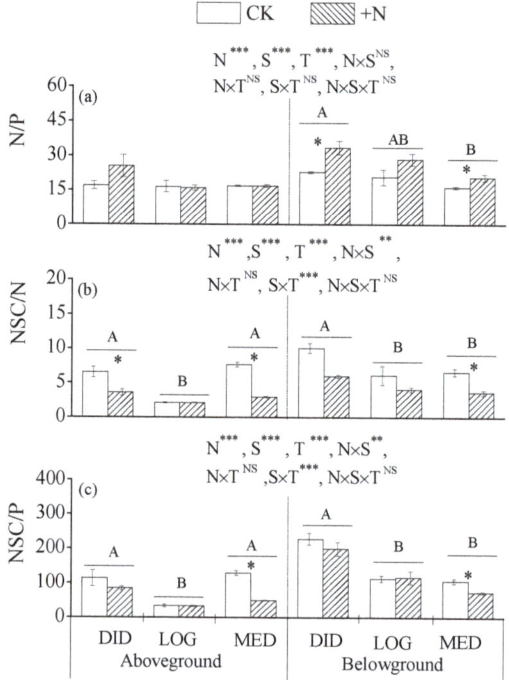

Figure 3. N/P (**a**), NSC/N (**b**), NSC/P (**c**) in aboveground and belowground tissues of the three understory plants in the Chinese fir forest plantation treated by nitrogen addition (+N) or in the control (CK). Note: Values shown are the means $\pm$ SE ($n = 4$). The asterisks (*) indicate significant differences at the $p < 0.05$ level between the control and N-treatment plots within a single species. Different capital letters indicate significant differences ($p < 0.05$) among the three species within the same tissue based on Duncan's multiple range test. N = nitrogen, P = phosphorus, NSC = nonstructural carbohydrates; N = nitrogen addition, S = species, T = tissues. ^{NS}, not significant; * $p < 0.05$; ** $p < 0.01$; *** $p < 0.001$. DID: *D. dichotoma*, LOG: *L. gracile*, MED: *M. dodecandrum*.

3.4. Linkages between Plant Tissue Acquirement and Rhizosphere Soil Supply

The tissue N concentration was generally and positively correlated with rhizosphere soil NH_4^+-N and/or NO_3^--N for each of the three understory plants (Table 3). The belowground tissue P concentration was significantly and positively correlated with rhizosphere soil available P for *D. dichotoma* and *L. gracile* but not for *M. dodecandrum*, while the aboveground tissue P concentration was significantly and positively correlated with rhizosphere soil NO_3^--N for *L. gracile* and *M. dodecandrum* but not for *D. dichotoma* (Table 3). The belowground tissue NSC concentration was negatively correlated with rhizosphere soil NH_4^+-N for *M. dodecandrum* and *D. dichotoma* and positively correlated with rhizosphere soil available P for *L. gracile*, while the aboveground tissue NSC concentration was positively correlated with rhizosphere soil NH_4^+-N and NO_3^--N for *L. gracile* and negatively correlated with rhizosphere soil NH_4^+-N for *M. dodecandrum* (Table 3). Likewise, the correlations among the ratios of N/P, NSC/N, NSC/P and soil available nutrients varied with plant species and tissues (Table 3).

Table 3. The correlation coefficients ($n = 8$) between plant tissue and rhizosphere soil nutrients for each of the three understory plant species in the Chinese fir plantation.

Plant Variables	N	P	NSCs	N/P	NSCs/N	NSCs/P
			Belowground Tissue			
Dicranopteris dichotoma						
NH_4^+-N	0.88 **	−0.15 ns	−0.74 *	0.81 *	−0.89 **	−0.41 ns
NO_3^--N	0.93 **	0.18 ns	−0.53 ns	0.61 ns	−0.81 *	−0.53 ns
Available P	0.29 ns	0.93 **	0.22 ns	−0.53 ns	−0.04 ns	−0.66 ns
Lophatherum gracile						
NH_4^+-N	0.95 **	0.04 ns	−0.04 ns	0.60 ns	−0.50 ns	0.10 ns
NO_3^--N	0.64 ns	0.77 *	0.16 ns	−0.30 ns	−0.12 ns	−0.54 ns
Available P	−0.33 ns	0.76 *	0.73 *	0.81 *	0.01 ns	−0.84 **
Melastoma dodecandrum						
NH_4^+-N	0.68 ns	0.66 ns	−0.79 *	0.60 ns	−0.78 *	−0.83 *
NO_3^--N	0.25 ns	0.11 ns	−0.62 ns	0.25 ns	−0.55 ns	−0.59 ns
Available P	0.06 ns	0.15 ns	−0.16 ns	−0.12 ns	−0.17 ns	0.03 ns
			Aboveground Tissue			
Dicranopteris dichotoma						
NH_4^+-N	0.75 *	0.16 ns	−0.57 ns	0.57 ns	−0.76 *	−0.39 ns
NO_3^--N	0.90 **	0.16 ns	−0.60 ns	0.71 *	−0.83 *	−0.42 ns
Available P	0.51 ns	−0.16 ns	−0.15 ns	−0.32 ns	−0.02 ns	0.56 ns
Lophatherum gracile						
NH_4^+-N	0.59 ns	0.61 ns	0.71 *	−0.10 ns	0.11 ns	−0.06 ns
NO_3^--N	0.84 **	0.75 *	0.86 **	0.03 ns	−0.31 ns	−0.06 ns
Available P	−0.20 ns	0.37 ns	−0.14 ns	0.33 ns	−0.55 ns	−0.53 ns
Melastoma dodecandrum						
NH_4^+-N	0.85 **	0.89 **	−0.88 **	−0.30 ns	−0.92 **	−0.94 **
NO_3^--N	0.77 *	0.83 *	−0.45 ns	−0.38 ns	−0.70 ns	−0.72 *
Available P	0.52 ns	0.51 ns	−0.64 ns	−0.58 ns	−0.56 ns	0.10 ns

NSCs = nonstructural carbohydrates; ns, not significant, $p > 0.05$; * $p < 0.05$; ** $p < 0.01$.

4. Discussion

4.1. Links between Plant Tissue Acquirement and Rhizosphere Soil Supply

Similar to the results in many previous studies [3,30], soil available N, including NH_4^+-N and NO_3^--N, increased with N addition in the form of NH_4NO_3. Consequently, in our study, the N concentrations in aboveground and belowground tissues were found to be higher in response to N addition for the three studied species. This result was consistent with those from previous studies in forest ecosystems [31–33]. Gurmesa found that plants could take up excessive N even in a N-saturated forest [34], which supported our results, as our plantation is located in a N-rich ecosystem within a region with extreme N deposition [35–37]. Thus, the higher soil N supply increased the N concentrations in aboveground and belowground tissues in response to N addition.

Surprisingly, the available P concentration and activities of both enzymes (NAG and AP) in rhizosphere and bulk soils of the three species generally did not change with N addition. Dong et al. (2015) also found the soil total P did not alter by N addition, while organic matter and total N increased due to three years' N addition of 100 kg hm^{-2} $year^{-1}$ in the same site of this study [26]. These results indicated that the intensity and duration of the simulated N addition might still be within the buffer range of the soil P supply in the studied plantation [3]. Thus, unsurprisingly, the P concentration in plant tissues was generally unaffected by N addition in our study. However, the P concentration in aboveground tissue of *M. dodecandrum* did increase with N addition. In general, the effect of N addition on plant P may be altered via P supply, P uptake and P resorption [38–40]. In our study, soil available P and belowground tissue P showed minimal responses to N addition. Some studies suggested that N addition decreased P uptake by reducing root biomass and inhibiting mycorrhiza growth [10,41]. Other studies found that N addition decreased the extraradical hyphae of Chinese fir mycorrhiza, but did not alter fine root biomass in a subtropical plantation forest [42]. Thus, we speculate that P resorption might be a potential mechanism increasing aboveground P concentration in *M. dodecandrum*. Compared with the other two species (a sun fern and a neutral grass), *M. dodecandrum* is a creeping shrub with traits characteristic of shade plants and is thus better suited to growing in an understory environment (Figure A1). To meet the requirements for growth, *M. dodecandrum* (the only studied species that did not exhibit decreased biomass with N addition) might improve P resorption to overcome the P insufficiency in soils and increase its aboveground P concentration to maintain N/P stoichiometric homeostasis.

As expected, the ratio of mineral N to available P in soils and the N/P ratio in belowground tissues generally increased with N addition, which partially supported our first hypothesis that N addition alters the balance between N and P in soils and understory plants. However, the N/P ratio in the aboveground tissue was not affected by N addition in any species, which seems counter to our first hypothesis of an imbalance between N and P in understory plants driven by N addition. The divergent patterns of the N/P ratio between aboveground and belowground tissues indicated that the understory plants might maintain N/P homeostasis in aboveground tissue to meet the requirement for leaf photosynthesis and plant growth [13,43], regardless of the alteration of the N/P ratio in plant belowground tissues due to N addition.

To achieve N/P homeostasis in aboveground tissue, the three understory species might use different mechanisms, such as the resorption in *M. dodecandrum* mentioned above, which should be further studied. The N/P ratio in plant tissue has been widely used as a diagnostic tool for evaluating nutrient limitation in terrestrial ecosystems [44,45]. Previous studies observed that the N/P ratio in tree leaves decreased with N deposition, and this indicated that N deposition might aggravate P limitation of plant growth in forest ecosystems [46]. In our study, the N/P ratio of aboveground tissue was generally less than 16, except for in *D. dichotoma* treated by N addition, and the N/P ratio in aboveground tissue was unchanged by N addition. These results suggested that N addition might not aggravate the P limitation of understory plant growth in this subtropical plantation.

4.2. Effects of N Addition on Understory Plant Growth and NSC Allocation

Understory plant growth might be altered by N addition through a direct effect of nutrient supply and an indirect effect of the overstory canopy via competition for resources such as light and water [3,14,31]. In our study, the biomass of two dominant understory species (*D. dichotoma* and *L. gracile*) significantly decreased with N addition. A recent study found that a moderate supply of N stimulated the understory vegetation growth in a tropical forest because N inputs satisfied plant demands for N [14]. However, some research suggested that understory vegetation productivity was mainly limited by light in a boreal forest, and N addition increased shading by the tree canopy (thus, less light was available to understory plants) [12]. Some other studies also found that plant growth was inhibited by N addition due to the aggravation of P limitation, with a mismatch in N and P stoichiometry [47,48].

Based on the responses of soil and plant nutrients to N addition discussed above, N addition likely did not aggravate the P limitation of understory plant growth in this subtropical plantation. Therefore, the observed decline in understory plant biomass may be most likely caused by increased shading by stimulated overstory tree growth due to N addition. Compared with overstory trees, understory plants are more easily limited by light [10]. Our data also indicated that N addition promoted the growth and litter production of Chinese fir (the canopy tree) (Figure 1) and thus could lead to a decrease in the light available to understory vegetation. The decrease in light availability might help explain the negative effect on understory vegetation growth in the N addition plots.

The NSCs in plant tissue are the products of photosynthesis and the main energy sources for plant growth [29]. Because NSCs can reflect plant response to light and nutrient availability [49,50], they may provide an effective way to reveal the underlying mechanism of N addition effects on understory plants driven mainly by light or nutrient resources. The storage of NSCs is lower in shade than in sun environments because carbohydrate synthesis is often limited by lower light availability [29]. In our study, NSCs, especially SS, concentrations in understory plants significantly decreased with N addition. However, previous studies found that carbohydrate reserves increased with elevated N supply when N inputs satisfied plant demand for N and increased the photosynthetic capacity [16,17]. There are several potential reasons for the reduction in NSC concentration of understory plants in response to N addition in our study. First, N addition may have increased the plant growth and decreased the accumulation of NSCs because the exogenous N supply stimulated the synthesis of amino acids and amide compounds to suppress the accumulation of carbohydrates for protein synthesis when N was in excess [49]. Second, an excessive concentration of N in aboveground tissue may have resulted in inorganic N toxicity, which may have downregulated the photosynthetic capacity [14]. Third, the light available to the understory plants may have significantly decreased because the tree canopy cover increased under the elevated N supply [51,52]. Our study found that the soil available N was strongly correlated with the N concentration in the tissues of the three understory species, which indicated that the deposited N might not have been excessive or toxic. We also found negative correlations between soil NH_4^+-N and tissue NSCs in *D. dichotoma* and *M. dodecandrum*, but both soil NH_4^+-N and NO_3^--N were positively correlated with aboveground tissue NSCs in *L. gracile*. Therefore, our results further indicated that the growth of understory species might be limited by light availability but not dominated by N (excess or toxicity).

In addition, the ratio of NSCs to nutrients reflects the relationship between the nutrients and the production of NSCs and their use efficiencies [53]. In our study, the ratios of NSCs to nutrients (NSC/N and NSC/P) in plant tissues generally decreased with N addition, implying that each unit of N and each unit of P produced fewer NSCs. The decrease in the NSC/N and NSC/P ratios in tissues also suggested that light availability may limit the photosynthetic rate and growth of understory plants in response to N addition [14,31].

Our results revealed that the three understory species were not consistent in their response to N addition, although they grow well in acid soils. *D. dichotoma* was a high light-demanding species, and *M. dodecandrum* was a shade tolerance species, but *L. gracile* was a duality species that can survive in low light and high light environment. Nitrogen addition has a direct influence on overstory vegetation that changes the light available reaching the understory. It is the reason that explains the different responses to N addition among the three understory species. First, decreased in the biomass of high light-demanding *D dichotoma* due to N addition is a consequence of lower light availability induced by the promoted growth of the Chinese fir tree canopy. This result is consistent with that of other research in that N fertilization has a significant influence on forest tree canopies, which can significantly reduce the light available to the understory plants [51,52]. Second, N addition promoted the growth of the Chinese fir tree canopy, leading to *L. gracile* being light limited as well as increasing the belowground nutrient accumulation and microbial activity of *L. gracile*. Compared with the sun and neutral plants (*D. dichotoma* and *L. gracile*), the shade plant (*M. dodecandrum*) showed a stronger capacity to synchronously increase N and P levels and decrease SS and ST pools in its tissues to maintain productivity and adapt to the shadier environment in the N addition treatment. In contrast, P and NSCs in the sun and neutral plants showed a weaker response to the N addition treatment, which led to a decrease in plant biomass with N addition. Furthermore, the correlations among soil-available nutrients, tissue nutrients, and NSCs differed among species, which further indicated species-specific mechanisms in response to N addition due to the differences in nutrient- and light-related traits among the three understory species.

5. Conclusions

Nitrogen addition did not lead to a mismatch in N and P stoichiometry in the aboveground tissues of understory plants, although N addition increased N availability in soils and plants. However, N addition decreased NSCs, especially SS, in two of the three studied species and the NSC/N and NSC/P ratios of all three species. Meanwhile, the aboveground biomass of *D. dichotoma* and *L. gracile* significantly decreased after four years of simulated N addition, and the biomass of *M. dodecandrum* did not respond to N addition. These results suggested that N addition might inhibit the growth of understory plants through decreasing the nonstructural carbohydrates and light availability indirectly rather than by changing nutrients and N/P stoichiometry directly, although species-specific responses to N addition occurred in the Chinese fir plantation. The limitation of available light to understory species through the facilitation of N addition on overstory canopy growth may be the underlying mechanism, and thinning should therefore be used to improve understory vegetation biomass and other potential functions to mitigate the adverse effects caused by N deposition on understory plant species.

Author Contributions: F.C. was responsible for funding acquisition and resources. F.C. and G.G.W. conceptualized the study. F.W. performed the data curation and investigation. F.C. and H.W. participated in the design of the study. X.F. and W.B. supervised the experimental process. F.W. wrote the original draft. R.M. reviewed and edited the manuscript. All authors read and approved the final manuscript.

Funding: This research was funded by [the National Natural Science Foundation of China] grant numbers [31730014 and 31870427] and [Jiangxi Provincial Department of Science and Technology] grant numbers [20153BCB22008, 20165BCB19006, and 20181ACH80006]. The APC was funded by [2011 Collaborative Innovation Center of Jiangxi Typical Trees Cultivation and Utilization, Jiangxi Agricultural University].

Acknowledgments: We thank Xiu-Lan Zhang, Zhang-Min Li and Gao-Yang Wu for their field work and sample analysis.

Conflicts of Interest: The authors have declared that no competing interests exist.

Appendix A

Table A1. General properties of the Chinese fir forest plantation before experimental treatment in 2011.

Variables	Control	Nitrogen Addition	*T*-test
Soil			
Bulk density (g cm^{-3})	1.22 ± 0.03	1.25 ± 0.03	ns
pH	4.43 ± 0.04	4.40 ± 0.08	ns
Organic carbon (g kg^{-1})	21.20 ± 2.2	22.41 ± 1.62	ns
Total nitrogen (g kg^{-1})	1.29 ± 0.13	1.26 ± 0.09	ns
Total phosphorus (g kg^{-1})	0.29 ± 0.02	0.31 ± 0.03	ns
Stand			
Density (ha^{-1})	2250 ± 45	2150 ± 62	ns
Average DBH (cm)	12.2 ± 0.2	11.9 ± 0.3	ns
Average height (m)	8.5 ± 0.2	8.7 ± 0.2	ns

Data are cited from our previous study [3]; ns indicates not significant between the control and the nitrogen addition treatment at the $p < 0.05$ level. Values are the means $\pm$ SE ($n = 4$). DBH = diameter at breast height.

Table A2. Girth growth rates at breast height in the Chinese fir forest plantation in the nitrogen addition treatment and control from 2012 to 2015.

Variables	Control	Nitrogen Addition	*T*-test
Girth growth rate			
1-year (mm year^{-1})	10.89 ± 0.69	13.69 ± 1.14	$p < 0.05$
2-year (mm year^{-1})	15.13 ± 0.70	18.64 ± 1.18	$p < 0.05$
3-year (mm year^{-1})	9.50 ± 0.57	10.66 ± 0.60	ns
4-year (mm year^{-1})	13.72 ± 1.15	15.23 ± 1.14	ns
Average (mm year^{-1})	12.31 ± 0.61	14.56 ± 0.88	$p < 0.05$

ns indicates not significant between the control and the nitrogen addition treatment at the $p < 0.05$ level. Values are the means $\pm$ SE ($n = 4$).

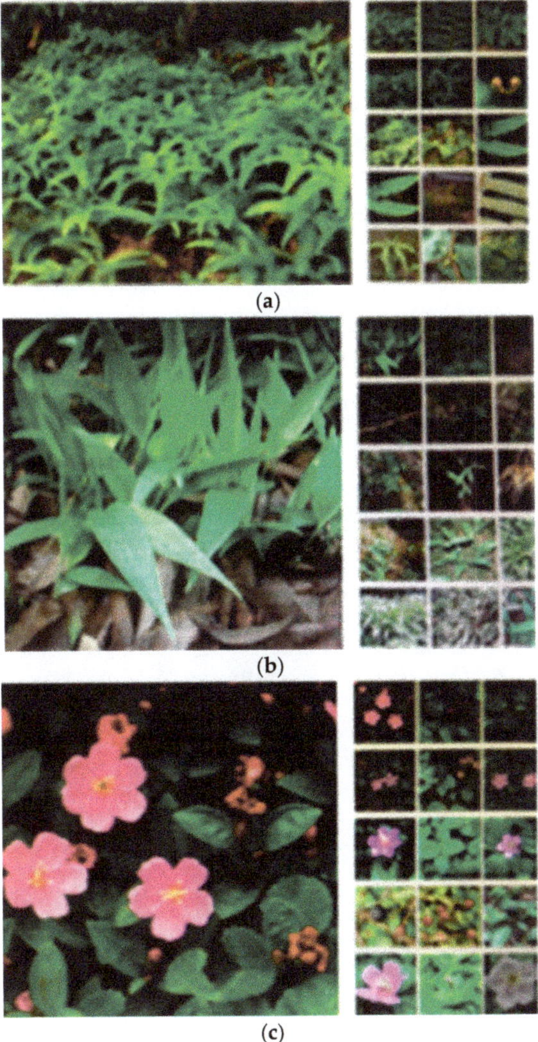

Figure A1. The morphological characteristics of the three species of understory plants in our study. (**a**) *Dicranopteris dichotoma* [54] (**b**) *Lophatherum gracile* [55] (**c**) *Melastoma dodecandrum* [56].

References

1. Galloway, J.N.; Townsend, A.R.; Erisman, J.W.; Bekunda, M.; Cai, Z.; Freney, J.R.; Sutton, M.A. Transformation of the nitrogen cycle, recent trends, questions, and potential solutions. *Science* **2008**, *320*, 889–892. [CrossRef] [PubMed]
2. Liu, X.; Zhang, Y.; Han, W.; Tang, A.; Shen, J.; Cui, Z.; Fangmeier, A. Enhanced nitrogen deposition over China. *Nature* **2013**, *494*, 459–462. [CrossRef] [PubMed]
3. Chen, F.S.; Niklas, K.J.; Liu, Y.; Fang, X.M.; Wan, S.Z.; Wang, H. Nitrogen and phosphorus additions alter nutrient dynamics but not resorption efficiencies of Chinese fir leaves and twigs differing in age. *Tree Physiol.* **2015**, *35*, 1106–1117. [CrossRef] [PubMed]
4. Lu, X.; Mao, Q.; Gilliam, F.S.; Luo, Y.; Mo, J. Nitrogen deposition contributes to soil acidification in tropical ecosystems. *Glob. Chang. Biol.* **2014**, *20*, 3790–3801. [CrossRef] [PubMed]

5. Wang, M.; Zhang, W.W.; Li, N.; Liu, Y.Y.; Zheng, X.B.; Hao, G.Y. Photosynthesis and growth responses of *Fraxinus mandshurica* Rupr. seedlings to a gradient of simulated nitrogen deposition. *Ann. For. Sci.* **2018**, *75*, 1. [CrossRef]

6. Kumar, P.; Chen, H.Y.; Thomas, S.C.; Shahi, C. Linking resource availability and heterogeneity to understory species diversity through succession in boreal forest of Canada. *J. Ecol.* **2018**, *106*, 1266–1276. [CrossRef]

7. Bobbink, R.; Hicks, K.; Galloway, J.; Spranger, T.; Alkemade, R.; Ashmore, M.; Emmett, B. Global assessment of nitrogen deposition effects on terrestrial plant diversity: A synthesis. *Ecol. Appl.* **2010**, *20*, 30–59. [CrossRef]

8. Gilliam, F.S.; Hockenberry, A.W.; Adams, M.B. Effects of atmospheric nitrogen deposition on the herbaceous layer of a central Appalachian hardwood forest. *J. Torry. Bot. Soc.* **2006**, *133*, 240–254. [CrossRef]

9. Lu, X.; Mo, J.; Gilliam, F.S.; Zhou, G.; Fang, Y. Effects of experimental nitrogen additions on plant diversity in an old-growth tropical forest. *Glob. Chang. Biol.* **2010**, *16*, 2688–2700. [CrossRef]

10. Gilliam, F.S. Response of the herbaceous layer of forest ecosystems to excess nitrogen deposition. *J. Ecol.* **2006**, *94*, 1176–1191. [CrossRef]

11. Hurd, T.M.; Brach, A.R.; Raynal, D.J. Response of understory vegetation of Adirondack forests to nitrogen additions. *Can. J. For. Res.* **1998**, *28*, 799–807. [CrossRef]

12. Strengbom, J.; Näsholm, T.; Ericson, L. Light, not nitrogen, limits growth of the grass *Deschampsia flexuosa* in boreal forests. *Can. J. Bot.* **2004**, *82*, 430–435. [CrossRef]

13. Kou, L.; Chen, W.; Jiang, L.; Dai, X.; Fu, X.; Wang, H.; Li, S. Simulated nitrogen deposition affects stoichiometry of multiple elements in resource-acquiring plant organs in a seasonally dry subtropical forest. *Sci. Total. Environ.* **2018**, *624*, 611–620. [CrossRef] [PubMed]

14. Mao, Q.; Lu, X.; Mo, H.; Gundersen, P.; Mo, J. Effects of simulated N deposition on foliar nutrient status, N metabolism and photosynthetic capacity of three dominant understory plant species in a mature tropical forest. *Sci. Total. Environ.* **2018**, *610*, 555–562. [CrossRef] [PubMed]

15. Elser, J.J.; Bracken, M.E.; Cleland, E.E.; Gruner, D.S.; Harpole, W.S.; Hillebrand, H.; Smith, J.E. Global analysis of nitrogen and phosphorus limitation of primary producers in freshwater, marine and terrestrial ecosystems. *Ecol. Lett.* **2007**, *10*, 1135–1142. [CrossRef] [PubMed]

16. Ibrahim, M.H.; Jaafar, H.Z.; Rahmat, A.; Rahman, Z.A. The relationship between phenolics and flavonoids production with total non-structural carbohydrate and photosynthetic rate in *Labisia pumila* Benth. under high CO_2 and nitrogen fertilization. *Molecules* **2011**, *16*, 162–174. [CrossRef] [PubMed]

17. Xiao, L.; Liu, G.; Li, P.; Xue, S. Nitrogen addition has a stronger effect on stoichiometries of non-structural carbohydrates, nitrogen and phosphorus in *Bothriochloa ischaemum* than elevated CO_2. *Plant Growth Regul.* **2017**, *83*, 325–334. [CrossRef]

18. Lü, X.T.; Reed, S.; Yu, Q.; He, N.P.; Wang, Z.W.; Han, X.G. Convergent responses of nitrogen and phosphorus resorption to nitrogen inputs in a semiarid grassland. *Glob. Chang. Biol.* **2013**, *19*, 2775–2784. [CrossRef]

19. Knox, K.J.E.; Clarke, P.J. Nutrient availability induces contrasting allocation and starch formation in resprouting and obligate seeding shrubs. *Funct. Ecol.* **2005**, *19*, 690–698. [CrossRef]

20. Xiong, Y.; Xu, X.; Zeng, H.; Wang, H.; Chen, F.; Guo, D. Low Nitrogen Retention in Soil and Litter under Conditions without Plants in a Subtropical Pine Plantation. *Forests* **2015**, *6*, 2387–2404. [CrossRef]

21. Kou, L.; Zhang, X.; Wang, H.; Yang, H.; Zhao, W.; Li, S. Nitrogen additions inhibit nitrification in acidic soils in a subtropical pine plantation: Effects of soil pH and compositional shifts in microbial groups. *J. For. Res.* **2017**, *29*, 1–10. [CrossRef]

22. Lü, C.; Tian, H. Spatial and temporal patterns of nitrogen deposition in China: Synthesis of observational data. *J. Geophys. Res. Atmos.* **2007**, *112*, 10–15. [CrossRef]

23. Hu, X.F.; Chen, F.S.; Wine, M.L.; Fang, X.M. Increasing acidity of rain in subtropical tea plantation alters aluminum and nutrient distributions at the root-soil interface and in plant tissues. *Plant Soil* **2017**, *417*, 261–274. [CrossRef]

24. Woodruff, D.R.; Meinzer, F.C. Water stress, shoot growth and storage of non-structural carbohydrates along a tree height gradient in a tall conifer. *Plant Cell Environ.* **2011**, *34*, 1920–1930. [CrossRef] [PubMed]

25. Allen, S.E. *Chemical Analysis of Ecological Materials*; Blackwell Scientific Publications: Oxford, UK, 1989.

26. Dong, W.Y.; Zhang, X.Y.; Liu, X.Y.; Fu, X.L.; Chen, F.S.; Wang, H.M.; Wen, X.F. Responses of soil microbial communities and enzyme activities to nitrogen and phosphorus additions in Chinese fir plantations of subtropical China. *Biogeosciences* **2015**, *12*, 5537–5546. [CrossRef]

27. Dubois, M.; Gilles, K.A.; Hamilton, J.K.; Rebers, P.T.; Smith, F. Colorimetric method for determination of sugars and related substances. *Anal. Chem.* **1956**, *28*, 350–356. [CrossRef]

28. Liu, J.F.; Arend, M.; Yang, W.J.; Schaub, M.; Ni, Y.Y.; Gessler, A.; Li, M.H. Effects of drought on leaf carbon source and growth of European beech are modulated by soil type. *Sci. Rep.* **2017**, *7*, 42462. [CrossRef]

29. Song, X.; Peng, C.; Zhou, G.; Gu, H.; Li, Q.; Zhang, C. Dynamic allocation and transfer of non-structural carbohydrates, a possible mechanism for the explosive growth of Moso bamboo (*Phyllostachys heterocycla*). *Sci. Rep.* **2016**, *6*, 25908. [CrossRef] [PubMed]

30. Zhu, B.; Panke-Buisse, K.; Kao-Kniffin, J. Nitrogen fertilization has minimal influence on rhizosphere effects of smooth crabgrass (Digitaria ischaemum) and bermudagrass (Cynodon dactylon). *J. Plant Ecol.* **2014**, *8*, 390–400. [CrossRef]

31. Mao, Q.; Lu, X.; Wang, C.; Zhou, K.; Mo, J. Responses of understory plant physiological traits to a decade of nitrogen addition in a tropical reforested ecosystem. *For. Ecol. Manag.* **2017**, *401*, 65–74. [CrossRef]

32. Nordin, A.; Näsholm, T.; Ericson, L. Effects of simulated N deposition on understorey vegetation of a boreal coniferous forest. *Funct. Ecol.* **1998**, *12*, 691–699. [CrossRef]

33. Talhelm, A.F.; Pregitzer, K.S.; Burton, A.J. No evidence that chronic nitrogen additions increase photosynthesis in mature sugar maple forests. *Ecol. Appl.* **2011**, *21*, 2413–2424. [CrossRef] [PubMed]

34. Gurmesa, G.A.; Lu, X.; Gundersen, P.; Mao, Q.; Zhou, K.; Fang, Y.; Mo, J. High retention of ^{15}N-labeled nitrogen deposition in a nitrogen saturated old-growth tropical forest. *Glob. Chang. Bio.* **2016**, *22*, 3608–3620. [CrossRef] [PubMed]

35. Jia, Y.; Yu, G.; He, N.; Zhan, X.; Fang, H.; Sheng, W.; Wang, Q. Spatial and decadal variations in inorganic nitrogen wet deposition in China induced by human activity. *Sci. Rep.* **2014**, *4*, 3763. [CrossRef] [PubMed]

36. Ma, Z.; Zhang, X.; Zhang, C.; Wang, H.; Chen, F.; Fu, X.; Lei, Q. Accumulation of residual soil microbial carbon in Chinese fir plantation soils after nitrogen and phosphorus additions. *J. For. Res.* **2018**, *29*, 953–962. [CrossRef]

37. Tang, Y.; Zhang, X.; Li, D.; Wang, H.; Chen, F.; Fu, X.; Yu, G. Impacts of nitrogen and phosphorus additions on the abundance and community structure of ammonia oxidizers and denitrifying bacteria in Chinese fir plantations. *Soil. Biol. Biochem.* **2016**, *103*, 284–293. [CrossRef]

38. Jones, A.G.; Power, S.A. Field-scale evaluation of effects of nitrogen deposition on the functioning of heathland ecosystems. *J. Ecol.* **2012**, *100*, 331–342. [CrossRef]

39. Kou, L.; Jiang, L.; Fu, X.; Dai, X.; Wang, H.; Li, S. Nitrogen deposition increases root production and turnover but slows root decomposition in *Pinus elliottii* plantations. *New. Phytol.* **2018**, *218*, 1450–1461. [CrossRef] [PubMed]

40. Yang, H. Effects of nitrogen and phosphorus addition on leaf nutrient characteristics in a subtropical forest. *Trees* **2018**, *32*, 383–391. [CrossRef]

41. Nakaji, T.; Fukami, M.; Dokiya, Y.; Izuta, T. Effects of high nitrogen load on growth, photosynthesis and nutrient status of *Cryptomeria japonica* and *Pinus densiflora* seedlings. *Trees* **2001**, *15*, 453–461.

42. Schulze, E.D. Air pollution and forest decline in a spruce (*Picea abies*) forest. *Science* **1989**, *244*, 776–783. [CrossRef] [PubMed]

43. Li, L.; McCormack, M.L.; Chen, F.; Wang, H.; Ma, Z.; Guo, D. Different responses of absorptive roots and arbuscular mycorrhizal fungi to fertilization provide diverse nutrient acquisition strategies in Chinese fir. *For. Ecol. Manag* **2019**, *433*, 64–72. [CrossRef]

44. Güsewell, S. N: P ratios in terrestrial plants: Variation and functional significance. *New Phytol.* **2004**, *164*, 243–266. [CrossRef]

45. Kang, H.; Xin, Z.; Berg, B.; Burgess, P.J.; Liu, Q.; Liu, Z.; Liu, C. Global pattern of leaf litter nitrogen and phosphorus in woody plants. *Ann. For. Sci.* **2010**, *67*, 811. [CrossRef]

46. Li, Y.; Niu, S.; Yu, G. Aggravated phosphorus limitation on biomass production under increasing nitrogen loading: A meta-analysis. *Glob. Chang. Biol.* **2016**, *22*, 934–943. [CrossRef] [PubMed]

47. Yuan, Z.Y.; Chen, H.Y. Decoupling of nitrogen and phosphorus in terrestrial plants associated with global changes. *Nat. Clim. Chang.* **2015**, *5*, 465. [CrossRef]

48. Yue, K.; Fornara, D.A.; Yang, W.; Peng, Y.; Li, Z.; Wu, F.; Peng, C. Effects of three global change drivers on terrestrial C: N: P stoichiometry: A global synthesis. *Glob. Chang. Biol.* **2017**, *23*, 2450–2463. [CrossRef]

49. Ai, Z.M.; Xue, S.; Wang, G.L.; Liu, G.B. Responses of non-structural carbohydrates and C: N: P Stoichiometry of *Bothriochloa ischaemum* to nitrogen addition on the Loess Plateau, China. *J. Plant Growth Regul.* **2017**, *36*, 714–722. [CrossRef]

50. Ai, Z.; He, L.; Xin, Q.; Yang, T.; Liu, G.; Xue, S. Slope aspect affects the non-structural carbohydrates and C: N: P stoichiometry of *Artemisia sacrorum* on the Loess Plateau in China. *Catena* **2017**, *152*, 9–17. [CrossRef]

51. Hedwall, P.O.; Nordin, A.; Brunet, J.; Bergh, J. Compositional changes of forest-floor vegetation in young stands of Norway spruce as an effect of repeated fertilisation. *For. Ecol. Manag.* **2010**, *259*, 2418–2425. [CrossRef]

52. Hedwall, P.O.; Strengbom, J.; Nordin, A. Can thinning alleviate negative effects of fertilization on boreal forest floor vegetation? *For. Ecol. Manag.* **2013**, *310*, 382–392. [CrossRef]

53. Li, M.H.; Xiao, W.F.; Wang, S.G.; Cheng, G.W.; Cherubini, P.; Cai, X.H.; Zhu, W.Z. Mobile carbohydrates in Himalayan treeline trees I. Evidence for carbon gain limitation but not for growth limitation. *Tree Physiol.* **2008**, *28*, 1287–1296. [CrossRef] [PubMed]

54. Dicranopteris dichotoma. Available online: http://frps.eflora.cn/frps/Dicranopteris%20dichotoma (accessed on 12 September 2018).

55. Lophatherum gracile. Available online: http://frps.eflora.cn/frps/Lophatherum%20gracile (accessed on 12 September 2018).

56. Melastoma dodecandrum. Available online: http://frps.eflora.cn/frps/Melastoma%20dodecandrum (accessed on 12 September 2018).

Article

Immediate Changes in Organic Matter and Plant Available Nutrients of Haplic Luvisol Soils Following Different Experimental Burning Intensities in Damak Forest, Hungary

Jack M. Bridges [1],*, George P. Petropoulos [2] and Nicola Clerici [3]

[1] Geography and Earth Sciences, University of Aberystwyth, Wales SY23 3DB, UK
[2] Department of Soil & Water Resources, Hellenic Agricultural Organization (HAO) "Demeter" (former NAGREF), 1 Theofrastou St., 41335 Larissa, Greece; petropoulos.george@gmail.com
[3] Biology Program, Faculty of Natural Sciences and Mathematics, Universidad del Rosario, Carrera 26 # 63B-48, Bogotá 111221, Colombia; nicola.clerici@urosario.edu.co
* Correspondence: bridgesj874@gmail.com; Tel.: +44-741-156-5456

Received: 3 May 2019; Accepted: 23 May 2019; Published: 24 May 2019

Abstract: One of the major pedological changes produced by wildfires is the drastic modification of forest soil systems properties. To our knowledge, large research gaps are currently present concerning the effect of such fires on forest Haplic Luvisols soils in Central Europe. In this study, the effects of experimental fires on soil organic matter and chemical properties at different burning intensities in a Central European forest were examined. The study was conducted at Damak Forest, in Hungary, ecosystem dominated by deciduous broadleaf trees, including the rare Hungarian oak *Quercus frainetto* Ten. The experimental fires were carried out in nine different plots on Haplic Luvisol soils transferred from Damak Forest to the burning site. Three types of fuel load were collected from the forest: litter layer, understorey and overstorey. Groups of three plots were burned at low (litter layer), medium intensity (litter and understorey) and high intensity (litter, understorey and overstorey). Pre-fire and post-fire soil samples were taken from each plot, analysed in the laboratory and statistically compared. Key plant nutrients of organic matter, carbon, potassium, calcium, magnesium and phosphorus were analysed from each sample. No significant differences in soil organic matter and carbon between pre- and post-fire samples were observed, but high intensity fires did increase soil pH significantly. Calcium, magnesium and phosphorus availability increased significantly at all fire intensity levels. Soil potassium levels significantly decreased (ca. 50%) for all intensity treatments, in contrast to most literature. Potassium is a key nutrient for ion transport in plants, and any loss of this nutrient from the soil could have significant effects on local agricultural production. Overall, our findings provide evidence that support the maintaining of the current Hungarian fire prevention policy.

Keywords: soil properties; experimental fires; nutrients; UV-spectroscopy analysis; thermal infrared thermometer

1. Introduction

Wildfires affect agricultural and forest land covers more than any other land cover type [1]. Prescribed fires involve basically controlled burning of forests as a mean to reduce fire fuel levels. This practice minimises the extent, severity and danger of potential wildfires [2]. In Europe, prescribed fire use has disappeared from many countries due to the intensification of agriculture and socio-economical changes [3]. In most European countries prescribed burning is prohibited. Where permitted (e.g., in UK, Germany, France, Spain, Slovenia), there are strict controls on the timing and location of the burning extension. European fire management policies are still aimed at fire suppression instead of using prescribed fires as a preventative measure. At present, only 10,000 ha year^{-1} of the Mediterranean basin is prescribed burnt; for comparison, this is about 3% of the wildfire extent in Spain, France and Portugal alone [4,5].

The use of prescribed fires in Europe is mostly an 'unnatural' event for both biotic and abiotic systems. Ecosystem alterations can arise as a consequence of fires, such as niche alteration for native species [6], invasion of alien species [7], and drastic changes to forest soil systems properties [2]. Climate change is an additional threat to wildfires frequency, due to projected higher mean temperatures, increase in the number of summer days and the decline of average precipitation. Climate change models suggest that by 2100, temperatures in central Europe will increase up to about 3 °C [8]. This increase in temperatures will cause soils to become drier for longer periods and therefore would extend the wildfire season globally. Southern Europe has already experienced record summer temperatures in 2017. The Intergovernmental Panel on Climate Change's (IPCC) Fifth Report (AR5) on forest fire risk shows as a general trend that each region, below the 55th parallel north, will either increase in forest fire risk or stay at the same level [9]. Temperate forests are particularly sensitive to the effects of climate change due to their long lifespan preventing rapid adaptation. The large amount of C in forest soils means that any deviation of the level will have a significant effect on global C balance and climate change [10]. The future IPCC projections support that the effects of wildfires will be an increasing concern for both the scientific community and the governmental agencies.

A major factor influencing the effects of fires on soil properties is fire severity, a parameter primarily controlled by combustion and site factors. Combustion factors include quantity, moisture and type of fuel, whereas site factors include topography, wind direction, air temperature and humidity [2]. There are two key components of fire behaviour that influence fire severity: intensity and duration. Intensity is defined as the rate of thermal production in the fire, whereas duration is the length of time soil is exposed to fire. The most varied change to forest soils during burning is the loss of soil organic matter (OM) and plant available phosphorous and nitrogen in mineral soils. Fire severity plays a significant role in the effect on this important carbon pool. Substantial consumption of OM occurs between 200 and 250 °C and is complete at ca. 460 °C [11]. Fire does not maintain a consistent temperature in the soil. Some parts of the burn area might experience over 460 °C, whereas other areas may only experience less than 200 °C. This means that the OM may only experience minor volatilisation at low temperatures compared to complete oxidation at high temperatures. There is consensus about combustion causing reduction or total removal of the forest floor [12]. The reduction of OM is not uniform with depth: for example, some studies have found that concentrations of OM decreases significantly by the heating effect at 1 cm depth, but not at 2 and 3 cm depth in a pine forest ecosystem, North-East Spain [13].

Studies have found differences between prescribed burning and wildfire effects on soil 10 years following the fire event: i.e., lower carbon content following prescribed fire and higher soil C content following wildfire, due to the accumulation of charcoal and the encroachment of post-fire N-fixing vegetation [14]. Studies support these findings, describing, e.g., OM returning to normal after one year [13]. They also showed that fire has a fertilising effect, due to the dissolution of ashes and the mineralisation of charcoal enriching the soil.

The key nutrients within forest soils, mainly contained in its organic matter component (OM) are: nitrogen (N), available phosphorus (P), exchangeable potassium (K), exchangeable calcium (Ca) and exchangeable magnesium (Mg). Each nutrient reacts differently to fire, depending on its individual volatilisation threshold [15]. The combustion of nutrients bound in vegetation and soil organic matter add inorganic forms of K, Ca, Mg, P and N to the soil [16]. Each nutrient has its own response to burning, e.g., early studies have found that concentrations of K, Ca and Mg ions can increase, whereas N and S often decrease [17]. Fire intensity is directly linked to the temperature an object experiences during fire, which in turn impacts the type of nutrient and the amount volatilised. Fire acts as a rapid mineralising agent [18]. This mineralisation has been studied extensively for N and P, because they are the two key macronutrients for plants. Soil nutrients that have low volatilisation thresholds are the first to be mineralised. There is an immediate reduction in soil organic N due to volatilisation [19]. Substantial proportions of soil organic N survive low intensity fires, however, moderate to high intensity fires convert most soil organic nitrogen to inorganic forms [2]. The effects of fire on properties of forest soils have been extensively reviewed and found that generally, losses of P through volatilisation are low [2]. However, the combustion of vegetation and litter cause major modifications on the P cycle. Micronutrients are also affected immediately after burning; e.g., total content and reducible forms of Mn increase significantly following fire due to the ash produced [20].

Soil pH has been also found to increase post fire. This increase has been attributed to the addition of base cations in ash, and to organic acid denaturing due to the heating of the soil [2]. However, significant increases only occur at temperatures >450 °C [21]; this temperature links with the total combustion of fuel as the ash produced has the capacity to neutralise soil acidity. Through analysing mixed forest soils, topsoil pH could increase as much as three units immediately after burning due to the production of K and Na oxides [22]. This significant increase can only be found in non-calcareous soils; calcareous soils are already alkaline, and in them the pH increases induced by fire are often negligible.

Currently, very little is known about the effect fires have on forest soils in Central Europe. Most research in Europe comes from Mediterranean forest studies, while there are few investigations about the areas that, according to the IPCC climate change scenarios, will have in the future climate similar to that of the Mediterranean region [9]. Soil in Mediterranean forests have low moisture and available nutrients, and plants are adapted by co-existing to prevent competition in resource scare environments [23]. Field research on fires, to our knowledge, is limited in Hungary in particular. Primary research is generally on the effects to biodiversity, mainly in grasslands and based on questionnaires to fire departments [24,25]. In particular, little research has been done on the effects of fire on Haplic Luvisols in Central Europe. Luvisols occupy ca. 5% of the total continental land area on Earth, and are found mainly in west-central Russia, the United States, Central Europe, the Mediterranean and southern Australia [26].

In purview of the above, the present study objectives were to: (1) characterise the effect of different fire intensities in a burn experiment on the organic matter, pH, potassium, phosphorus, calcium and magnesium of forest Haplic Luvisol through laboratory analysis; (2) examine the effectiveness of the experimental burning design at measuring changes in soil properties, and, (3) provide recommendations for fire authorities in Hungary about potential risks to the dominant soil Haplic Luvisol from both prescribed fires and wildfires. For this purpose, a field experiment of three fire treatments on forest Haplic Luvisol in Damak forest, NE Hungary was implemented.

2. Materials and Methods

2.1. Study Area

The study was carried out in Damak forest (48°31′ N, 20°82′ E), in the county of Borsod-Abaúj-Zemplén, Hungary (Figure 1). The area is characterised by a temperate continental climate. The mean annual temperature of the area is 9.5 °C; warmest period ranges from May to September (20.0 °C in July). Maximum precipitation within this warm period is 82 mm (July); mean annual precipitation is 567 mm. Forest typology is classified as mixed broadleaf and coniferous woodland (Figure 2). Dominant species are Hungarian oak, common nettle *Urtica dioica* L., black locust *Robinia pseudoacacia* L., Scots pine *Pinus sylvestris* L. and hornbeam *Carpinus betulus* L.

Soils in the studied area are of the Haplic Luvisol type [27,28]. This is a non-calcareous soil, with neutral pH and high in organic matter. In stark contrast to the Hungarian Plains (80 km South-West, higher precipitation and lower temperatures in Damak forest has led to the development of this soil type. As this is a forest soil, there is an organic horizon, comprised primarily of leafy material that is separated from the mineral horizon [27]. For this study, we identified a Haplic Luvisol that to our knowledge was negligibly affected by human disturbance, at about 250 m from the experimental site.

In Damak Forest, where soil samples were collected, there have been no recent significant disturbances that could affect this study, in particular prescribed or wild fires.

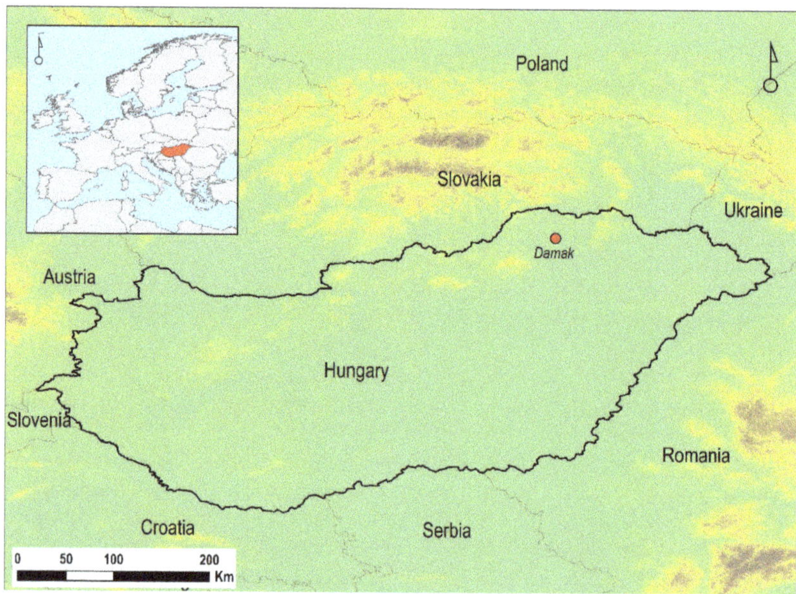

Figure 1. Study area: Damak Forest (red point; geographical coordinates 48°31′ N, 20°82′ E) in Borsod-Abaúj-Zemplén county, Hungary. Inset map: Hungary (red) in Europe.

Figure 2. Characteristic vegetation and strata in Damak forest: (**a**) soil to litter layer interface, (**b**) litter layer, (**c**) understorey and (**d**) overstorey. Photo: Jack M. Bridges.

2.2. Experimental Design

Experimental burning is used to simulate fire conditions in a safe area, where the effects of fire can be accurately observed. Studies using experimental burning are usually carried out between late spring and early autumn [29,30]. This is the optimum period where soil is dry and fuel is burnable. Experimental burns should recreate the same conditions as a real forest fires. One of the problems with the literature of experimental burning is that there are no universally established fuel load amounts for low, medium and high intensity fires. Studies have recommended for high and medium intensity fires, to use 40 and 20 t ha^{-1} of fuel, respectively [31]. For smaller scale experimental burning plots this equates to a minimum of 4 kg m^{-2} of fuel for high intensity and 2 kg m^{-2} for medium intensity.

The experimental approach is built along the lines of previously published studies [29,30,32] measuring fire effects on soil properties on the basis of setting up experimental fires of different burning intensities. The experimental burning area in our study is located in a 1500 m^2 paddock, about 200 m from the source of forest soil, providing a safe area to conduct the experiment. A total of nine 0.5 × 0.5 m plots and 0.225 m^3 of forest soil was transported to the area. The litter layer from the soil was removed and the upper 10 cm of mineral soil was collected. Soil was then distributed evenly to form the nine 0.1 m depth plots (Figure 3) to recreate the conditions in the forest. To prevent factors such as slope, drainage and texture affecting our results, all the soil was taken from the exact same location within Damak forest. This area was representative of the whole forest, and no single species dominated the under and overstorey, while the litter layer contained similar material to most of the forest floor.

Figure 3. Characteristics of the experiment burn plots: (**a**) 0.1 m depth, (**b**) 0.5 m width, (**c**) plots (*n* = 9), and (**d**) three plots per fuel load. Photo: Jack M. Bridges.

Three sets of three plots were burned close to the soil samples collection site with three distinct fire intensity treatments; low (F1), medium (F2) and high (F3). Three different fuel layers were used to achieve these fire intensities (Table 1), to represent degrees of heterogeneity in forest structure. The plots were ignited by a 30 s exposure from a propane torch in a circular motion around the plot. Pre-fire samples (F0) were taken from each plot for comparison. The vegetation used as the fuel load came from the same area as the soil, creating realistic conditions of a wildfire/prescribed fire. The first treatment (F1), involved the addition of ca. 4 kg m^{-2} of litter layer to three plots. The second treatment (F2), ca. 12 kg m^{-2} of litter layer and understory fuel was added to another three plots. In the third treatment (F3), ca. 40 kg m^{-2} of litter layer understorey and overstorey fuel was added to three plots (Table 2). Allocation of fire treatment to the plots was randomised to prevent pseudo-replication in a Latin square design [33].

Table 1. One-way ANOVA mean difference between low (F1), medium (F2) and high (F3) fire intensity treatments, and pre-fire (F0) for pH (soil), soil organic matter (OM), carbon (C), calcium (Ca), potassium (K), magnesium (Mg) and phosphorus (P). $n = 9$ samples for F0 and $n = 3$ samples for F1, 2 and 3 treatments. ** Significant difference at $p < 0.01$. * Significant difference at $p < 0.05$.

Treatment	F0						
	pH (soil)	OM (%)	C (mg/kg)	Ca (mg/kg)	K (mg/kg)	Mg (mg/kg)	P (mg/kg)
F1	0.03	0.53	0.50	242.00 **	190.00 **	6.00 **	0.25
F2	0.44 *	0.14	0.70	1272.00 **	190.33 **	22.00 **	33.25 **
F3	0.45 *	0.82	4.10	1452.00 **	187.33 **	13.74 **	39.7 **

Table 2. Weight of fuel for each plot.

Subplot	Weight of Litter Layer (kg)	Weight of Understorey Fuel (kg)	Weight of Overstorey Fuel (kg)	Total Fuel Weight (kg)
1A	1	0	0	1
1B	1	0	0	1
1C	1	0	0	1
2A	1	1.6	0	2.6
2B	1	1.7	0	2.7
2C	1	1.4	0	2.4
3A	1	1.2	7.2	9.4
3B	1	1.5	6.1	8.6
3C	1	1.2	6.9	9.1

Subplot 1 (A, B, C) = Low intensity fire. Subplot 2 (A, B, C) = Medium intensity fire. Subplot 3 (A, B, C) = High intensity fire.

During fire, temperatures on the surface of the soil were measured by a thermal infrared thermometer. Plot temperatures were recorded at 2 min intervals until the maximum temperature fell below 100 °C. This value was chosen as it has been shown to be the threshold temperature for the most relevant changes in soil [29,31,34].

2.3. Sample Collection and Laboratory Analysis

The remaining litter was discarded and the whole ash layer was collected and weighed from each plot; sample mean ash weight from F1 plots was 9.00 g, 36.00 g from F2 plots and 200.33 g from F3 plots. Prior to soil sampling, the litter layer was removed. An effective and representative method of soil sampling was employed. This involved taking 10 cm cores in a zigzag path across the plot and mixing to form an average sample; this was repeated for each plot [35]. The samples were taken using a trowel, and this was cleaned after each sample was taken to prevent cross-contamination. The samples were stored in a Kraft wet-strength paper bag labelled with the sample subplot and whether it was pre-fire or post-fire.

Samples were then air dried and sieved to isolate and keep the <2 mm diameter soil particles, which were then used for the analysis of chemical properties. One plot had three laboratory samples made up to improve the accuracy of the instrumentation (sample 3A).

Chemical properties analysis was then carried out in a laboratory. Organic matter content (%OM) was determined by Loss-on-Ignition (LOI). Samples were properly heated to ensure the complete loss of moisture from the soil [36,37]. F0, F1, F2 and F3 samples were then weighed (Figure A2, Appendix A). The samples were then heated to 430 °C in a muffle furnace for 16 h and weighed. This removed most of the carbon and OM [38]. By comparing the two weights before and after ignition, LOI can be calculated, which is the value for OM. C makes up ca. 50% of the OM value [39,40], therefore C was estimated by multiplying LOI by 0.5.

An acetic acid extraction (Figure A1, Appendix A) provided an estimate of the plant available content of the macro and micro nutrients contained in the samples. Using atomic absorption spectrophotometry (AAS) [41], elements Ca, Mg and K were measured (Figure A3, Appendix A). UV-visible Spectroscopy [42] was used to determine the value of P (Figure A4, Appendix A). The pH of the ash layer, F0, F1, F2 and F3 samples were measured using a Whatman pHA wet-bulb pH meter, pH obtained with water and a proportion of 1:2.5 soil to liquid (Figure A5, Appendix A).

An in-house standard reference material (ABS3) validated the precision of the sample measurements. This standard is soil based and was stored in a sealed container in cool, dark storage prior to use. All dry samples were weighted with the same mass balance to ensure consistency. Sample 3A was repeated three times and a control (blank) was used to ensure the reliability of all methods. Analysis of variance (ANOVA), at $\alpha = 0.05$, between F0 and the three post-fire samples (F1, F2 and F3) was performed to detect any significant statistical differences among results.

3. Results

3.1. Burn Characteristics

Temperatures at the soil surface during the low intensity fires (F1) averaged 228 °C. The highest temperature recorded was 400 °C. Mean burn time >100 °C was 0.79 min (Figure 4). Weight of ash averaged 11 g ($n = 3$ for all values). Due to the infrared thermometer's maximum temperature threshold of 420 °C, a true average of the soil surface temperature could not be calculated for the medium and high intensity fires. For this reason, we measured the percentage of time >420 °C. This resulted an average of 43% of the total burn time for medium intensity fires (F2). Highest temperature recorded was ≥420 °C. Mean burn time >100 °C was 15.98 min (Figure 4). Weight of ash averaged 36 g ($n = 3$ for all values). Temperatures at the soil surface during the high intensity fires (F3) averaged >420 °C for 85% of the total burn time. Highest temperature recorded was ≥420 °C. Mean burn time >100 °C was 190.67 min (Figure 4). Weight of ash averaged 200.6 g ($n = 3$ for all values).

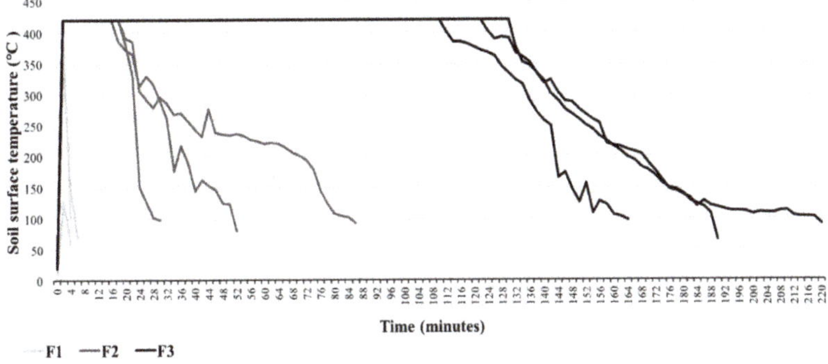

Figure 4. The burn time and soil surface temperatures of each subplot for F1, F2 and F3 treatments.

3.2. Fire Intensity Treatment Effects on Haplic Luvisol Properties

3.2.1. Soil and Ash pH

Results obtained for the pH of soil showed a slight decrease for F1 (low intensity fire treatment; Figure 5). This was followed by an increase in pH after F2 and F3 (medium and high fire treatment, respectively). F3 was significantly increased at $p > 0.05$ compared with the F0 measured pH (Table 1), F1 and F2 were not significantly different. This shows that immediately after the burn, medium and high intensity fires can change the Haplic Luvisol pH from slightly acidic to neutral. The pH of ash increased by ca. 1.0 for each treatment: F1 8.42, F2 9.79, F3 10.33. This is a dramatic change as the pH scale is logarithmic (Table 3). The coefficient of variation for F1 and F2 was 5%, whereas it was 14% for F3.

Table 3. Mean values of pH (soil and ash), soil organic matter (OM), carbon (C), calcium (Ca), potassium (K), magnesium (Mg) and phosphorus (P) in low (F1), medium (F2) and high (F3) fire intensity treatments, and pre-fire (F0). $n = 9$ samples for F0 and $n = 3$ samples for F1, F2 and F3.

Variable	F0	F1	F2	F3	Reference Material Accuracy
pH (soil)	6.39 ± 0.34	6.38 ± 0.12	7.17 ± 0.63	7.01 ± 0.36	N/A
pH (ash)	N/A	8.42 ± 0.55	9.79 ± 0.43	10.33 ± 1.54	N/A
OM (%)	6.60 ± 1.70	6.07 ± 1.88	6.74 ± 1.45	5.78 ± 1.15	N/A
C (mg/kg)	33.00 ± 8.50	33.50 ± 9.40	33.70 ± 7.25	28.90 ± 5.75	N/A
Ca (mg/kg)	2048.00 ± 39.00	2290.00 ± 47.00	3320.00 ± 78.00	3500.00 ± 12.00	97.00
K (mg/kg)	402.48 ± 0.44	212.48 ± 0.03	212.15 ± 0.03	215.15 ± 0.05	85.00
Mg (mg/kg)	25.67 ± 0.02	31.67 ± 0.05	47.67 ± 0.05	39.41 ± 0.02	59.00
P (mg/kg)	13.30 ± 2.08	13.55 ± 1.96	46.55 ± 4.75	53.00 ± 4.00	80.00

3.2.2. Organic Matter and Carbon

Results obtained for organic matter (OM) content varied greatly among the plots (Table 3). F1 and F3 lowered the OM% by a small amount (0.53% and 0.82% respectively), though not significantly (Table 1), whereas, F2 increased the amount of OM by only 0.14%. The coefficient of variation for all treatments was around 20% (Table 3). These results show that organic matter in Haplic Luvisol is not significantly affected by fire.

The carbon content of the soil samples (Table 3) remained at similar levels for F1 and F2 treatments compared to F0, 33 mg/kg. For F3, the amount of carbon decreased by 13%, from 33.00 to 29 mg/kg (Table 3). Carbon levels in the soil samples did not significantly change after any of the treatments (Table 1). Coefficient of variation was 30% for all treatments. Carbon levels in Haplic Luvisol after a fire event of any intensity are not affected.

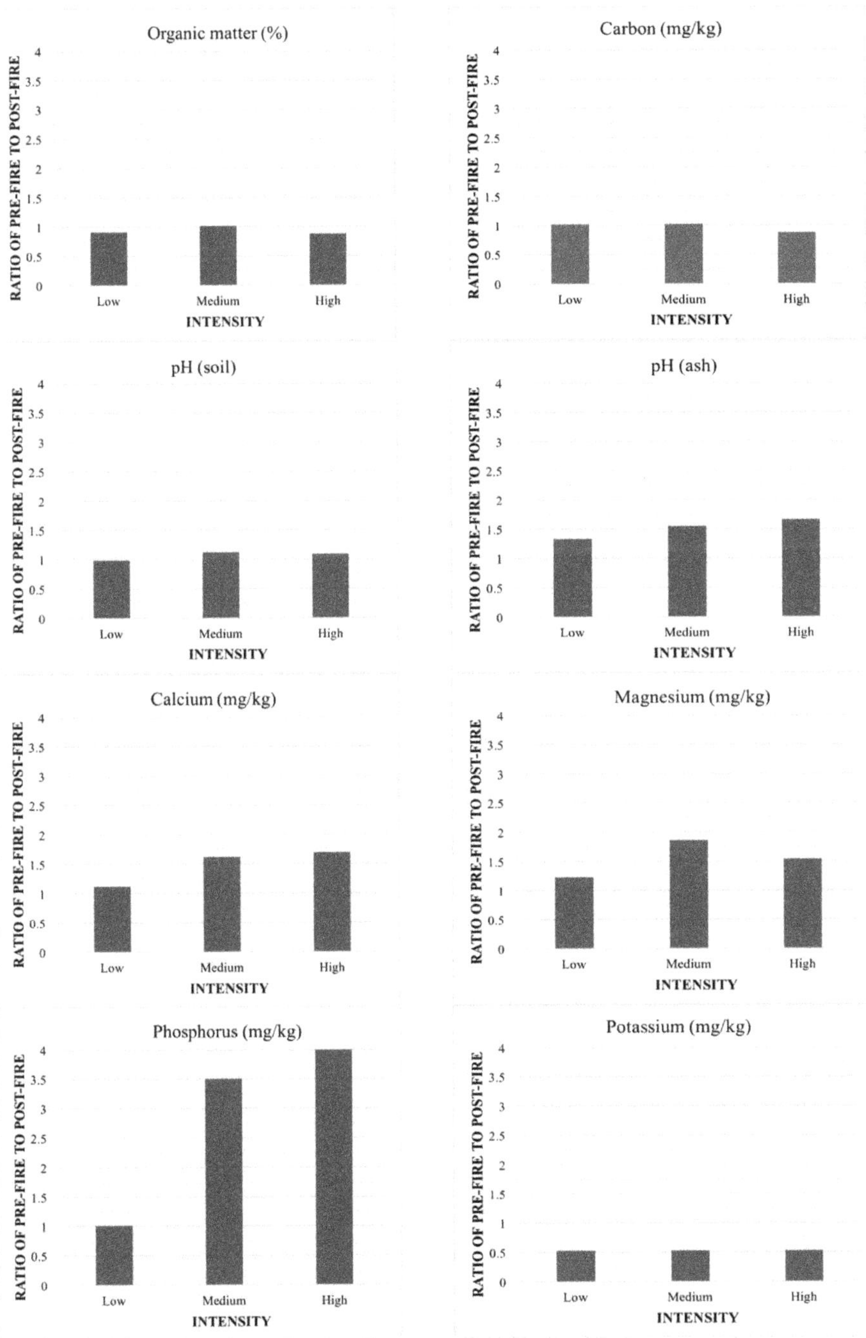

Figure 5. Ratio of pre-fire to post-fire change for pH (soil and ash), organic matter, carbon, calcium, magnesium, potassium and phosphorus. >1 = positive change, 1 = no change, <1 = negative change. Note the different scales on each graph.

3.2.3. Available calcium

Available calcium in soil increases with fire intensity (Figure 5), higher intensity fires result in more available calcium present. There was a noticeable increase after F2 and F3, over 1000 mg/kg more calcium measured than after F1 (Table 3). Each fire treatment showed a significant increase at $p < 0.01$ in comparison to F0 (Table 1). The coefficient of variation for calcium was 2% for F0, F1 and F2; while F3 was 0.3%. The reference material accuracy was 97% for calcium. This was the highest out of all the parameters measured (Table 3), meaning the accuracy of the Atomic Absorption Spectroscopy was very high. Fire therefore, has a significant effect on the amount of calcium present in Haplic Luvisol immediately after a fire event of any intensity.

3.2.4. Available Potassium

The potassium levels in the soil decreased significantly for F1, F2 and F3 at a significance of $p < 0.01$ (Table 1). The negative correlation has shown that higher intensity fires decrease the amount of potassium present in the soil immediately after the burn. The decreasing ratio is constant for all three treatments (Figure 5), dropping from 402.48 mg/kg (F0) to 212.48, 212.15 and 215.15 mg/kg (F1, F2 and F3, respectively). The coefficient of variation was; 0.11% for F0 and 0.01% for F1, F2, F3 (Table 3). Reference material accuracy was calculated at 85%, showing a high accuracy of the AAS instrumentation. Therefore, the level of potassium in Haplic Luvisol can be significantly affected by fire of any intensity immediately after the fire event.

3.2.5. Available Magnesium

Magnesium levels varied across the three treatments (Figure 5); F2 had the highest level at 47.67 mg/kg followed by F3 (39.41 mg/kg) then F1 (31.67 mg/kg) in comparison with the pre-fire level of 25.67 mg/kg. All treatments showed a significant increase in plant available magnesium at $p > 0.01$ (Table 1). The coefficient of variation was; 0.08% for F0, 0.16% for F1, 0.10% for F2 and 0.05% for F3 (Table 3). The reference material accuracy was 59%, which shows a medium accuracy of the AAS instrumentation for magnesium parameter.

3.2.6. Available Phosphorous

The higher the fire intensity, the higher the phosphorus availability. (Figure 5). The amount of P stays constant between F0 and F1, at 13 mg/kg, rising to 46.55 and 53.00 mg/kg after F2 and F3, respectively. These increases in P are significant at $p > 0.01$ (Table 1). The coefficient of variation was; 15.6% for F0, 14.5% for F1, 10.2 % for F2 and 7.5% for F3. Reference material accuracy was calculated at 80% (Table 3), this is a high accuracy for the UV-visible spectroscopy.

4. Discussion

4.1. Burn Characteristics

Recreating conditions of a natural burn, whether prescribed fire or wildfire, is one of the biggest challenges of experimental burning. The amount of biomass and type of species can dictate the intensity of the fire; the challenge is that no forest composition is the same. There are other dominating factors such as weather and topography that influence fire behaviour; weather conditions determine whether the fuel load is 'available' to burn. Weather and topography cannot be controlled in this study so therefore the focus was the effect of different fuel load and type of fuel.

For this experiment, extensive research was carried out to produce the best experimental burn design as detailed in the Materials and Methods section. Some studies [43,44] that used the same recommendations of fuel load for medium and high intensity burns, 20 and 40 t ha^{-1} respectively, found that medium intensity fires burned for ca. 17 min and high intensity fires burned for ca. 37 min [45]. The Damak experiment followed the same fuel load recommendations; however, the mean burn time >100 °C for F2 and F3 were 55.67 and 190.67 min, respectively. An explanation for this is could be attributed to the size of burn area; the area for one study was 105 m^2 [44] and another was 80 m^2 [43]. A larger burn area would create a more severe fire and therefore, burn at a quicker rate. These differences are possibly due to different instrumentation, being that these studies used thermocouples that measured temperature directly within the soil. Our experiment used an infrared thermometer that measured the soil surface temperature. The burning material on the soil surface would make both the burn time and temperature a higher reading compared to studies that use thermocouples due to the soils insulating properties.

4.2. Effect of Fire Intensity on Soil Properties

The pH of noncalcareous soils (such as Haplic Luvisol) should increase after burning, due to the effect of organic acid denaturisation [2] and the release of alkaline cations. Below the threshold of medium fire intensity, burning does not affect soil pH. This is likely to be due to some of the plots for F1 being at a lower pH originally compared to the average pH of all the plots pre-fire. This would mean that the F1 pH would appear to decrease compared to the pH of the plots before burning (F0). pH is not consistent in soil. Substances such as decomposing leaf litter and plant materials can have a higher pH than the underlying soil. Therefore, when F1 was applied to the soil, some of the leaf material is likely to have mixed with the soil. This would produce a higher pH for F1 plots compared to the F0. The fact that the soil sample bags were left for two months before laboratory analysis, might have been enough time for the plant material to change the original soil pH. Studies have found a decrease in pH after low intensity followed by an increase at higher intensity burn sites [32]. It is important to note that the results for F1 pH were not significantly different to F0.

The pH of F2 was higher than F0 but not significantly. However, F3 pH was significantly higher than F0. This shows that the pH after high intensity fires is significantly affected, supported by other studies [2]. Studies have found that significant increases in soil pH only occurred at temperatures >450–500 °C [21], this supports our F3 pH data. However, this shows that the maximum temperature for F2 cannot have exceeded 450 °C. Many authors report increased pH values after a prescribed fire [46].

The pH of ash above the soil for F1, F2 and F3 were significantly higher than both the original soil pH and each other. This was to be expected as the carbonates and oxides that are produced from combustion are highly alkaline. This is supported by studies that found that topsoil pH (including the ash layer), under a mixed conifer forest, increased three units immediately after the burning [22]. Our high intensity burn plots experienced an increase of six pH units within the ash layer. It is likely that ash leaching, chemicals draining from the ash into the underlying soil by the action of percolating liquid (usually rainwater), could turn the soil from slightly acidic to more alkaline. This would possibly have an effect on tree species, especially in forests like Damak. Common pine *Pinus sylvestris* and nettle would likely cope with these changes to the soil, because the latter prefers high levels of phosphate. However, species like hornbeam can only grow in neutral soils, so in the case the pH becomes too alkaline it will not be suitable for its growth. There would possibly be a profound effect on the native Hungarian oak, as they are specially adapted to heavy acidic soils. Additionally, following a forest fire in the region, invasive species such as black locust would probably pioneer alkaline rich soil: this species has a tolerable pH range of 4.6–8.2, but prefers dry alkaline soils.

The varied results for the OM and C post-fire in our experiment are supported by other studies [47]. The effect of fire on soil OM is highly dependent on a number of factors that are seen in our study including; fire intensity, varied distribution of organic matter in the soil, nature of burned vegetation. Effects have been highly variable in many studies [15,48] with effects ranging from complete combustion of organic matter to increases of up to 30%. This is a prominent discussion in soil research as there is no generalised trend between organic matter and burning, it is variable controlled. For example, soil heated under laboratory conditions have frequently reported significant losses in OM and C of up to 100% [49], whereas after a wildfire there have been records of an increase [10]. Another study suggests this could be due to the addition of necromass in forests (dead wood and organisms) after a fire event which is not as prevalent in a laboratory experiment [50]. One study produced a comprehensive dataset of soils effected by wildfire or prescribed fire. Their study concluded that lower C content was found after prescribed fires while higher C content was found after wildfires [14]. This was attributed to the accumulation of charcoal and encroachment of post-fire N-fixing vegetation. We do, however, acknowledge the presence of some limitations due to the application of the LOI method to determine OM and OC. LOI is based on two assumptions: (1) LOI equals OM and (2) the ratio OM/OC is the same for all treatments. LOI incorporated more than just OM, for example hydrated covers of clays, which are expected to be partially lost when a thermal shock occurs, therefore LOI was not exactly equal to OM in this case. In addition, as different temperatures were reached among fire treatments, the OM/OC ratio would not be the same in all of them, likely: OM/OC ratio decreases with the thermal treatment. For these reasons, what we found in our results related to OM and OC can be mainly attributed to using a broad non-specific technique of measuring OM and OC, with its inherent limitations. Further research should imply a more precise method to determine OM and OC, like a modern dichromate oxidation method. This would ensure a precise quantification of OM and OC.

Calcium results suggest that all fire intensities significantly affect calcium levels in Haplic Luvisol. This finding is supported by other studies examining the effect of fire on Ca levels in soil [2]. As the volatilisation threshold temperature of Ca occurs at 1484 °C [15], it is relatively insensitive to fire (temperature of burning woody fuel ca. 1100 °C). Therefore, it was not expected in this study to observe any reductions in the amount of Ca. The reason why Ca increases significantly after fire is thought to have come from the mineralisation of the organic matter [32]. Studies suggest that the exchange capacity of surface horizons is depleted after fire which increases the number of cations that are not volatised (Ca and Mg) [51].

The soil samples measured after the burnings would contain Ca released from the overlying biomass. This could be evidence as to why there was a major increase in Ca between F1 and F2/F3. The increased biomass (ca. 2 and 8 kg more for F2 and F3, respectively) would contribute much more Ca cations than 1 kg of litter (F1). Another possibility is that the temperatures in F2 and F3 reached the volatilisation threshold of K and P, ca. 750 °C [52].

The potassium (K) level after all fire treatments was significantly lower than the pre-fire levels, ca. 50% reduction. This is not supported by much of the literature measuring K levels post-fire events [2,53]. As the volatilisation of K occurs at ca. 750 °C, this reduction would be expected in very high fire intensity events, however, we have observed the reduction in F1 (low intensity) so the reduction in K cannot be attributed to volatilisation. The instrumentation accuracy was high for K (87%), so the reduction cannot be an instrumentation error. There is a possibility that being left in the soil sample bags for 2 months the K cations bound to inorganic substances; however, similar reductions would be observed in F0 if this was the case. A probable reason is that the available potassium was immobilised within mineral structures driven by thermal treatment.

There are very limited studies exploring the effect of fire on K levels in Haplic Luvisol. The studies that have been carried out support our findings that K levels are reduced, after a fire event, in Haplic Luvisol [54]. One study measured significantly lower K levels in the 0–5 cm soil layer post-fire for Umbric leptosols, attributing the loss to erosion and leaching [55]. Studies suggest available K should increase after low intensity prescribed fires [46]. However, there were no erosional or leaching processes occurring in our experiment therefore, the reason why there is a reduction of K in Haplic Luvisol is likely to be due to the composition of the soil. Another explanation could be the moisture level of the soil. It has been found that soil moisture significantly affects the availability of K [56]. The Damak soil was dry (observed in fieldwork) which could have reduced the level of K recorded. However, dry soil would be present in F0 measurements so the K values should be similar to F1, F2 and F3 K values. The results from this study show that fire was the cause of the reduction in K. The effect fire has on K levels in Haplic Luvisol is a therefore a key finding.

Mg levels post-fire were all significantly increased. This increase has been found in many post-fire soil analysis studies [2]. The volatilisation of Mg occurs at ca. 1107 °C [15] therefore there should be no reduction in its availability post-fire. As with Ca, the increase in available Mg is likely to come from the mineralisation of the organic matter in the soil.

Following a prescribed fire, we were expecting increased available phosphorus [46]. The fact that available phosphorous increases with fire intensity can be attributed to the conversion of organic P to orthophosphate [57]. This occurs during the combustion of organic matter and is dependent on the type of soil, vegetation species and leaching [58]. We observed an increase in P for Haplic Luvisol underlying a temperate deciduous forest with no leaching (as samples were taken immediately post-fire). This increase confirms other findings in this area of soil science [59]. The general trend is that P increases significantly dependant mainly on the fire intensity [2]. Small amounts of P can be lost through volatilisation. The threshold temperature for P is ca. 774 °C [60], therefore, losses would be expected in F2 and F3 as the burn time and temperatures could reach >774 °C. However, it is important to note these losses would be very low in comparison to the converted P. In this experiment, there is a clear positive relationship between available P and fire intensity. This result has been found in many levels [2].

In relation to the changes observed in soil OM and C, were somehow inconclusive. This is likely because soil was taken at a depth of 10 cm. Therefore, there would be a dilution effect on the impact of fire on the upper soil (organic matter is mainly affected in the upper 2 cm layer). It could also depend on the amount of OM in the soil pre-fire and soil type. This is supported by other studies [10]. However, in a forest setting, organic matter has shown to return to pre-fire levels after one year [13]. Therefore, any effect of fire on the OM and C would be recoverable in the long term.

One limitation of this study was the instrumentation used to measure soil temperature. This can be observed in the burn time (Figure 4), being much longer than studies that used thermocouples [43,44]. It is likely that the infrared thermometer measured flames from the burning vegetation instead of the soil surface, thus giving the higher temperature (flame). An accurate measurement of the soil surface temperature could not be measured. Therefore, assumptions had to be made in relation to whether certain cations reached their volatilisation threshold. Ideally, a thermocouple would be used that could be inserted into the soil surface and measure the temperature. In addition to that, this study was limited to 0.5 m^2 plots through health and safety restrictions. With accessible funding and Government approval, larger plots could be used, as is the case in Spain, where researchers have had access to 80 m^2 plots [29,31]. A larger scale plot would be a better way to recreate the conditions of a prescribed fire/wildfire. Furthermore, the time between soil sample collection and analysis would ideally be as short as possible, however for our experiment, the gap between collection and analysis was two months. Samples were stored in dark and dry conditions; nevertheless, several processes could still take place within the soil. For example, the exchangeable cations released by the fire treatments could bind to inorganic substances and affect the results for that cation. The soil samples contained root systems and small amounts of vegetation. Their decomposition over two months could lower the pH through the release of carbon dioxide. However, in our study, the processes described above would have had negligible effect on our samples; pH increased in all samples and the change, for all cations measured, was significantly increased. Lastly, the same prescribed fire experiment could have also been implemented at other locations of Hungary with the same Haplic Luvisol soil type, to provide more conclusive results on the burning effect on forest Haplic Luvisol.

The immediate changes to forest soils following burning have been discussed. This study contributes to the concept that burning temperate forests in Central Europe can have a direct impact on soil fertility and forest management, whether that is from prescribed burning or wildfires. The use of small scale experimental fires to measure this effect has mirrored larger experimental burning and wildfire studies.

5. Conclusions

The main purpose of this study was to measure and analyse the immediate effect of fire on forest Haplic Luvisol. This was carried out using representative small scale fire plots that enables fire effect on soil to be measured without damaging large areas of land. These experimental fires carried out on soil matter in Damak Forest, North East Hungary, produced significant changes to the Haplic Luvisol organic matter and chemical properties. Increased changes were most obvious in exchangeable Ca, followed by available P and exchangeable Mg, with pH only significant in the high intensity fire treatment. Significant decrease was observed in all the fire intensities for exchangeable K. This is a key finding of this study, as it is one of the few studies that have found losses in exchangeable K post-fire. This contrasts with the general behaviour of exchangeable K in soils [2,46]. As expected, a higher intensity fire resulted in a greater change in nutrient levels, with the exception of Mg.

Another relevant finding of this study is the increase of pH, which was significant in the high intensity fires. Through leaching, the Haplic Luvisol will likely have a much higher pH, which can directly affect pH sensitive vegetation (e.g., native rare species such as *Quercus frainetto* Ten.).

This study indicates that there are potential negative effects on soil fertility based on the decrease of potassium in the experiment. Potassium is a key nutrient for the ion transport within vegetation. Any loss of this from the soil would have major influences on the agricultural sector of the Hungarian economy. With an increased forest fire risk across Europe, from the 2041–2070 A1B emission scenario [61], nutrient rich soil, such as Haplic Luvisol, will be adversely affected by losses in K. As such, this study provides a reference point for further investigation into the effect of fire intensity on Haplic Luvisol and for any soil type that could be subject to burning. To enable stronger conclusions to be drawn in future investigations, we recommend conducting a larger scale of experimental burning, to follow burned soils over a recovery period and to evaluate impacts on productivity. Nevertheless, we believe our findings provide useful evidence for Haplic Luvisol changes under fire disturbance, while also supporting the maintaining of the current Hungarian fire prevention policy.

Author Contributions: Conceptualisation, J.M.B. and G.P.P.; methodology, J.M.B. and G.P.P.; formal analysis, J.M.B.; investigation, J.M.B., G.P.P. and N.C.; writing—original draft preparation, J.M.B., G.P.P. and N.C.; writing—review and editing, J.M.B., G.P.P., and N.C.; visualisation, J.M.B.; supervision, G.P.P. and N.C.; project administration and funding acquisition J.M.B.

Funding: GPP's contribution to this work was supported by NERC's Newton Fund RCUK project Towards a Fire Early Warning System for Indonesia (ToFEWSI).

Acknowledgments: We thank János and Katalin Tóth, who granted permission for the experiment to be carried out on their land. We are grateful to Roxána Filetóth for facilitating the arrangement and in particular for her help with the fieldwork. Ian Saunders in the DGES Laboratory at Aberystwyth University who helped with the analysis of the soil samples. Andrew Thomas is acknowledged for insightful discussions on the manuscript.

Conflicts of Interest: The authors declare no conflict of interest. The funders had no role in the design of the study; in the collection, analyses, or interpretation of data; in the writing of the manuscript, or in the decision to publish the results.

Appendix A

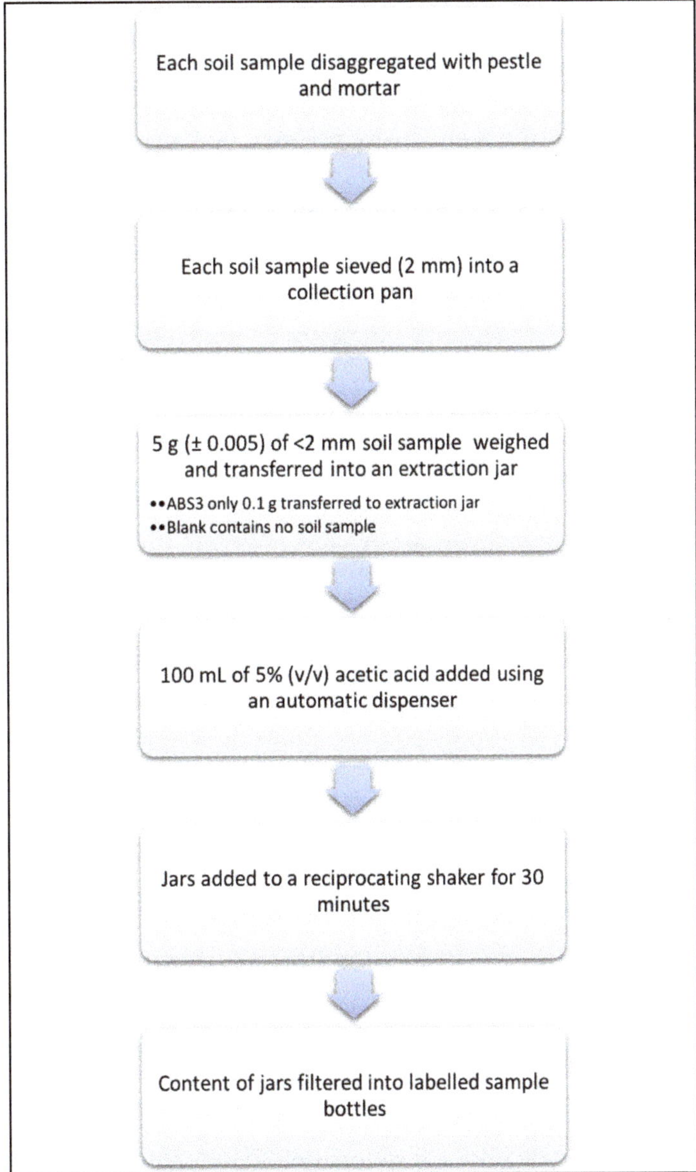

Figure A1. Method used to prepare soil samples for AAS and UV-spectroscopy analysis for plant available concentration.

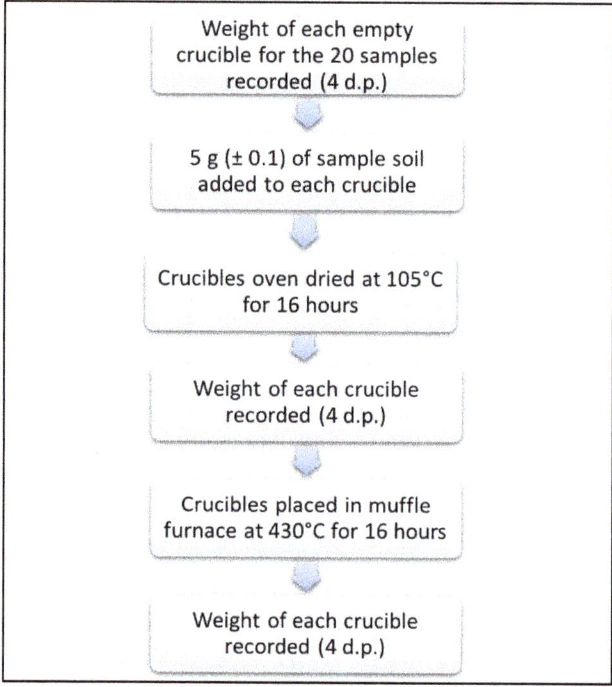

Figure A2. Method used for organic matter (OM) and C analysis of all samples.

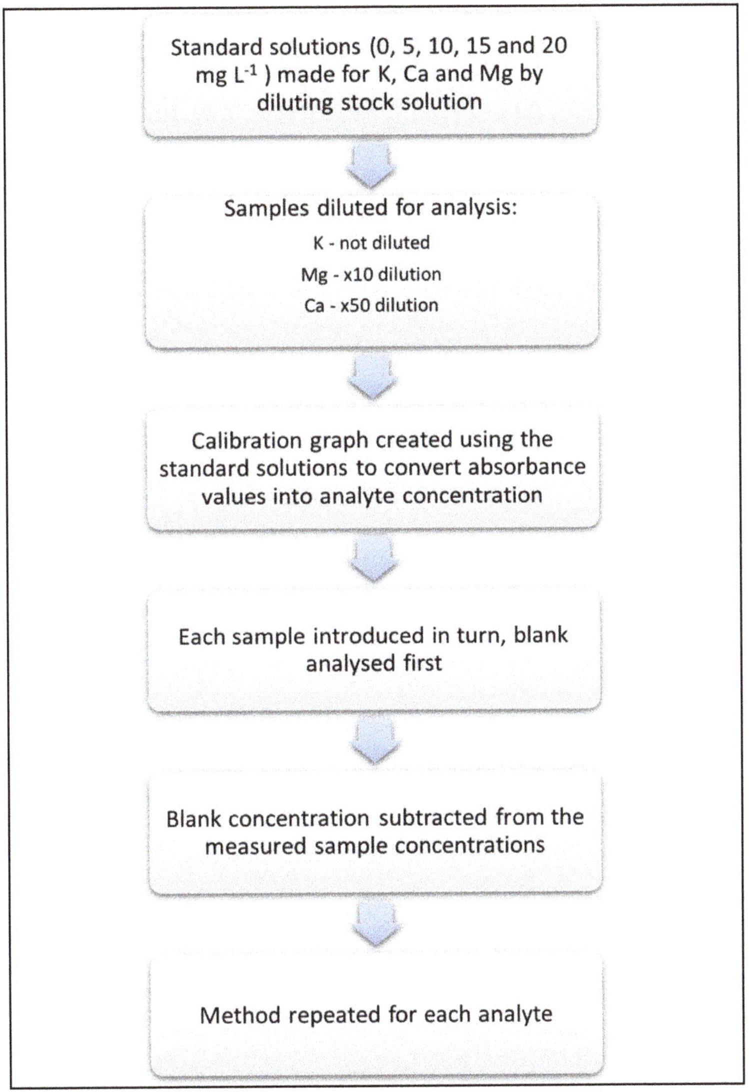

Figure A3. Method used for Ca, K and Mg analysis of all samples.

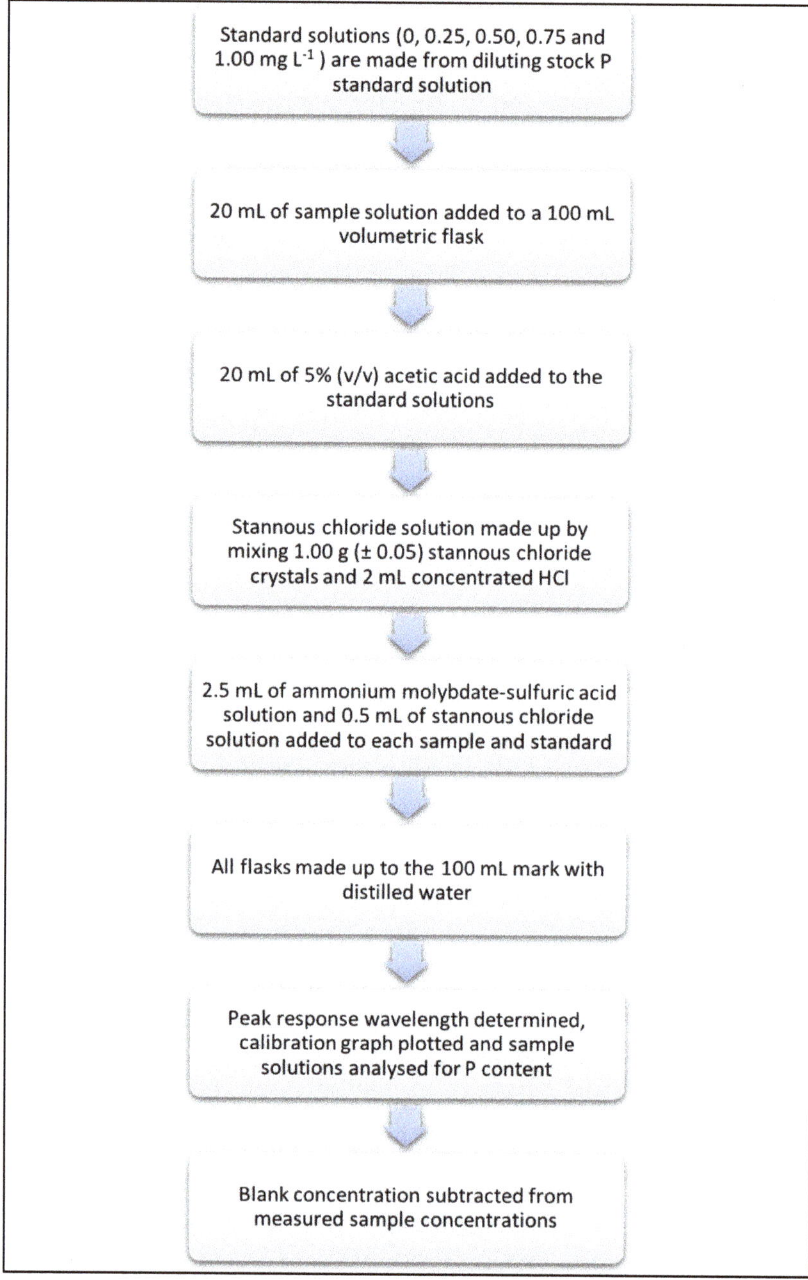

Figure A4. Methods used for P analysis of all samples.

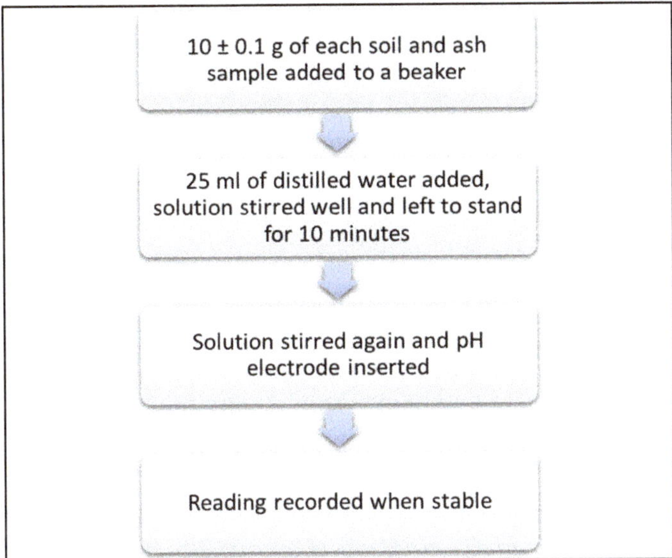

Figure A5. Method used for soil and ash pH analysis.

References

1. Barros, A.M.G.; Pereira, J.M.C. Wildfire Selectivity for Land Cover Type: Does Size Matter? *PLoS ONE* **2014**, *9*, e84760. [CrossRef] [PubMed]
2. Certini, G. Effects of fire on properties of forest soils: A review. *Oecologia* **2005**, *143*, 1–10. [CrossRef] [PubMed]
3. Castellnou, M.; Kraus, D.; Miralles, M. Prescribed Burning and Suppression Fire Techniques. In *Best Practices of Fire Use—Prescribed Burning and Suppression Fire Programmes in Selected Case-Study Regions in Europe*; Montiel, C., Kraus, D., Eds.; European Forest Institute: Joensuu, Finland, 2010; pp. 3–17.
4. Ascoli, D.; Bovio, G. Prescribed burning in Italy: Issues, advances and challenges. *iForest—Biogeosci. For.* **2013**, *6*, 79–89. [CrossRef]
5. Fernandes, P.; Davies, G.; Ascoli, D.; Fernández, C.; Moreira, F.; Rigolot, E.; Stoof, C.; Vega, J.; Molina, D. Prescribed burning in southern Europe: Developing fire management in a dynamic landscape. *Front. Ecol. Environ.* **2013**, *11*, 4–14. [CrossRef]
6. Tilman, D.; Lehman, C. Human-caused environmental change: Impacts on plant diversity and evolution. *Proc. Natl. Acad. Sci. USA* **2001**, *98*, 5433–5440. [CrossRef]
7. Moritz, M.A.; Batllori, E.; Bradstock, R.A.; Gill, A.M.; Handmer, J.; Hessburg, P.F.; Leonard, J.; McCaffrey, S.; Odion, D.C.; Schoennagel, T.; et al. Learning to coexist with wildfire. *Nature* **2014**, *515*, 58–66. [CrossRef]
8. Christensen, J.H.; Hewitson, B.; Busuioc, A.; Chen, A.; Gao, X.; Held, R.; Jones, R.; Kolli, R.K.; Kwon, W.K.; Laprise, R.; et al. Regional climate projections. In *Climate Change 2007: The Physical Science Basis. Contribution of Working Group I to the Fourth Assessment Report of the Intergovernmental Panel on Climate Change*, 1st ed.; Solomon, S., Qin, D., Manning, M., Chen, Z., Marquis, M., Averyt, K.B., Tignor, M., Miller, H.L., Eds.; Cambridge University Press: Cambridge, UK, 2007; pp. 847–940.
9. IPCC. *Fifth Assessment Report*; Cambridge University Press: Cambridge, UK, 2014.
10. González-Pérez, J.; González-Vila, F.; Almendros, G.; Knicker, H. The effect of fire on soil organic matter—A review. *Environ. Int.* **2004**, *30*, 855–870. [CrossRef]
11. Giovannini, G.; Lucchesi, S.; Giachetti, M. Effect of heating on some physical and chemical parameters related to soil aggregation and erodibility. *Soil Sci.* **1988**, *146*, 255–261. [CrossRef]
12. Simard, D.G.; Fyles, J.W.; Paré, D.; Nguyen, T. Impacts of clearcut harvesting and wildfire on soil nutrient status in the Quebec boreal forest. *Can. J. Soil Sci.* **2001**, *81*, 229–237.

13. Badía, D.; Martí, C.; Aguirre, A.; Aznar, J.; González-Pérez, J.; De la Rosa, J.; León, J.; Ibarra, P.; Echeverría, T. Wildfire effects on nutrients and organic carbon of a Rendzic Phaeozem in NE Spain: Changes at cm-scale topsoil. *Catena* **2014**, *113*, 267–275. [CrossRef]

14. Johnson, D.W.; Curtis, P.S. Effects of forest management on soil C and N storage: Meta analysis. *For. Ecol. Manag.* **2001**, *140*, 227–238. [CrossRef]

15. DeBano, L.F. The role of fire and soil heating on water repellency in wildland environments: A review. *J. Hydrol.* **2000**, *231*, 195–206. [CrossRef]

16. Schlesinger, W.H.; Dietze, M.C.; Jackson, R.B.; Phillips, R.P.; Rhoades, C.C.; Rustad, L.E.; Vose, J.M. Forest biogeochemistry in response to drought. *Glob. Chang. Biol.* **2016**, *22*, 2318–2328. [CrossRef] [PubMed]

17. Hough, D. *Long Corner Creek Hydrologic Project: Aspects of the Geology, Physiography and Soils*; Soil Conservation Authority: Victoria, Australia, 1981.

18. John, T.V.S.; Rundel, P.W. The role of fire as a mineralizing agent in a Sierran coniferous forest. *Oecologia* **1976**, *25*, 35–45. [CrossRef]

19. Binkley, D.; Fisher, R. *Ecology and Management of Forest Soils*, 4th ed.; Wiley-Blackwell: Chichester, UK, 2013.

20. Parra, J.; Rivero, V.; Lopez, T. Forms of Mn in soils affected by a forest fire. *Sci. Total Environ.* **1996**, *181*, 231–236. [CrossRef]

21. Arocena, J.M.; Opio, C. Prescribed fire-induced changes in properties of sub-boreal forest soils. *Geoderma* **2003**, *113*, 1–16. [CrossRef]

22. Ulery, A.L.; Graham, R.C.; Amrhein, C. Wood-ash composition and soil pH following intense burning. *Soil Sci.* **1993**, *156*, 358–364. [CrossRef]

23. Sardans, J.; Peñuelas, J. Plant-soil interactions in Mediterranean forest and shrublands: Impacts of climatic change. *Plant Soil* **2013**, *365*, 1–33. [CrossRef]

24. Deák, B.; Valkó, O.; Török, P.; Végvári, Z.S.; Hartel, T.; Schmotzer, A.; Kapocsi, I.; Tóthmérész, B. Grassland fires in Hungary—Experiences of nature conservationists on the effect of fire on biodiversity. *Appl. Ecol. Environ. Res.* **2014**, *12*, 267–283. [CrossRef]

25. Végvári, Z.; Valkó, O.; Deák, B.; Török, P.; Konyhás, S.; Tóthmérész, B. Effects of land use and wildfires on the habitat selection of Great Bustard (Otis tarda L.)—implications for species conservation. *Land Degrad. Dev.* **2016**, *27*, 910–918. [CrossRef]

26. FAO—UNESCO. *Soil Map World*, 5th ed.; UNESCO: Paris, France, 1981.

27. FAO. *World Reference Base for Soil Resources 2006: A Framework for International Classification, Correlation and Communication*; FAO: Rome, Italy, 2006.

28. European Soil Bureau Network and the Hungarian Soil Science Society. The Soils of Hungary. 2011. Available online: http://enfo.agt.bme.hu/drupal/sites/default/files/Soils%20of%20Hungary.pdf (accessed on 23 September 2017).

29. Campo, J.; Gimeno-García, E.; Andreu, V.; González-Pelayo, O.; Rubio, J. Aggregation of under canopy and bare soils in a Mediterranean environment affected by different fire intensities. *Catena* **2008**, *74*, 212–218. [CrossRef]

30. Rubio, J.L.; Forteza, J.; Andreu, V.; Cerni, R. Soil profile characteristics influencing runoff and soil erosion after forest fire: A case study (Valencia, Spain). *Soil Technol.* **1997**, *11*, 67–78. [CrossRef]

31. Gimeno-Garcia, E.; Andreu, V.; Rubio, J. Changes in organic matter, nitrogen, phosphorus and cations in soil as a result of fire and water erosion in a Mediterranean landscape. *Eur. J. Soil Sci.* **2000**, *51*, 201–210. [CrossRef]

32. Giovannini, G.; Lucchesi, S. Modifications induced in soil physico-chemical parameters by experimental fires at different intensities. *Soil Sci.* **1997**, *162*, 479–486. [CrossRef]

33. Hurlbert, S.H. Pseudoreplication and the design of ecological field experiments. *Ecol. Monogr.* **1984**, *54*, 187–211. [CrossRef]

34. Whelan, R.J. *The Ecology of Fire*, 2nd ed.; Cambridge University Press: Cambridge, UK, 1995.

35. Jackson, M.L. *Soil Chemical Analysis*; Prentice-Hall: Englewood Cliffs, NJ, USA, 1958.

36. Schulte, E.E.; Hopkins, B.G. Estimation of soil organic matter by weight loss-on-ignition. In *Soil Organic Matter: Analysis and Interpretation, SSSA Special Publication*, 46th ed.; Magdoff, F., Tabatabai, M.A., Hanlon, E.A., Eds.; Soil Society of America: Madison, WI, USA, 1996; pp. 21–31.

37. Combs, S.M.; Nathan, M.V. Soil Organic Matter. In *Recommended Chemical Soil Test. Procedures*, 1st ed.; Warncke, D., Brown, J., Eds.; Missouri Agricultural Experiment Station SB 1001: Columbia, MO, USA, 1998; pp. 53–59.

38. Zhang, H.; Wang, J.J. Loss on Ignition Method. In *Soil Test Methods from the Southeastern United States*; Sikora, F.J., Moore, K.P., Huluka, G., Kissel, D.E., Miller, R., Mylavarapu, R., Oldham, J.L., Eds.; Southern Cooperative Series Bulletin: Stillwater, OK, USA, 2014; pp. 155–158.

39. Pribyl, D.W. A critical review of the conventional SOC to SOM conversion factor. *Geoderma* **2010**, *156*, 75–83. [CrossRef]

40. Ball, D.F. Loss-on-ignition as an estimate of organic matter and organic carbon in non-calcareous soil. *Eur. J. Soil Sci.* **1964**, *15*, 84–92. [CrossRef]

41. Welz, B.; Sperling, M. *Atomic Absorption Spectrometry*, 3rd ed.; John Wiley & Sons: New York, NY, USA, 2008.

42. Rossel, R.V.; Walvoort, D.J.J.; McBratney, A.B.; Janik, L.J.; Skjemstad, J.O. Visible, near infrared, mid infrared or combined diffuse reflectance spectroscopy for simultaneous assessment of various soil properties. *Geoderma* **2006**, *131*, 59–75. [CrossRef]

43. González-Pelayo, O.V.; Andreu, J.; Campo, E.; Gimeno-García, E.; Rubio, J.L. Hydrological properties of a Mediterranean soil burned with different fire intensities. *Catena* **2006**, *68*, 186–193. [CrossRef]

44. Granged, A.; Jordán, A.; Zavala, L.; Muñoz-Rojas, M.; Mataix-Solera, J. Short-term effects of experimental fire for a soil under eucalyptus forest (SE Australia). *Geoderma* **2011**, *167*, 125–134. [CrossRef]

45. Etienne, M.; Legrand, C. A non-destructive method to estimate shrubland biomass and combustibility. In Proceedings of the 2nd International Conference on Forest Fire Research, Comissão de Coordenação da Região Centro Coimbra, Coimbra, Portugal, 1994; pp. 425–434.

46. Alcañiz, M.; Outeiro, L.; Francos, M.; Úbeda, X. Effects of prescribed fires on soil properties: A review. *Sci. Total Environ.* **2018**, *613*, 944–957. [CrossRef] [PubMed]

47. González-Pérez, J.A.; González-Vila, F.J.; Polvillo, O.; Almendros, G.; Knicker, H.; Salas, F.; Costa, J.C. *Wildfire and Black Carbon in Andalusian Mediterranean Forest*; Viegas, D.X., Ed.; Millpress: Rotterdam, The Netherlands, 2000; pp. 1–7.

48. Chandler, C.; Cheney, P.; Thomas, P.; Trabaud, L.; Williams, D. *Fire in Forestry Volume 1: Forest Fire Behavior and Effects*; Wiley: New York, NY, USA, 1983.

49. Fernández, I.; Cabaneiro, A.; Carballas, T. Organic matter changes immediately after a wildfire in an Atlantic forest soil and comparison with laboratory soil heating. *Soil Biol. Biochem.* **1997**, *29*, 1–11. [CrossRef]

50. Rashid, G.H. Effects of fire on soil carbon and nitrogen in a Mediterranean oak forest of Algeria. *Plant Soil* **1987**, *103*, 89–93. [CrossRef]

51. Smith, D.W. Concentrations of soil nutrients before and after fire. *Can. J. Soil Sci.* **1970**, *50*, 17–29. [CrossRef]

52. Neary, D.G.; Ryan, K.C.; DeBano, L.F. Wildland Fire in Ecosystems: Effects of Fire on Foils and Water. General Technical Report. 2005. Available online: https://www.fs.fed.us/rm/pubs/rmrs_gtr042_4.pdf (accessed on 7 October 2017).

53. DeBano, L.F.; Ffolliott, P.; Neary, D. *Fire's Effects on Ecosystems*; John Wiley & Sons: New York, NY, USA, 1998.

54. Ene, C.; Rogobete, G.; Adam, I.; Ciobanu, V.D. Influence of fires on soils of Berzasca and Moldova Noua forestries (wouth-west of Romania). In Proceedings of the 12th WSEAS international conference on Mathematics and computers in biology, business and acoustics, Braşov, Romania, 11–13 April 2011; pp. 114–121.

55. Fonseca, F.; De Figueiredo, T.; Nogueira, C.; Queirós, A. Effect of prescribed fire on soil properties and soil erosion in a Mediterranean mountain area. *Geoderma* **2017**, *307*, 172–180. [CrossRef]

56. Kuchenbuch, R.; Claassen, N.; Jungk, A. Potassium availability in relation to soil moisture. *Plant Soil* **1986**, *95*, 233–243. [CrossRef]

57. Cade-Menun, B.; Berch, S.; Preston, C.; Lavkulich, L. Phosphorus forms and related soil chemistry of Podzolic soils on northern Vancouver Island. I. A comparison of two forest types. *Can. J. For. Res.* **2000**, *30*, 1714–1725. [CrossRef]

58. Kutiel, P.; Shaviv, A. Effects of soil type, plant composition and leaching on soil nutrients following a simulated forest fire. *For. Ecol. Manag.* **1992**, *53*, 329–343. [CrossRef]

59. Schaller, J.; Tischer, A.; Struyf, E.; Bremer, M.; Belmonte, D.U.; Potthast, K. Fire enhances phosphorus availability in topsoils depending on binding properties. *Ecology* **2015**, *96*, 1598–1606. [CrossRef]

60. Raison, R.J.; Khanna, P.K.; Woods, P.V. Mechanisms of element transfer to the atmosphere during vegetation fires. *Can. J. For. Res.* **1985**, *15*, 132–140. [CrossRef]
61. Lung, T.; Lavalle, C.; Hiederer, R.; Dosio, A.; Bouwer, L.M. A multi-hazard regional level impact assessment for Europe combining indicators of climatic and non-climatic change. *Glob. Environ. Chang.* **2013**, *23*, 522–536. [CrossRef]

Article

Fine Root Biomass Mediates Soil Fauna Community in Response to Nitrogen Addition in Poplar Plantations (*Populus deltoids*) on the East Coast of China

Haixue Bian [1,2], Qinghong Geng [1,2], Hanran Xiao [1,2], Caiqin Shen [3], Qian Li [4], Xiaoli Cheng [5], Yiqi Luo [6], Honghua Ruan [1,2] and Xia Xu [1,2,*]

1   College of Biology and the Environment, Nanjing Forestry University, Nanjing, Jiangsu 210037, China; haixue_bian@163.com (H.B.); gengqh0127@163.com (Q.G.); xiaobaisky123@163.com (H.X.); hhruan@njfu.edu.cn (H.R.)
2   Co-Innovation Centre for Sustainable Forestry in Southern China, Nanjing Forestry University, Nanjing, Jiangsu 210037, China
3   Dongtai Forest Farm, Yancheng, Jiangsu 224200, China; shencaiqin123@163.com
4   Advanced Analysis and Testing Center, Nanjing Forestry University, Nanjing, Jiangsu 210037, China; liqian.1987@outlook.com
5   Key Laboratory of Aquatic Botany and Watershed Ecology, Wuhan Botanical Garden, Chinese Academy of Sciences, Wuhan 430074, China; chengxiaoli@wbgcas.cn
6   Center for Ecosystem Science and Society, Northern Arizona University, Flagstaff, AZ 86011, USA; yiqi.luo@nau.edu
*   Correspondence: xuxia.1982@njfu.edu.cn; Tel.: +86-25-8542-7086

Received: 4 January 2019; Accepted: 31 January 2019; Published: 3 February 2019

Abstract: Soil fauna is critical for maintaining ecosystem functioning, and its community could be significantly impacted by nitrogen (N) deposition. However, our knowledge of how soil-faunal community composition responds to N addition is still limited. In this study, we simulated N deposition (0, 50, 100, 150, and 300 kg N ha^{-1} year^{-1}) to explore the effects of N addition on the total and the phytophagous soil fauna along the soil profile (0–10, 10–25, and 25–40 cm) in poplar plantations (*Populus deltoids*) on the east coast of China. Ammonium nitrate (NH_4NO_3) was dissolved in water and sprayed evenly under the canopy with a backpack sprayer to simulate N deposition. Our results showed that N addition either significantly increased or decreased the density (D) of both the total and the phytophagous soil fauna (D_{total} and D_p) at low or high N addition rates, respectively, indicating the existence of threshold effects over the range of N addition. However, N addition had no significant impacts on the number of groups (G) and diversity (H) of either the total or the phytophagous soil fauna (G_{total}, G_p and H_{total}, H_p). With increasing soil depth, D_{total}, D_p, G_{total}, and G_p largely decreased, showing that the soil fauna have a propensity to aggregate at the soil surface. H_{total} and H_p did not significantly vary along the soil profile. Importantly, the threshold effects of N addition on D_{total} and D_p increased from 50 and 100 to 150 kg N ha^{-1} year^{-1} along the soil profile. Fine root biomass was the dominant factor mediating variations in D_{total} and D_p. Our results suggested that N addition may drive changes in soil-faunal community composition by altering belowground food resources in poplar plantations.

Keywords: soil fauna; N addition; soil profile; community structure; food resources; poplar plantations

---

1. Introduction

Nitrogen (N) deposition, which is predicted to continuously increase, has become a major global concern [1–3]. In terrestrial ecosystems, most of the deposited N eventually dissipates into soils [4].

N accumulation in soils could change the community structure of soil fauna and could thereby impact ecosystem functioning [5]. Collectively, the soil fauna is a major consumer and decomposer and is therefore an essential component of the forest ecosystem [6,7]. As the "engineer of the soil ecosystem", soil fauna plays an important role in dissolving residues and altering biogeochemical cycles [8,9]. However, our understanding of the responses of soil fauna community to increased levels of N input is still limited.

Increasing evidence has demonstrated that N addition that mimics natural N deposition can undoubtedly have profound impacts on soil fauna communities. Despite much research, however, no consensus exists regarding the impacts of N addition in soil fauna communities. For instance, N addition significantly increases the densities of both phytophagous soil fauna and total soil fauna, which could be attributed to enhanced ammonium (NH_4^+) production and nitrification processes from increased root exudation and organic matter input to the system [10]. In addition, Raub, et al. [11] confirmed the positive effect of N-rich food resources on the abundance of soil fauna. Conversely, a long-term N addition study reports that continuous N addition significantly reduces the density and taxa richness of the soil fauna due to the decreased allocation of carbon (C) to leaf and fine root litter, especially C from the roots associated with changes in rhizodeposition [12–14]. Additionally, some studies find no significant effects of N addition on the density of soil fauna due to reduced competitive exclusion of soil fauna and a short observation period [15]. A lack of impacts of N addition on the diversity of soil fauna has also been reported; results from the demands for N are similar among various soil fauna [16,17] or elevated intraguild predation [11]. The different responses of soil fauna communities to N addition—whether positive, negative, or non-significant—may primarily depend on the N addition rate [4]. The impact of N addition on soil fauna communities could change directionally and dramatically at a critical concentration [18,19], suggesting a threshold effect. The study of soil fauna community responses to N addition at a range of rates is therefore of great importance.

N addition could affect the vertical distribution of soil fauna, leading to major variations in the soil fauna community with increasing soil depth [20]. The density of soil fauna, for example, could appreciably increase in the topsoil layer due to N addition while exhibiting no significant changes in deep soil layers [21]. Soil depth is one of the most important factors influencing soil fauna communities due to variations in the soil's physical and chemical properties along the depth gradient [22,23]. Soil fauna may be more abundant in the subsoil, where they take refuge to avoid the high N concentrations in the upper soil layers under N addition [24]. Moreover, thresholds, if present, could shift from high to low along the soil profile [24]. Although N addition is reported to affect the soil fauna community, field-based studies of the responses of a soil fauna community to soil depth are still lacking, making predictions about soil food web changes with N deposition in plantations difficult.

The soil fauna community's responses to N addition are mediated by the quantity and quality of the food resources available, such as roots, litter, and fungi [25,26]. Typically, soil fauna density is correlated with root biomass [23], though previous studies in various ecosystems have shown that N addition can have a variety of impacts on fine root biomass, either stimulating the production of fine root biomass [27] or retarding it [28]. Moreover, root chemical traits are key factors in shaping soil fauna communities, and do so via mediating species-specific interactions and affecting the soil fauna at multiple trophic levels [29]. It is well established that plant root-derived nutrition, such as N, may strongly influence a soil fauna community [30]. Litter quantity and quality are two additional factors that have a major influence on the activity and composition of soil fauna [31]. High quantities of litter can stimulate an increase in soil fauna density, mainly due to concomitant increases in the food supply and the habitat space of the ecosystem [11,32]. In addition, the density and community parameters of soil fauna also largely depend on the chemical and physical qualities of the litter [31]. A previous study showed that the density of soil fauna could increase as the litter C:N ratio decreased [33]. Furthermore, some soil fauna prefer feeding on older, more decomposed litter, and this preference strongly depends on the structure of the litter, e.g., tensile strength and palatability [26]. Recent studies, however, have suggested that root-derived food resources can be much more effective in fueling belowground food webs than those that come from leaf litter (e.g., [30,34]). Roots as a food source for soil fauna are considered to be an essential component of the underground food web [35]. The quantity and quality of food resources are likely to change in response to N addition, mediating the effects of N addition on soil fauna communities.

China grows more than 7 million ha of poplar plantations. Biomass production and C fixation by poplars (*Populus*) thus play important roles in mediating global climate change [36]. Compared to the natural forest ecosystem, changes in the soil fauna community in response to N addition in poplar plantations are less studied, especially for the whole soil fauna community [19]. Over the decades, many studies have evaluated the impact of N addition on particular species of soil fauna [37,38], while few studies have reported whole soil fauna community responses to N addition. Studies focused on one or two soil fauna species are not sufficiently indicative of the changes in whole soil fauna community [39]. It is thus necessary to study the response of the whole soil fauna to N addition in poplar plantations. In this study, we aimed to: (1) explore the effects of N addition and soil depth on the whole soil fauna community, and (2) determine the primary factors controlling the responses of the whole soil fauna community across N treatments and the soil profile in a poplar plantation at the Dongtai Forest Farm, eastern China.

2. Materials and Methods

2.1. The Experimental Site and Design

Our experimental site is located at the Dongtai Forest Farm in Yancheng, Jiangsu Province, eastern China (120°49′ E, 32°52′ N). The farm is close to the Yellow Sea State Forest Park (on the coast of the Yellow Sea) and has a climate classified as Cfa (Humid subtropical climate) according to Köppen [40]. The mean annual temperature (MAT) is 13.7 °C and the mean annual precipitation (MAP) is 1051 mm. The soil of the forest farm is a desalting meadow and sandy soil with a pH value ~8 [41].

Our N addition experiment was established in May 2012. We chose 12-year-old pure poplar plantations (*Populus deltoids cv. 'I-35'*) with uniform site conditions and management measures as our plots. We used a randomized block design with a gradient of five levels of N addition (0, 50, 100, 150, and 300 kg N ha^{-1} year^{-1}) in three replicate blocks (25 × 190 m). Each N treatment subplot was 25 × 30 m with a 10 m buffer zone between any two adjacent subplots. The distance between any two adjacent blocks was at least 500 m. We chose a range of N addition rates because the ambient N deposition rate is about 50 kg N ha^{-1} year^{-1} for this area [42] and Liu et al. [18] estimated that the critical loads for N deposition in Jiangsu province could be more than 200 kg N ha^{-1} year^{-1}. In each month of the growing season (May through October), one-sixth of the yearly amount of NH_4NO_3 was dissolved in 20 L water and sprayed evenly under the canopy by backpack sprayer to simulate natural N deposition. Each control subplot received 20 L of water.

2.2. Microclimate

Soil temperature (0–15 cm) and moisture (0–20 cm) were measured once or twice a month using a Delta-T WET-2 (UK) from January to December 2016. Litter was collected in each subplot from November to December 2016 with a 2 m × 2 m litter trap constructed of 2 mm mesh nylon cloth. We oven-dried the litter at 65 °C for 48 h and weighed it to calculate litter mass.

2.3. Measurement of C and N in Litter and Soil Samples

Litter samples for measuring C and N content were collected in October 2016, oven-dried, ground, filtered by sieving (0.5 mm), and analyzed with an elemental analyzer (PerkinElmer 2400 II, Waltham, MA, USA). Soil cores of 4 cm in diameter were taken in October 2016 from the 0–15 cm soil layer in each subplot. Soil samples were transferred to our lab, air-dried, ground, and dipped into 0.5 M hydrochloric acid (HCl) to remove carbonates [43,44]. We then measured soil organic C (SOC) and total N (TN) by combustion with an elemental analyzer (Elementar, Vario EL III, Elementar Analysen Systeme GmbH, Elementar Analysensysteme GmbH, Hanau, Germany).

2.4. Estimation of Fine Root Biomass

Soil cores were taken from the top 45 cm of soil in each subplot with polyvinyl chloride (PVC) tubes (with an inside diameter of 4.6 cm) in June 2016 to estimate the fine root biomass [45–47]. Soil cores were numbered and then transported to our laboratory at Nanjing Forestry University, where they were frozen at −20 °C before analyzing. The cores were separated into three sections by depth (0–10, 10–25, and 25–40 cm), then carefully washed by wet sieving (0.5 mm) under gently flowing water to remove attached soil and dark-brown/black debris. Collected root samples were separated into coarse (>2 mm in diameter) and fine (≤2 mm in diameter, live and dead) roots, oven-dried at 65 °C for 48 h and weighed to calculate the live and dead fine root biomass.

2.5. Soil Fauna Sampling and Identification

Soil samples were collected from the 0–10, 10–25, and 25–40 cm soil layers using a soil coring with a diameter of 4 cm. Different layers represent organic horizon (O horizon, 0–10 cm), eluvial horizon (A horizon, 10–25 cm), and deposition horizon (B horizon, 25–40 cm), respectively [48,49]. In June 2016, four soil cores were collected from each 25 × 30 m subplot and pooled together as a replicate sample. The soil samples were immediately shipped back to our laboratory. The soil fauna was then collected from each soil sample using Tullgren extractors (Tullgren Funnel Unit, BURKARD, BURKARD SCIENTIFIC Ltd., Uxbridge, UK) [50–53]. All collected fauna samples were preserved in 75% ethanol and then sorted under a dissecting microscope (Nikon Eclipse E200, Nikon Instech Co., Ltd., Tokyo, Japan). Soil fauna was identified according to Yin [54,55].

2.6. Statistical Analysis

To determine the structure of the soil fauna in our study site, we chose density, the number of groups, and the Shannon-Wiener diversity index to describe the characteristics and composition of the soil fauna of each sample. The soil fauna density (D) was calculated as the following:

$$D = \frac{N}{V};\tag{1}$$

where D is the density of soil fauna, N is the total number of individuals, and V is the volume of the soil samples. The number of groups (G) was estimated as the number of groups of the fauna in the same order. The Shannon-Wiener diversity index (H) of each sample was estimated according to Whittaker [56]:

$$H = -\sum_{i=1}^{n} P_i \ln P_i \tag{2}$$

where H is the Shannon-Wiener diversity index of each sample, n is the number of groups of soil fauna, and P_i is the proportion of the number of individuals in the ith order to the total number of individuals, which could be calculated with the following equation:

$$P_i = \frac{N_i}{N} \tag{3}$$

where N_i is the number of individuals in the ith order.

Two-way ANOVA was used to examine the impact of N addition, soil layers, and their interactions on the fine root biomass (including live, dead, and total fine roots), density, number of groups, and biodiversity of phytophagous and total soil fauna. Repeated-measures ANOVA was performed to test the responses of litter mass, soil temperature, and moisture to N addition. One-way ANOVA was used to examine the responses of the litter C:N ratio, soil TN, and SOC to N addition. Linear regression analyses were performed to examine the relationships between soil fauna and fine root biomass (live, dead, and total), SOC, TN, mean annual soil temperature and between moisture and the soil temperature in June. All statistical analyses were performed using SPSS 22.0 for Windows (SPSS Inc., Chicago, IL, USA).

3. Results

3.1. Microclimate, Litter C and N, Soil Samples, and Fine Root Biomass

Soil temperature and moisture showed remarkable seasonal variations. Soil temperature increased from January to August and then decreased from August to December (Table 1, Figure S1a) and soil moisture was low in summer and high in spring and fall (Table 1, Figure S1b). N addition did not affect the C:N ratio of the litter, SOC, TN, soil temperature or moisture (all $p > 0.05$, Tables 1 and 2). While N addition had no significant effect on live, dead, or total fine root biomass (all $p > 0.05$), root biomass in general decreased with increasing soil depth (all $p < 0.01$, Table 3, Figure S2).

Table 1. Results of repeated-measures ANOVA for seasonal responses of litter mass, soil temperature, and moisture to nitrogen (N) addition (0, 50, 100, 150, and 300 kg N ha^{-1} year^{-1}). Statistical analysis was performed on log-transformed data. p levels are indicated beside the F values by the following: *: <0.05, **: <0.01, ***: <0.001, and no asterisks indicate $p > 0.05$. df represents degree of freedom.

Variables	N Addition (N)		Time (T)		N × T	
	df	F, p	df	F, p	df	F, p
Litter mass	4	0.41	1	311.77 ***	4	0.58
Soil temperature	4	0.49	15	1582.24 ***	60	0.29
Soil moisture	4	0.67	15	233.25 ***	60	0.38

Table 2. Results of one-way ANOVA for responses of litter carbon (C): N ratio, soil N (TN), and soil organic C (SOC) to N addition. Statistical analysis was performed on log-transformed data. p levels are indicated beside the F values by the following: *: <0.05, **: <0.01, ***: <0.001, and no asterisks indicate $p > 0.05$.

Variables	N Addition (N)	
	df	F, p
Litter C:N ratio	4	1.68
TN	4	0.96
SOC	4	1.67

Table 3. Results of two-way ANOVA for responses of the density (*D*), number of groups (*G*), Shannon-Wiener diversity index (*H*) of soil fauna and fine root biomass (live root biomass, dead root biomass, and total root biomass) to nitrogen (N) addition and soil depth (0–10, 10–25, and 25–40 cm). Statistical analysis was performed on log-transformed data. *P* levels are indicated beside the F values by the following: *: <0.05, **: <0.01, ***: <0.001, and no asterisks indicate $p > 0.05$.

Variables		N Addition (N)		Soil Depth (depth)		N × Depth	
		df	F, p	df	F, p	df	F, p
	D_{total}	4	8.11 **	2	308.52 ***	8	2.77 *
Total soil fauna	G_{total}	4	1.91	2	10.63 *	8	2.19
	H_{total}	4	1.33	2	6.35	8	1.78

Table 3. *Cont.*

Variables		N Addition (N)		Soil Depth (depth)		N × Depth	
		df	*F, p*	*df*	*F, p*	*df*	*F, p*
Phytophagous soil fauna	D_p	4	5.91 *	2	118.38 ***	8	1.75
	G_p	4	1.74	2	8.13 *	8	2.79*
	H_p	4	2.28	2	4.94	8	2.81
Fine root biomass	Live root biomass	4	1.84	2	13.42 ***	8	0.72
	Dead root biomass	4	2.54	2	8.9 **	8	1.39
	Total root biomass	4	1.64	2	17.32 ***	8	0.88

3.2. Soil Fauna Community as Affected by N Addition

In general, N addition and soil depth both played important roles in altering the density of the soil fauna. With increasing N addition rate, the density of the total soil fauna (D_{total}) significantly increased first, then declined to the levels of the controls ($p < 0.01$, Table 3, Figure 1a). In the 25–40 cm soil layer, for example, D_{total} increased from N_0 to N_3 and then declined from N_3 to N_4 (Figure 1a). In addition, D_{total} significantly decreased with increasing soil depth ($p < 0.001$, Table 3, Figure 1a). The interaction between N addition and soil depth had a significant impact on D_{total} ($p < 0.05$, Table 3). D_{total} increased from N_0 to N_1 and then declined with higher N addition (from N_1 to N_4) in the topsoil layer (0–10 cm). In the deep soil layers (10–25 and 25–40 cm), however, declines in D_{total} started at N_2 and N_3, respectively, compared to the topsoil layer. The density of phytophagous soil fauna (D_p) showed similar responses to N addition along the soil profile (Table 3, Figure 1b).

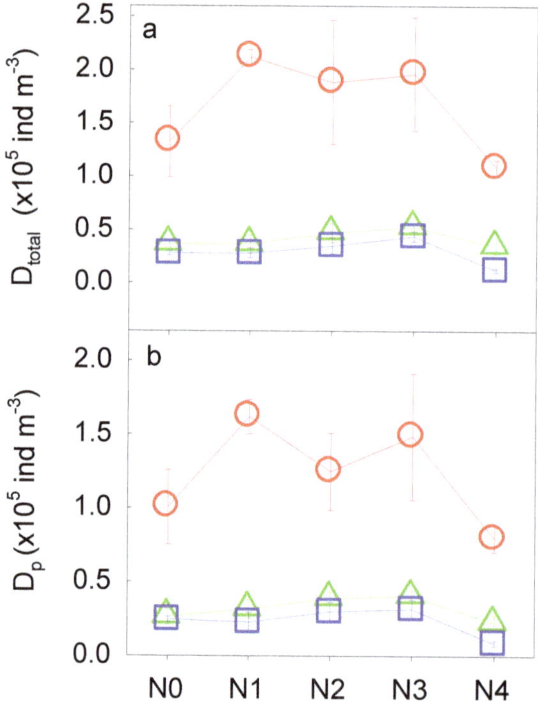

Figure 1. The density of total soil fauna (D_{total}, **a**) and phytophagous soil fauna (D_p, **b**) along the N addition gradient and the soil profile (red for 0–10 cm, green for 10–25 cm, and blue for 25–40 cm). N_0, N_1, N_2, N_3, N_4 indicate the gradient of N additions (N_0: 0 kg N ha^{-1} year^{-1}; N_1: 50 kg N ha^{-1} year^{-1}; N_2: 100 kg N ha^{-1} year^{-1}; N_3: 150 kg N ha^{-1} year^{-1}; N_4: 300 kg N ha^{-1} year^{-1}).

N addition had no significant impact on either the number of groups of total soil fauna (G_{total}) or the number of groups of phytophagous fauna (G_p) (all $p > 0.05$, Table 3, Figure 2). With increasing soil depth, in general, we found significant decreases in G_{total} and G_p (all $p < 0.05$, Table 3, Figure 2). Abrupt decreases in G_{total} and G_p at N_2 were observed in the topsoil layer (0–10 cm). We also found interactive effects between N addition and soil layer on G_p ($p < 0.05$, Table 3, Figure 2b). Along the soil profile, for instance, G_p reached its highest values at N_1, N_2, and N_3. For the Shannon-Wiener diversity index of the soil fauna, neither N addition nor soil depth had significant effects on the diversity of total soil fauna (H_{total}) or the diversity of phytophagous soil fauna (H_p) (all $p > 0.05$, Table 3, Figure 3).

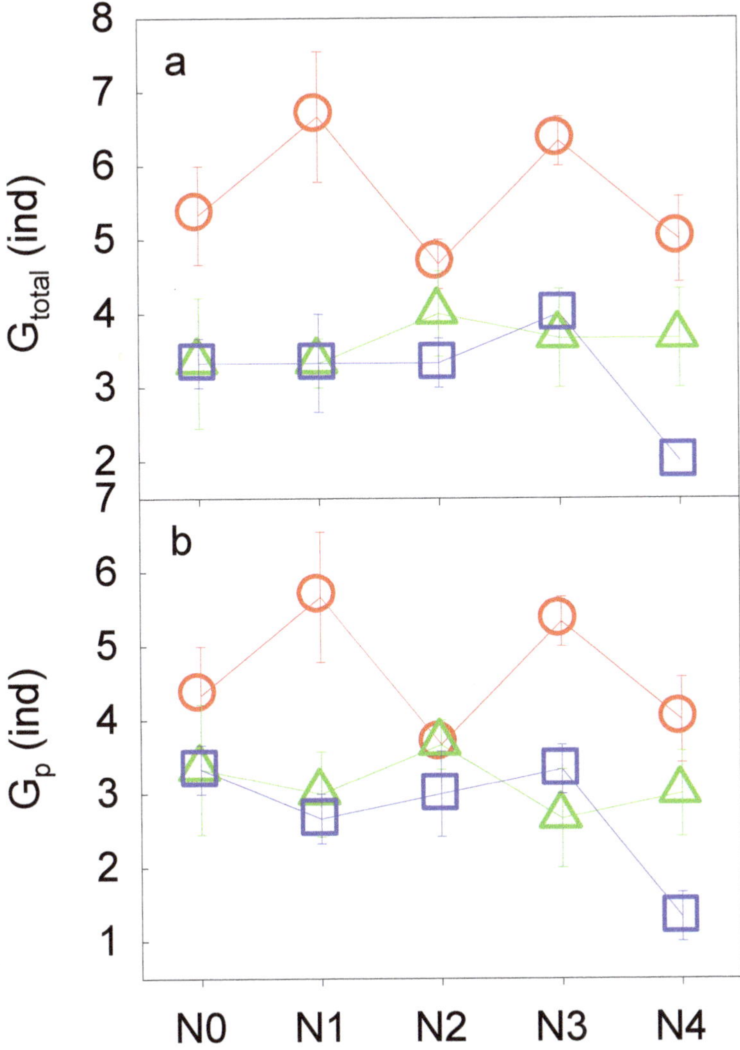

Figure 2. The number of groups of total soil fauna (G_{total}, **a**) and phytophagous soil fauna (G_p, **b**) along the N addition gradient and the soil profile. See Figure 1 for abbreviations.

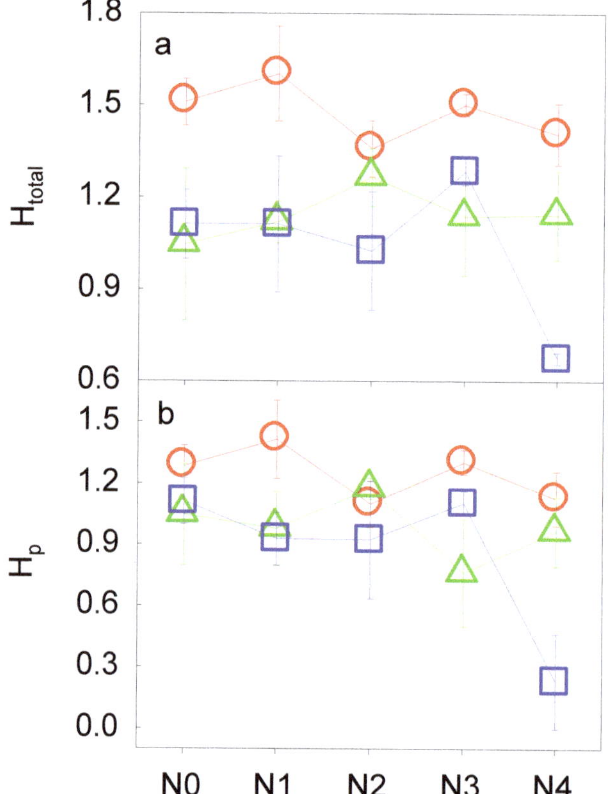

Figure 3. The Shannon-Wiener diversity index of total soil fauna (H_{total}, **a**) and phytophagous soil fauna (H_p, **b**) along the N addition gradient and the soil profile. See Figure 1 for abbreviations.

3.3. Fine Root Biomass Regulation of Soil Fauna Density

Variations in soil fauna densities (both D_p and D_{total}) were consistently regulated by fine root biomass (Table 4). We found that variations in D_p and D_{total} across N addition treatments and soil profile were positively correlated with the amount of live, dead, and total fine roots (all $p < 0.01$, Figures 4 and 5). The positive relationships remained even without the "outliers" — the very small values (both in fine root biomass on the *X*-axis as well as in the soil fauna density on the *Y* axis, all $p < 0.01$, Figures S3 and S4). Multifactor linear regression showed that the dominant factors controlling variations in D_p and D_{total} were the total and the dead fine root biomass across N addition treatments (all $p < 0.001$, Table 4).

Table 4. Results of multi-regression analysis of the density of phytophagous soil fauna (D_p) and total soil fauna (D_{total}) with fine root biomass (live, dead, and total), SOC, TN, mean annual soil temperature and moisture, and the soil temperature and moisture in June across N addition treatments. ***: $p < 0.001$.

Dependent	Model	Variables	Regression	r^2 and p
D_p	A-1	Total root	D_p = 2.79 + 0.72 total root	0.48 ***
	A-2	Total root, dead root	D_p = 2.95 + 0.63 total root+0.11 dead root	0.53 ***
D_{total}	B-1	Total root	D_{total} = 3.03 + 0.67 total root	0.42 ***
	B-2	Total root, dead root	D_{total} = 3.20 + 0.58 total root+0.12 dead root	0.49 ***

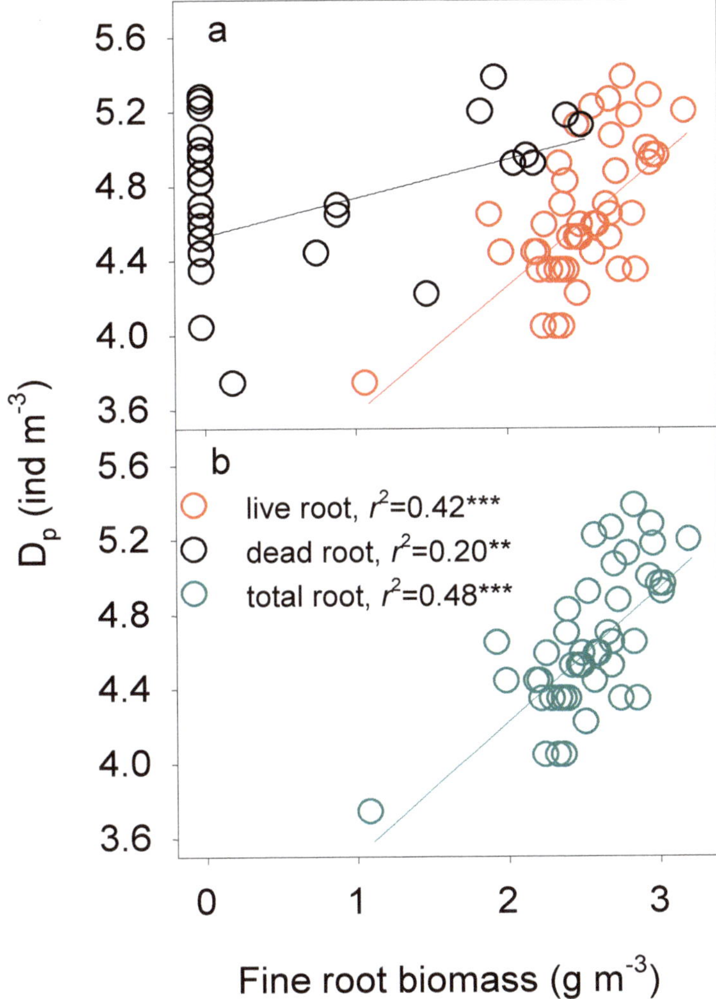

Figure 4. Log-log regulation of fine root biomass ((**a**) live and dead root biomass; (**b**) total root biomass) on the density of phytophagous soil fauna (D_p). **: $p < 0.01$, ***: $p < 0.001$.

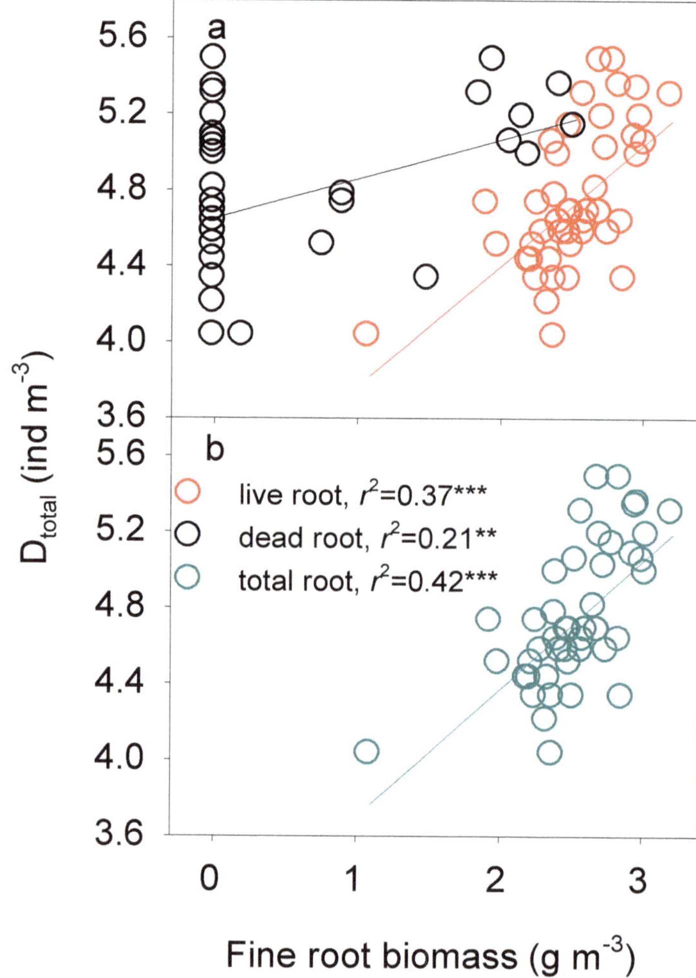

Figure 5. Log-log regulation of fine root biomass ((**a**) live and dead root biomass; (**b**) total root biomass) on the density of total soil fauna (D_{total}). **: $p < 0.01$, ***: $p < 0.001$.

4. Discussion

4.1. N-addition Impacts on Soil Fauna Density

Although N addition, up to a certain rate, could have positive impacts on soil fauna, high rates of N input appears to negatively affect the fauna community. The increases in D_{total} at low N addition rates are in line with previous findings, which show that N addition, to a certain extent, can be beneficial to soil fauna (e.g., [57–59]). The increases in D_{total} under N addition may have resulted from (1) increases in available N in the soil [60]; (2) increases in the food sources (soil bacteria and fungi) for the soil fauna [5,61]; (3) increases in rhizodeposition-C accessible to soil fauna [14,62]; and (4) more hospitable soil conditions for soil fauna (e.g., decreases in soil pH associated with N addition) [63]. Interestingly, D_{total} dramatically decreased from N_3 to N_4, exhibiting a negative response to high N addition. Decreases in the density of soil fauna at high levels of N addition may be attributed to at least two reasons. First, excessive acidic substances resulted from N addition may negatively affect soil fauna [64,65]. For instance, soil acidification [66] significantly restricts the number of soil nematodes and could thus decrease D_{total} [67,68]. Second, N may cause soil eutrophication due to the increase in nitrate (NO_3^-) and NH_4^+ concentrations in soil, and thereby has a negative influence on soil fauna [19,67]. Similar to the variations observed in D_{total}, D_p increased from N_0 to N_3 and decreased from N_3 to N_4. The decline in D_p is consistent with previous studies, which found that N addition that increases NH_4^+ concentration significantly decreases D_p, as the NH_4^+ suppresses phytophagous soil fauna [13,67,69]. Additionally, we found that D_p positively responds to low N addition rates, which is similar to the trend observed for D_{total}, probably due to the large proportion (~75% across all treatments) of phytophagous soil fauna to total soil fauna in our study. Since few studies have reported any positive impacts of N addition on D_p, the mechanisms remain to be explored.

A threshold effect of N addition on soil fauna is usually widely found [19,70]. In our study, we detected a threshold in the effects of N addition on soil fauna at approximately 150 kg N ha^{-1} year^{-1} for both D_{total} and D_p. N addition within a certain concentration range enables competitive soil fauna to coexist with less competitive soil fauna, leading to an increase in density, while excessive N addition is a threat to the less competitive soil fauna, as it negatively alters the soil's physical and chemical properties [2]. When an ecosystem reaches N saturation, excessive exogenous N might increase leaching of some basic ions, such as calcium ion (Ca^{2+}), aluminum ion (Al^{3+}), and magnesium ion (Mg^{2+}), from the system [24], consequently destroying the structure and function of the ecosystem, causing soil acidification and nutrient imbalances, breaking intraspecific competition dynamics, and eroding the system's resistance to external disturbances [71]. Differences in N thresholds among ecosystems probably result from a quantitative variation in the primary available N. N addition may increase the quality and quantity of organic matter in young, developing ecosystems while having a toxic effect on the soil fauna in old ecosystems that reached N saturation from N addition [12]. Thus, the threshold effects of N deposition in different ecosystems and their potential impacts on soil fauna are worthy of further investigation from a community structure perspective.

4.2. N-addition Impacts on the Number of Groups and Diversity of Soil Fauna

N addition has either positive or negative effects on the number of groups and Shannon-Wiener diversity index of soil fauna in various ecosystems (e.g., [58,59,72,73]). However, a non-significant impact of N addition on those variables has been widely reported (e.g., [16,17,19]), supporting our findings in this study. The lack of N addition effects may be explained as follows: (1) N addition did not significantly alter the proportion of soil fauna in each group to the total number. Statistically, we found that the number of soil fauna in each group increased evenly with N addition (Table S1, Figures S5 and S6); (2) litter, as one of the main food sources for soil fauna [26], had no significant response to N addition either in terms of quantity (litter mass) or quality (C:N ratio) (all $p > 0.05$, Tables 1 and 2); (3) N addition had no significant influence on soil microclimate (soil temperature and moisture) (all $p > 0.05$, Table 1). The abrupt decreases in G_{total} and G_p at N_2 may have been related to a decrease in food resources, e.g., fine root biomass, at N_2. Additionally, the discrepancy in the effects of N addition on the number of groups and the diversity of soil fauna between our study and the previous studies may have resulted from the differences in soils, as our experimental site is located in an alkaline coastal area.

4.3. Increasing N-Addition Thresholds along Soil Profile

N addition did not affect the "surface aggregation" of soil fauna but increased the thresholds with increasing soil depth. The density and the number of groups of soil fauna usually decrease along soil profile [48,74], and this holds true under N addition in our study. Surface soil is more hospitable to soil fauna, as it always has a higher substrate content and air circulation rate [53,75]. Higher N addition thresholds for soil fauna may be attributed to the fact that topsoil could be more sensitive to environmental stress (e.g., N addition) than deeper soil layers [75]. Soil fauna move from the topsoil downwards to the deeper soil layers when high amounts of N are added to escape the depression of high N addition in the topsoil [76].

4.4. Regulation of Fine Root Biomass on Soil Fauna

Responses of soil fauna to N addition at the site level may be influenced by many factors, such as soil temperature and moisture [77], pH value [78], and resource supply [15]. Based on the multifactor regression analyses, we found that fine root biomass was the dominant factor controlling variations in the density of soil fauna for both D_{total} and D_p. Fine roots and their exudates may positively influence soil biota through stimulated hyphal growth with increasing quantities of fine root biomass [79]. A growing body of evidence suggests that soil food webs rely heavily on "root C" because fine roots can be a source of readily available C for both soil microorganisms and phytophagous soil fauna [80]. Likewise, many studies have shown that phytophagous soil fauna are largely controlled by root-derived nutrition [30,81], just as aboveground herbivore populations are strongly linked to net primary production [82].

5. Conclusions

Our study in poplar plantations on the east coast of China suggests that N addition could have substantial effects on soil fauna along the soil profile. Across the N addition treatments, low N addition rates had positive effects on soil fauna densities, likely due to improved physical and chemical soil conditions. With increasing N addition rates, responses of soil fauna community returned to the control level. Soil fauna responded differently to N addition in different soil layers, which may largely relate to changes in the soil substrate content with N addition. Multifactor linear regressions showed that fine root biomass mediated N induced changes in the soil fauna density. The global N deposition rate is projected to continuously increase in the future [3,83], which will undoubtedly affect soil fauna and further the biogeochemical cycles. Therefore, it is imperative to explore impacts of N addition on soil fauna communities to understand the way an ecosystem functions under the global N deposition.

Supplementary Materials: The following are available online at http://www.mdpi.com/1999-4907/10/2/122/s1, Figure S1: Variations in soil temperature (a) and moisture (b) from January to December along the nitrogen (N) addition gradient in 2016. N0, N1, N2, N3, N4 indicate different N additions (N0: 0 kg N ha^{-1} year^{-1}; N1: 50 kg N ha^{-1} year^{-1}; N2: 100 kg N ha^{-1} year^{-1}; N3: 150 kg N ha^{-1} year^{-1}; N4: 300 kg N ha^{-1} year^{-1}), Figure S2: The live (a), dead (b), and total (c) fine root biomass along the N addition gradient and the soil profile (red for 0–10 cm, green for 10–25 cm, and blue for 25–40 cm). See Figure S1 for abbreviations, Figure S3: Regulation of fine root biomass (a: live and dead root biomass; b: total root biomass) on the density of phytophagous soil fauna (Dp), without the very small points (circled in the panels). **: $p < 0.01$, ***: $P < 0.001$, Figure S4: Regulation of fine root biomass (a: live and dead root biomass; b: total root biomass) on the density of total soil fauna (Dtotal), without the very small points (circled in the panels). **: $p < 0.01$, ***: $p < 0.001$, Figure S5: The proportion of each individual soil fauna over total soil fauna along the nitrogen (N) addition gradient and the soil profile. See Figure S1 for abbreviations, Figure S6: The proportions of macrofauna, mesofauna, and microfauna over total soil fauna along the nitrogen (N) addition gradient and the soil profile. See Figure S1 for abbreviations. Table S1: Results of two-way ANOVA for the responses of each individual soil fauna's proportion over total soil fauna to nitrogen (N) addition (0, 50, 100, 150, and 300 kg N ha^{-1} year^{-1}) and soil depth (0–10, 10–25, and 25–40 cm). P levels are indicated beside the F values by the following: *: <0.05, **: <0.01, ***: <0.001, and no asterisks indicate $p > 0.05$.

Author Contributions: Conceptualization, H.R. and X.X.; Data curation, H.B., Q.G. and H.X.; Formal analysis, H.B., Q.G. and X.X.; Funding acquisition, H.R. and X.X.; Investigation, H.B., Q.G. and H.X.; Resources, C.S., H.R. and X.X.; Supervision, H.R. and X.X.; Validation, H.X.; Writing–original draft, H.B. and X.X.; Writing–review & editing, Q.L., X.C., Y.L., H.R. and X.X.

Funding: This study is financially supported by the National Key Research and Development Program of China (2016YFD0600204), the Recruitment Program for Young Professionals, the Jiangsu Specially Appointed Professors Program, the Natural Science Key Fund for Colleges and Universities of Jiangsu Province of China (17KJA180006), the Six Talent Peaks Program of Jiangsu Province (JY-041 & TD-XYDXX-006), the Nanjing Forestry University Science Fund for Distinguished Young Scholars (the "5151" Talent Program), the Jiangsu Collegiate Science and Technology Fund for the Excellent Innovative Research Teams, and the Priority Academic Program Development of Jiangsu Higher Education Institutions (PAPD).

Acknowledgments: We thank many lab members for their help with the field work.

Conflicts of Interest: The authors declare no conflict of interest.

References

1. Penuelas, J.; Poulter, B.; Sardans, J.; Ciais, P.; van der Velde, M.; Bopp, L.; Boucher, O.; Godderis, Y.; Hinsinger, P.; Llusia, J.; et al. Human-induced nitrogen-phosphorus imbalances alter natural and managed ecosystems across the globe. *Nat. Commun.* **2013**, *4*, 2934. [CrossRef] [PubMed]
2. Niu, S.; Classen, A.T.; Dukes, J.S.; Kardol, P.; Liu, L.; Luo, Y.; Rustad, L.; Sun, J.; Tang, J.; Templer, P.H. Global patterns and substrate-based mechanisms of the terrestrial nitrogen cycle. *Ecol. Lett.* **2016**, *19*, 697–709. [CrossRef] [PubMed]
3. Schwede, D.B.; Simpson, D.; Tan, J.; Fu, J.S.; Dentener, F.; Du, E.; deVries, W. Spatial variation of modelled total, dry and wet nitrogen deposition to forests at global scale. *Environ. Pollut.* **2018**, *243*, 1287–1301. [CrossRef]
4. Xu, G.-L.; Mo, J.-M.; Zhou, G.-Y.; Fu, S.-L. Preliminary response of soil fauna to simulated N deposition in three typical subtropical forests. *Pedosphere* **2006**, *16*, 596–601. [CrossRef]
5. Sun, F.; Tariq, A.; Chen, H.; He, Q.; Guan, Y.; Pan, K.; Chen, S.; Li, J.; Zhao, C.; Wang, H.; et al. Effect of nitrogen and phosphorus application on agricultural soil food webs. *Arch. Agron. Soil Sci.* **2017**, *63*, 1176–1186. [CrossRef]
6. Xin, W.D.; Yin, X.Q.; Song, B. Contribution of soil fauna to litter decomposition in Songnen sandy lands in northeastern China. *J. Arid Environ.* **2012**, *77*, 90–95. [CrossRef]
7. Oliverio, A.M.; Gan, H.; Wickings, K.; Fierer, N. A DNA metabarcoding approach to characterize soil arthropod communities. *Soil. Biol. Biochem.* **2018**, *125*, 37–43. [CrossRef]
8. Li, X.; Yin, X.; Wang, Z.; Fan, W. Litter mass loss and nutrient release influenced by soil fauna of Betula ermanii forest floor of the Changbai Mountains, China. *Appl. Soil Ecol.* **2015**, *95*, 15–22. [CrossRef]
9. Lu, P.; Dai, N.H.; Sun, X.Y.; Zhang, G.H.; Xu, D.R.; Zhan, Y.H. Composition and structure of soil fauna community in the Dexing Copper Mine tailings pool after revegetation. *Turk. J. Zool.* **2018**, *42*, 307–315.
10. Lin, H.; He, Z.; Hao, J.; Tian, K.; Jia, X.; Kong, X.; Akbar, S.; Bei, Z.; Tian, X. Effect of N addition on home-field advantage of litter decomposition in subtropical forests. *For. Ecol. Manag.* **2017**, *398*, 216–225. [CrossRef]
11. Raub, F.; Scheuermann, L.; Hoefer, H.; Brandl, R. No bottom-up effects of food addition on predators in a tropical forest. *Basic Appl. Ecol.* **2014**, *15*, 59–65. [CrossRef]
12. Gan, H.; Zak, D.R.; Hunter, M.D. Chronic nitrogen deposition alters the structure and function of detrital food webs in a northern hardwood ecosystem. *Ecol. Appl.* **2013**, *23*, 1311–1321. [CrossRef] [PubMed]
13. Eisenhauer, N.; Cesarz, S.; Koller, R.; Worm, K.; Reich, P.B. Global change belowground: Impacts of elevated CO_2, nitrogen, and summer drought on soil food webs and biodiversity. *Glob. Chang. Biol.* **2012**, *18*, 435–447. [CrossRef]
14. Preece, C.; Penuelas, J. Rhizodeposition under drought and consequences for soil communities and ecosystem resilience. *Plant Soil* **2016**, *409*, 1–17. [CrossRef]
15. Cole, L.; Buckland, S.M.; Bardgett, R.D. Influence of disturbance and nitrogen addition on plant and soil animal diversity in grassland. *Soil Biol. Biochem.* **2008**, *40*, 505–514. [CrossRef]
16. Jiang, M.; Wang, X.; Liusui, Y.; Sun, X.; Zhao, C.; Liu, H. Diversity and Abundance of Soil Animals as Influenced by Long-Term Fertilization in Grey Desert Soil, China. *Sustainability* **2015**, *7*, 10837–10853. [CrossRef]
17. Zhu, X.; Zhu, B. Diversity and abundance of soil fauna as influenced by long-term fertilization in cropland of purple soil, China. *Soil. Tillage Res.* **2015**, *146*, 39–46. [CrossRef]
18. Liu, X.; Duan, L.; Mo, J.; Du, E.; Shen, J.; Lu, X.; Zhang, Y.; Zhou, X.; He, C.; Zhang, F. Nitrogen deposition and its ecological impact in China: An overview. *Environ. Pollut.* **2011**, *159*, 2251–2264. [CrossRef]
19. Ochoa-Hueso, R.; Rocha, I.; Stevens, C.J.; Manrique, E.; Luciañez, M.J. Simulated nitrogen deposition affects soil fauna from a semiarid Mediterranean ecosystem in central Spain. *Biol. Fertil. Soils* **2014**, *50*, 191–196. [CrossRef]
20. Nkem, J.N.; Bruyn, L.A.L.D.; Hulugalle, N.R.; Grant, C.D. Changes in invertebrate populations over the growing cycle of an N-fertilised and unfertilised wheat crop in rotation with cotton in a grey Vertosol. *Appl. Soil Ecol.* **2002**, *20*, 69–74. [CrossRef]

21. Xu, G.L.; Schleppi, P.; Li, M.H.; Fu, S.L. Negative responses of Collembola in a forest soil (Alptal, Switzerland) under experimentally increased N deposition. *Environ. Pollut.* **2009**, *157*, 2030–2036. [CrossRef] [PubMed]

22. Yin, X.; Song, B.; Dong, W.; Xin, W.; Wang, Y. A review on the eco-geography of soil fauna in China. *J. Geogr. Sci.* **2010**, *20*, 333–346. [CrossRef]

23. Potapov, A.M.; Goncharov, A.A.; Semenina, E.E.; Korotkevich, A.Y.; Tsurikov, S.M.; Rozanova, O.L.; Anichkin, A.E.; Zuev, A.G.; Samoylova, E.S.; Semenyuk, I.I. Arthropods in the subsoil: Abundance and vertical distribution as related to soil organic matter, microbial biomass and plant roots. *Eur. J. Soil Biol.* **2017**, *82*, 88–97. [CrossRef]

24. Xu, G.L.; Mo, J.M.; Fu, S.L.; Gundersen, P.; Zhou, G.Y.; Xue, J.H. Response of soil fauna to simulated nitrogen deposition: A nursery experiment in subtropical China. *Acta Sci. Circumstantiae* **2007**, *19*, 603–609. [CrossRef]

25. Morrien, E. Understanding soil food web dynamics, how close do we get? *Soil Biol. Biochem.* **2016**, *102*, 10–13. [CrossRef]

26. Zieger, S.L.J. Trophic Structure of Soil Animal Food Webs of Deciduous Forests as Analyzed by Stable Isotope Labeling. Ph.D. Thesis, Universität Göttingen, Göttingen, Germany, 2016.

27. Leppälammi-Kujansuu, J.; Ostonen, I.; Strömgren, M.; Nilsson, L.O.; Kleja, D.B.; Sah, S.P.; Helmisaari, H.-S. Effects of long-term temperature and nutrient manipulation on Norway spruce fine roots and mycelia production. *Plant Soil* **2013**, *366*, 287–303. [CrossRef]

28. Li, W.; Jin, C.; Guan, D.; Wang, Q.; Wang, A.; Yuan, F.; Wu, J. The effects of simulated nitrogen deposition on plant root traits: A meta-analysis. *Soil Biol. Biochem.* **2015**, *82*, 112–118. [CrossRef]

29. Tsunoda, T.; Dam, N.M.V. Root chemical traits and their roles in belowground biotic interactions. *Pedobiologia* **2017**, *65*, 58–67. [CrossRef]

30. Eissfeller, V.; Beyer, F.; Valtanen, K.; Hertel, D.; Maraun, M.; Polle, A.; Scheu, S. Incorporation of plant carbon and microbial nitrogen into the rhizosphere food web of beech and ash. *Soil Biol. Biochem.* **2013**, *62*, 76–81. [CrossRef]

31. Sauvadet, M.; Chauvat, M.; Fanin, N.; Coulibaly, S.; Bertrand, I. Comparing the effects of litter quantity and quality on soil biota structure and functioning: Application to a cultivated soil in Northern France. *Appl. Soil Ecol.* **2016**, *107*, 261–271. [CrossRef]

32. Sayer, E.J.; Sutcliffe, L.M.E.; Ross, R.I.C.; Tanner, E.V.J. Arthropod Abundance and Diversity in a Lowland Tropical Forest Floor in Panama: The Role of Habitat Space vs. Nutrient Concentrations. *Biotropica* **2010**, *42*, 194–200. [CrossRef]

33. Ilieva-Makulec, K.; Olejniczak, I.; Szanser, M. Response of soil micro- and mesofauna to diversity and quality of plant litter. *Eur. J. Soil Biol.* **2006**, *42*, S244–S249. [CrossRef]

34. Gilbert, K.J.; Fahey, T.J.; Maerz, J.C.; Sherman, R.E.; Bohlen, P.; Dombroskie, J.J.; Groffman, P.M.; Yavitt, J.B. Exploring carbon flow through the root channel in a temperate forest soil food web. *Soil Biol. Biochem.* **2014**, *76*, 45–52. [CrossRef]

35. Melaniem, P.; Reinhard, L.; Stefan, S.; Mark, M. Compartmentalization of the soil animal food web as indicated by dual analysis of stable isotope ratios (15N/14N and 13C/12C). *Soil Biol. Biochem.* **2009**, *41*, 1221–1226.

36. Wang, S.; Tan, Y.; Fan, H.; Ruan, H.; Zheng, A. Responses of soil microarthropods to inorganic and organic fertilizers in a poplar plantation in a coastal area of eastern China. *Appl. Soil Ecol.* **2015**, *89*, 69–75. [CrossRef]

37. Liang, W.J.; Lou, Y.L.; Qi, L.; Shuang, Z.; Zhang, X.K.; Wang, J.K.; Coleman, D.; Fu, S.L.; Zou, X.M. Nematode faunal response to long-term application of nitrogen fertilizer and organic manure in Northeast China. *Soil Biol. Biochem.* **2009**, *41*, 883–890. [CrossRef]

38. Zhao, J.; Wang, F.; Li, J.; Zou, B.; Wang, X.; Li, Z.; Fu, S. Effects of experimental nitrogen and/or phosphorus additions on soil nematode communities in a secondary tropical forest. *Soil Biol. Biochem.* **2014**, *75*, 1–10. [CrossRef]

39. Boxman, A.W.; Blanck, K.; Brandrud, T.E.; Emmett, B.A.; Gundersen, P.; Hogervorst, R.F.; Kjonaas, O.J.; Persson, H.; Timmermann, V. Vegetation and soil biota response to experimentally-changed nitrogen inputs in coniferous forest ecosystems of the NITREX project. *For. Ecol. Manag.* **1998**, *101*, 65–79. [CrossRef]

40. WIKIPEDIA. Köppen Climate Classification. 1884. Available online: https://en.wikipedia.org/wiki/Köppen_climate_classification (accessed on 3 January 2019).

41. Wang, S.; Chen, H.Y.; Tan, Y.; Fan, H.; Ruan, H. Fertilizer regime impacts on abundance and diversity of soil fauna across a poplar plantation chronosequence in coastal Eastern China. *Sci. Rep.* **2016**, *6*, 20816. [CrossRef]

42. Zhu, J.; Wang, Q.; He, N.; Smith, M.D.; Elser, J.J.; Du, J.; Yuan, G.; Yu, G.; Yu, Q. Imbalanced atmospheric nitrogen and phosphorus depositions in China: Implications for nutrient limitation. *J. Geophys. Res. Biogeosci.* **2016**, *121*, 1605–1616. [CrossRef]

43. Ramnarine, R.; Voroney, R.P.; WagnerRiddle, C.; Dunfield, K.E. Carbonate removal by acid fumigation for measuring the δ13C of soil organic carbon. *Can. J. Soil Sci.* **2011**, *91*, 247–250. [CrossRef]

44. Wotherspoon, A.; Voroney, R.P.; Thevathasan, N.V.; Gordon, A.M. Comparison of Three Methods for Measurement of Soil Organic Carbon. *Commun. Soil Sci. Plant Anal.* **2015**, *46*, 362–374. [CrossRef]

45. Xu, X.; Niu, S.; Sherry, R.A.; Zhou, X.; Zhou, J.; Luo, Y. Interannual variability in responses of belowground net primary productivity (NPP) and NPP partitioning to long-term warming and clipping in a tallgrass prairie. *Glob. Chang. Biol.* **2012**, *18*, 1648–1656. [CrossRef]

46. Derner, J.D.; Briske, D.D. Does a tradeoff exist between morphological and physiological root plasticity? A comparison of grass growth forms. *Acta Oecol.* **1999**, *20*, 519–526. [CrossRef]

47. Gao, Y.Z.; Giese, M.; Lin, S.; Sattelmacher, B.; Zhao, Y.; Brueck, H. Belowground net primary productivity and biomass allocation of a grassland in Inner Mongolia is affected by grazing intensity. *Plant Soil* **2008**, *307*, 41–50. [CrossRef]

48. Yang, B.; Zhang, W.; Xu, H.; Wang, S.; Xu, X.; Fan, H.; Chen, H.Y.H.; Ruan, H. Effects of soil fauna on leaf litter decomposition under different land uses in eastern coast of China. *J. For. Res.* **2018**, *29*, 973–982. [CrossRef]

49. Wang, G.B.; Deng, F.F.; Xu, W.H.; Chen, H.Y.H.; Ruan, H.H. Poplar plantations in coastal China: Towards the identification of the best rotation age for optimal soil carbon sequestration. *Soil Use Manag.* **2016**, *32*, 303–310. [CrossRef]

50. Wallwork, J.A. The Distribution and Diversity of Soil Fauna. *Q. Rev. Biol.* **1977**, *52*, 319.

51. Wu, H.; Lu, M.; Lu, X.; Guan, Q.; He, X. Interactions between earthworms and mesofauna has no significant effect on emissions of CO_2 and N_2O from soil. *Soil Biol. Biochem.* **2015**, *88*, 294–297. [CrossRef]

52. Bokhorst, S.; Huiskes, A.; Convey, P.; Pmvan, B.; Aerts, R. Climate change effects on soil arthropod communities from the Falkland Islands and the Maritime Antarctic. *Soil Biol. Biochem.* **2008**, *40*, 1547–1556. [CrossRef]

53. Li, Y.; Chen, Y.; Xu, C.; Xu, H.; Zou, X.; Chen, H.Y.H.; Ruan, H. The abundance and community structure of soil arthropods in reclaimed coastal saline soil of managed poplar plantations. *Geoderma* **2018**, *327*, 130–137. [CrossRef]

54. Yin, W. *Subtropical Soil Animals in China (in Chinese)*; Science Press: Beijing, China, 2000.

55. Yin, W. *Chinese subtropical soil animals (in Chinese)*; Science Press: Beijing, China, 1992.

56. Whittaker, R.H. Evolution and Measurement of Species Diversity. *Taxon* **1972**, *21*, 213–251. [CrossRef]

57. Sjursen, H.; Michelsen, A.; Jonasson, S. Effects of long-term soil warming and fertilisation on microarthropod abundances in three sub-arctic ecosystems. *Appl. Soil Ecol.* **2005**, *30*, 148–161. [CrossRef]

58. Van, d.W.A.; Rhem, G.; Korevaar, H.; Schouten, A.J.; Gajmjagers, O.A.; Rutgers, M.; Mulder, C. Dissimilar response of plant and soil biota communities to long-term nutrient addition in grasslands. *Biol. Fertil. Soils* **2009**, *45*, 663–667.

59. Cusack, D.F.; Firestone, M.K. Changes in microbial community characteristics and soil organic matter with nitrogen additions in two tropical forests. *Ecology* **2011**, *92*, 621–632. [CrossRef] [PubMed]

60. Wall, D.H.; Bardgett, R.D.; Kelly, E. Biodiversity in the dark. *Nat. Geosci.* **2010**, *3*, 297–298. [CrossRef]

61. Kudrin, A.A.; Tsurikov, S.M.; Tiunov, A.V. Trophic position of microbivorous and predatory soil nematodes in a boreal forest as indicated by stable isotope analysis. *Soil Biol. Biochem.* **2015**, *86*, 193–200. [CrossRef]

62. Li, J.; Wen, Y.C.; Li, X.H.; Li, Y.T.; Yang, X.D.; Lin, Z.; Song, Z.Z.; Cooper, J.M.; Zhao, B.Q. Soil labile organic carbon fractions and soil organic carbon stocks as affected by long-term organic and mineral fertilization regimes in the North China Plain. *Soil Tillage Res.* **2018**, *175*, 281–290. [CrossRef]

63. Fenn, M.E.; Baron, J.S.; Allen, E.B.; Rueth, H.M.; Nydick, K.R.; Geiser, L.; Bowman, W.D.; Sickman, J.O.; Meixner, T.; Johnson, D.W. Ecological Effects of Nitrogen Deposition in the Western United States. *Bioscience* **2003**, *53*, 404–420. [CrossRef]

64. Fu, G.; Shen, Z.X. Response of alpine soils to nitrogen addition on the Tibetan Plateau: A meta-analysis. *Appl. Soil Ecol.* **2017**, *114*, 99–104. [CrossRef]

65. Lu, X.; Mo, J.; Gilliam, F.S.; Zhou, G.; Fang, Y. Effects of experimental nitrogen additions on plant diversity in an old-growth tropical forest. *Glob. Chang. Biol.* **2010**, *16*, 2688–2700. [CrossRef]

66. Xu, C.; Geng, Q.; Guo, L.; Li, Q.; Cheng, X.; Luo, Y.; Ruan, H.; Xu, X. Variation in Soil Carbon Content in Response to Long-Term Nitrogen Addition in Poplar Plantations (*Populus Deltoids*) on the East Coast of China. in preparation.

67. Wei, C.; Zheng, H.; Li, Q.; Lü, X.; Yu, Q.; Zhang, H.; Chen, Q.; He, N.; Kardol, P.; Liang, W. Nitrogen Addition Regulates Soil Nematode Community Composition through Ammonium Suppression. *PLoS ONE* **2012**, *7*, e43384. [CrossRef] [PubMed]

68. Guo, J.H.; Liu, X.J.; Zhang, Y.; Shen, J.L.; Han, W.X.; Zhang, W.F.; Christie, P.; Goulding, K.W.T.; Vitousek, P.M.; Zhang, F.S. Significant Acidification in Major Chinese Croplands. *Science* **2010**, *327*, 1008–1010. [CrossRef] [PubMed]

69. Collange, B.; Navarrete, M.; Peyre, G.; Mateille, T.; Tchamitchian, M. Root-knot nematode (Meloidogyne) management in vegetable crop production: The challenge of an agronomic system analysis. *Crop Prot.* **2011**, *30*, 1251–1262. [CrossRef]

70. Pinho, P.; Theobald, M.R.; Dias, T.; Tang, Y.S.; Cruz, C.; Martinsloução, M.A.; Máguas, C.; Sutton, M.; Branquinho, C. Critical loads of nitrogen deposition and critical levels of atmospheric ammonia for semi-natural Mediterranean evergreen woodlands. *Biogeosciences* **2012**, *9*, 1205–1215. [CrossRef]

71. Yang, J.-Y.; Fan, J. Review of study on mineralization, saturation and cycle of Nitrogen in forest ecosystems. *J. For. Res.* **2003**, *14*, 239–243.

72. Van Diepen, L.T.A.; Lilleskov, E.A.; Pregitzer, K.S.; Miller, R.M. Simulated Nitrogen Deposition Causes a Decline of Intra- and Extraradical Abundance of Arbuscular Mycorrhizal Fungi and Changes in Microbial Community Structure in Northern Hardwood Forests. *Ecosystems* **2010**, *13*, 683–695. [CrossRef]

73. Högberg, M.N.; Briones, M.J.I.; Keel, S.G.; Metcalfe, D.B.; Campbell, C.; Midwood, A.J.; Thornton, B.; Hurry, V.; Linder, S.; Näsholm, T. Quantification of effects of season and nitrogen supply on tree below-ground carbon transfer to ectomycorrhizal fungi and other soil organisms in a boreal pine forest. *New Phytol.* **2010**, *187*, 485–493. [CrossRef]

74. Doblasmiranda, E.; Sánchezpiñero, F.; Gonzálezmegías, A. Vertical distribution of soil macrofauna in an arid ecosystem: Are litter and belowground compartmentalized habitats? *Pedobiologia* **2009**, *52*, 361–373. [CrossRef]

75. Wei, H.; Liu, W.; Zhang, J.; Qin, Z. Effects of simulated acid rain on soil fauna community composition and their ecological niches. *Environ. Pollut.* **2016**, *220*, 460–468. [CrossRef]

76. Rusek, J.; Marshall, V.G. Impacts of Airborne Pollutants on Soil Fauna. *Annu. Rev. Ecol. Syst.* **2000**, *31*, 395–423. [CrossRef]

77. Wang, S.; Ruan, H. Effects of soil mesofauna and microclimate on nitrogen dynamics in leaf litter decomposition along an elevation gradient. *Afr. J. Biotechnol.* **2011**, *10*, 6732–6742.

78. Li, Q.; Bai, H.; Liang, W.; Xia, J.; Wan, S.; Wh, V.D.P. Nitrogen addition and warming independently influence the belowground micro-food web in a temperate steppe. *PLoS ONE* **2013**, *8*, e60441. [CrossRef] [PubMed]

79. Hishi, T.; Fujimaki, R.; McGonigle, T.P.; Takeda, H. Relationships among fine roots, fungal hyphae and soil microarthropods among different soil microhabitats in a temperate coniferous forest of Chamaecyparis obtusa. *Eur. J. Soil Biol.* **2008**, *44*, 473–477. [CrossRef]

80. Bradford, M.A. Re-visioning soil food webs. *Soil Biol. Biochem.* **2016**, *102*, 1–3. [CrossRef]

81. Pollierer, M.M.; Langel, R.; Körner, C.; Maraun, M.; Scheu, S. The underestimated importance of belowground carbon input for forest soil animal food webs. *Ecol. Lett.* **2007**, *10*, 729–736. [CrossRef] [PubMed]

82. Mcnaughton, S.J.; Oesterheld, M.; Frank, D.A.; Williams, K.J. Ecosystem-level patterns of primary productivity and herbivory in terrestrial habitats. *Nature* **1989**, *341*, 142–144. [CrossRef]

83. Du, E.; de Vries, W. Nitrogen-induced new net primary production and carbon sequestration in global forests. *Environ. Pollut.* **2018**, *242*, 1476–1487. [CrossRef]

MDPI

St. Alban-Anlage 66

4052 Basel

Switzerland

Tel. +41 61 683 77 34

Fax +41 61 302 89 18

www.mdpi.com

Forests Editorial Office

E-mail: forests@mdpi.com

www.mdpi.com/journal/forests